Complete
CHEMISTRY

RoseMarie Gallagher

Paul Ingram

OXFORD
UNIVERSITY PRESS

OXFORD
UNIVERSITY PRESS

Great Clarendon Street, Oxford OX2 6DP

Oxford University Press is a department of the University of Oxford.
It furthers the University's objective of excellence in research, scholarship,
and education by publishing worldwide in

Oxford New York

Auckland Cape Town Dar es Salaam Hong Kong Karachi
Kuala Lumpur Madrid Melbourne Mexico City Nairobi
New Delhi Shanghai Taipei Toronto

With offices in

Argentina Austria Brazil Chile Czech Republic France Greece
Guatemala Hungary Italy Japan Poland Portugal Singapore
South Korea Switzerland Thailand Turkey Ukraine Vietnam

Oxford is a registered trade mark of Oxford University Press
in the UK and in certain other countries

© R. Gallagher and P. Ingram 2000

The moral rights of the author have been asserted

Database rights Oxford University Press (maker)

First published 2000

Produced form GCSE Chemistry (ISBN: 019 914 1878) first published 1997
© R. Gallagher and P. Ingram 1997

British Library Cataloguing in Publication Data
Data available

ISBN : 978-0-19-914799-1

20 19 18 17

Additional setting and design work by AMR, Bramley, Basingstoke.

Typeset in Palatino

Printed in China

Acknowledgements

*The publisher would like to thank the following for their kind permission to
reproduce the following photographs:*

p6 Ronald Sheridan/Ancient Art & Architecture Collection, **p7** Science
Photo Library/P Plailly, **p9** Dominic Phography/C Ashmore, **p11** Zefa
Photo Library, **p12** Prestige, **p20** Fisons, **p26** Da Beers (top centre), Derby
Museum & Art Gallery (bottom), **p27** Zefa, **p28** Hunting Films (top),
Hulton Picture Library, **p29** Hulton Picture Library, **p32** Zefa (top),
p33 SPL/E Willard Spurr/US Library of Congress, **p34** Zefa (top), Britain
on View/Barry Hicks (right), **p42** BOC Ltd, **p50** P Brierly (centre), **p52** University of Liverpool
(centre), **p54** SPL/R Ressmeyer/Starlight, **p55** SPL/D Guyon, **p56** De Beers
(top), SPL/V Fleming (bottom), **p57** Lead Sheeting Association (left), Black
& Decker (centre), Hutchison Picture Library (right), **p58** Pilkington Glass,
SPL/P Ryan (right), **p62** Hulton Picture Library, **p66** S & R Greenhill
(bottom), **p76** SPL/S Stammers, **p77** Zefa (top), **p78** Aga (top), **p86** SPL/M
Bond, **p89** J Allan Cash, **p90** Barnabys (top), **p91** Baxi Heating (top), Chubb
Fire Ltd (bottom), **p95** J Allan Cash (bottom), **p104** Black & Decker,
p105 Central Electricity Board, **p111** Telegraph Colour Library, **p112** Severn
Trent Water (left), Weston Point Studio, **p118** Zefa (top centre), British Steel
(top right), Barnabys (bottom right), **p120** Hutchison Picure Library,
p127 Science Photo Library, **p128** BMW, **p129** BobKrist, Black Star (top
right), Johnson Matthey Technology Centre (bottom), **p132** SPL/C Winters,
p145 NHPZ/S Dalton (top), Hutchison Picture Library (bottom), **p149** Body
Shop (bottom), **p155** SPL/Hale Observatories, **p158** Oxford Scientific
Films/H Taylor, **p159** GeoScience Features, **p160** Dr A Waltham (all),
p162 Tony Stone Worldwide, **p163** GeoScience Features (top),
Telegraph Colour Library (bottom), **p165** Robert Harding
PictureLibrary, **p166** Telegraph Colour Library (top), SPL/D Taylor
(left), J Allan Cash (bottom left), **p167** SPL/M Bond, **p168** Dr A
Waltham, **p169** Telegraph Colour Library, **p170** Aluminium
Association, **p175** SPL/J Charmet, **pp176, 177** BOC Ltd (all),
pp178, 179 Dr A Waltham, **p180** SPL/Petit Format/Nestle (left),
Telegraph Colour Library (right), **pp181, 182** Robert Harding,
p183 SPL, **p184** Robert Harding, **p185** Permutit, **p187** Ecowater
Systems, **p190** J Allan Cash (left), Prestige (right), **p191** Barnabys
(left), **p194** Thermit Welding, **p196** Royal Mint, **pp201, 202** Dr A
Waltham, **p202** GeoScience Features, **p204** Unigate (top left), J Allan
Cash (centre bottom), **p206** Alcan Aluminium, Jonquiere Analytical
Centre (top right), SPL/NASA (bottom right), **p207** British Alcan
(top), **p208** Zefa, **p209** Camera Press (top), **p210** J Allan Cash (centre
and bottom right), **p211** Heinz (top), British Alcan (bottom),
p214 SPL/K Kent (centre), SPL/A Hart-Davis, **p215** SPL/H Pincis,
p216 Barnabys, **pp218, 219** ICI, **p220** SPL/A Hart-Davis (top), Dr A
Waltham (centre), ICI (bottom), **p224** SPL/M Dohrn (left), Oxford
Scientific Films (centre), Supersport (right), **p225** British Gas (left
and centre), S & R Greenhill (right), **p228** Sulphur Institute (top),
GeoScience Features (bottom), **p229** Friends of the Earth, **p230** ICI,
p232 Allsport (left), Hulton Picture Library (right), **p234** Dr A
Waltham (bottom), **p238** De Beers (top), Chubb Fire Ltd (bottom),
p239 M Holford (top), **p241** OSF/L Nielson (top), Dr A Waltham
(bottom), **p224** SPL/D Nunuk (top), Shell (bottom), **p245** J Allan
Cash, **p246** BP (top), Esso (bottom), **p255** Hulton Picture Library
(left), SPL/P Menzel (right), **p256** Tony Stone Worldwide,
pp265, 266, 267 Dr A Waltham, **p268** Dr A Waltham (left), OSF/S
Camazine (right), **pp272, 273, 272, 275, 277, 279, 280** Dr A Waltham,
p273 Barnaby's Picture Library, **p283** SPL (left), OSF/C Monteath
(right), **p284** SPL/P Menzel, **p287** Panos Pictures.**p292** Derby
Museums & Art Gallery (top), **p293** Corbis-Bettmann, **p295** UKAEA
Harwell, **p297** Andrew Lambert, **p299** Andrew Lambert,
p301 Andrew Lambert, Science Photo Library (centre), Tony
Waltham Geophotos (bottom), **p302** Andrew Lambert (top), Science
Photo Library (bottom), **p303** Science Photo Library, **p304** Science
Photo Library (top), Romil Ltd (bottom), **p305** GeoScience Features,
p306 Science Photo Library, **p308** Science Photo Library (top),
GeoScience Features (bottom) **p309** Bostik Limited (left), Andrew
Lambert (right).

Every effort has been made to contact copyright holders of material
reproduced in this book. Any omissions will be rectified in
subsequent printings if notice is given to the publisher.

Additional photography by **Chris Honeywell**, **Peter Gould**, and
Martin Sookias.

Illustrations are by **Michael Eaton**, **Jeff Edwards**, **David Graham**,
Nick Hawken, **AMR**, and **Illustra Graphics**.

The publisher would also like to thank: Anglian Water, BOC, British
Cement Association, British Steel, Buxton Lime Industries, ICI,
National Dairy Council, Unilever, Waste Watch, Zeneca, Dr Robert
Lenk and Romil Ltd, Waterbeach, Cambs.

Examination Questions

We are grateful to the following Examination Boards for permission
to reproduce examination paper questions: SEG examination
questions reproduced by permission of AQA; NEAB examination
questions reproduced by permission of Northern Examinations and
Assessment ... ed
by permiss...
Joint Educa...
for the solu...
necessarily ... ept
no respons... ... ing
in the answ...

Introduction

If you are taking chemistry at GCSE, IGCSE, or O-level, then this book is for you.

Which topics do you need to study? You may want to read all of this book from cover to cover – but you may not need to *study* it all. It all depends on your specific exam board syllabus. Sections 1–19 cover the core syllabus content. Section 20 contains additional topics required by some syllabuses. So find a copy of the syllabus, and use the contents list to tick the topics you are required to study.

Answering questions Answering questions helps you to get a good grasp of a topic. This book contains hundreds of questions. The short questions at the end of each unit are to check that you have understood the unit. Those at the ends of sections are exam level. Numerical answers are provided. If you get stuck on a question, ask your teacher for help.

Revision checklists These start on page 310. Use them to help you with your revision.

Exam questions These start on page 316. Working through these will help you be well prepared when exam day arrives. Numerical answers are provided.

When it's time to revise It is useful to revise all though your course, because it helps your long-term memory. But if you don't manage that, make sure you start in good time for exams!

Chemistry is an exciting and challenging subject. It will help you understand some of the issues that we face in our world today. We hope you will find this book easy to use, and that your studies are a great success.

RoseMarie Gallagher
Paul Ingram

Contents

Note: use the tick boxes ☐ to mark all material appropriate to your syllabus.

1.1 Everything is made of particles

A little history

Since earliest time, we humans have wondered what things were made of. Our ideas have developed slowly, and often gone off in the wrong direction for centuries. For example:

- The Greek philosopher Democritus (460–370 BC) and his followers believed that everything was made of solid particles called **atoms**, too small to see. All atoms were identical inside. But their shapes and sizes gave them different properties. Large round atoms tasted sweet, small sharp ones sour. Smooth ones were white, jagged ones black. They came in four colours: white, black, red, and green.
- Plato (428–348 BC) and Aristotle (384–270 BC) couldn't accept this. *They* believed that everything was made of four elements: earth, air, fire, and water, mixed in different quantities. A stone has a lot of earth but not much water. Break it up into smaller and smaller pieces, and each piece will keep the properties of the original stone, no matter how small you go.
- The **alchemists** (AD 1400–1650) also believed that everything was made of just a few elements: mercury, sulphur, and salt. They believed you could turn a 'base' metal such as lead into gold by adding the right amount of mercury. This quest kept them busy for several centuries. But along the way they discovered many important substances, including sulphuric, hydrochloric, and nitric acids, and phosphorus.

The alchemists got a bad name as cheats and liars. Some made a lot of money from rich men by promising to produce gold for them. They were eventually replaced by a new breed of honest **chemists**.

The Greek philosopher Aristotle (384–270 BC). A lot of thinking – but no experiments!

The start of modern ideas

In 1661 the Englishman Robert Boyle came up with a definition we still use today: **an element is any substance that can not be broken down into simpler substances**. In the centuries that followed, many elements were discovered. Air was found to be a *mixture* of elements and salt made of *two*. This finally put paid to earlier ideas. It was also found that elements combined with each other in fixed ratios.

In 1803 the English chemist John Dalton explained this by saying that elements were in fact made of indivisible particles. Guess what he called them? **Atoms**. He said that the atoms of one element could combine with those of another element in a fixed ratio. This idea caught on fast, because it made sense of so many discoveries.

But no one could *prove* that matter was made of separate particles, since they were too small to see. Until 1827, that is, when a Scottish botanist called Robert Brown noticed grains of pollen jiggling about in water. Something was obviously banging into them and knocking them about. It had to be **water particles**.

And so, the existence of particles in matter was proved. The erratic movement of the pollen grains became known as **Brownian motion**.

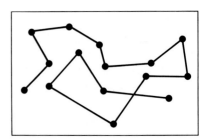

This shows the path of a pollen grain, bombarded by water particles. Smoke specks in a beam of light also show Brownian motion. They jerk all over the place because they are bombarded by air particles.

How *you* can prove the existence of particles

1 Place a crystal of potassium manganate(VII) in a beaker of water. The colour spreads because particles leave the crystal and mix through the water particles. The crystal **dissolves**.

2 Place an open gas jar of air upside down on an open gas jar of red-brown bromine vapour. The colour spreads upwards because particles of bromine mix through the particles of air.

In each case the particles mix because they collide with each other and bounce off in all directions. This mixing process is called **diffusion**. It couldn't happen if particles didn't exist.

What are these particles?

- The particles in some substances are separate atoms. Argon is the gas used in light bulbs. It consists of separate argon atoms.
- The particles in many substances consist of two or more atoms joined together. These particles are called **molecules**. Water, bromine, and the main gases in air are made up of molecules.
- The particles in other substances consist of atoms or groups of atoms that carry a charge. These particles are called **ions**. Potassium manganate(VII) is made of ions.

You'll find out more about all these particles in later sections.

A scanning electron micrograph. A scanning electron microscope scans the surface of a substance with a beam of electrons. The beam knocks electrons from the surface atoms, and these are directed onto a film or a fluorescent screen, giving an image of the surface. The green spheres represent carbon atoms. The bright splodge is a cluster of gold atoms sitting on top of the carbon.

'Seeing' the atom

Atoms, molecules, and ions are far too small to be seen with the naked eye.

But scientists have developed some powerful ways of looking at things. The **scanning electron miscroscope** can produce an image of a particle magnified millions of times. In 1979 it was used to 'see' single atoms of uranium and thorium.

1 What is *Brownian motion*? Give two examples of where you might see it.

2 Why does the purple colour spread when a crystal of potassium manganate(VII) is placed in water?

3 Bromine vapour is heavier than air. Even so, it spreads upwards in the experiment above. Why?

4 **a** What is *diffusion*? **b** Use the idea of diffusion to explain how the smell of perfume travels.

1.2 Solids, liquids, and gases

It is easy to tell the difference between a solid, a liquid and a gas:

A solid has a definite shape and a definite volume.

A liquid flows easily. It has a definite volume but no definite shape. Its shape depends on the container.

A gas has neither a definite volume nor a definite shape. It completely fills its container. It is much lighter than the same volume of solid or liquid.

Water: solid, liquid and gas

Water can be a solid (ice), a liquid (water) and a gas (water vapour or steam). Its state can be changed by heating or cooling:

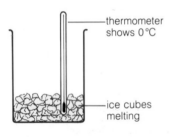

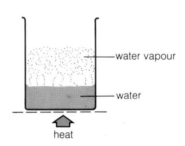

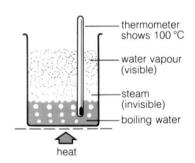

1 Ice slowly changes to **water**, when it is put in a warm place. This change is called **melting**. The thermometer shows 0 °C until all the ice has melted, so 0 °C is called its **melting point**.

2 When the water is heated its temperature rises, and some of it changes to **water vapour**. This change is called **evaporation**. The hotter the water gets, the more quickly it evaporates.

3 Soon bubbles appear. The water is **boiling**. Water vapour forms faster. It is now called **steam**. The thermometer shows 100 °C until all the water has changed to steam. 100 °C is the **boiling point** of water.

And when steam is cooled, the opposite changes take place:

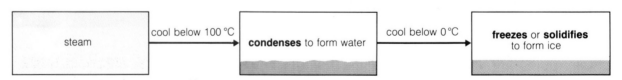

You can see that:
    condensing is the opposite of evaporating;
    freezing is the opposite of melting;
    the freezing point of water is the same as the melting point of ice, 0 °C.

Other things can change state too

Even iron and diamond can melt and boil! Some melting and boiling points are given in this table. Notice how different they can be.

Substance	Melting point/°C	Boiling point/°C
Oxygen	−219	−183
Ethanol	−15	78
Sodium	98	890
Sulphur	119	445
Iron	1540	2900
Diamond	3550	4832

A few substances change straight from solid to gas, without becoming liquid, when they are heated. This change is called **sublimation**. Carbon dioxide and iodine both sublime.

Melting and boiling points give useful clues

The melting and boiling points of a substance depend on the particles in it. This means that:
1 No two substances have the same melting and boiling points.
2 The melting and boiling point of a substance will change if you mix even a tiny amount of another substance with it.

So you can find out quite a lot from melting and boiling points:
1 You can use them to **identify** a substance.
2 You can use them to say whether it is a **pure** substance or whether it has something else mixed with it. For example:

This substance melts at 119 °C and boils at 445 °C. Can you identify it? (Look in the table above.)

You need some pure water for an experiment. This water freezes at about –2 °C and boils at about 101 °C. Is it pure?

The water in the beaker is obviously not pure. It has other things mixed with it. These are called **impurities**. And this is what happens:
1 An impurity *lowers* the freezing point of a substance and *raises* its boiling point.
2 The more impurity there is in a substance, the more its freezing and boiling points change.

In this play, the 'fog' was carbon dioxide, subliming from lumps on the stage floor.

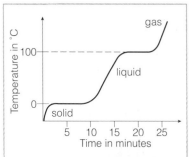

This graph is called a **heating curve**. It shows how the temperature of a substance changes when you heat it. The temperature stays constant while the solid melts, and while the liquid boils. What is the substance?

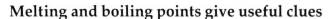

1 Write down two properties of a solid, two of a liquid, and two of a gas.
2 What word means the opposite of *boiling*?
3 Which has the lower freezing point, oxygen or ethanol?

4 What useful things can you tell from the melting and boiling points of a substance?
5 One sample of ethanol boils at around 79 °C. Another boils at around 81 °C. How can you tell they are not pure? Which one is *least* pure?

1.3 Particles in solids, liquids, and gases

You saw on page 9 that a substance can change from solid to liquid to gas. The individual *particles* of the substance are the same in each state. It is their *arrangement* that is different.

State	How the particles are arranged	Diagram of particles
Solid 	The particles in a solid are packed tightly in a fixed pattern. There are strong forces holding them together, so they cannot leave their positions. The only movements they make are tiny vibrations to and fro.	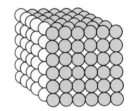
Liquid 	The particles in a liquid can move about and slide past each other. They are still close together but are not in a fixed pattern. The forces that hold them together are weaker than in a solid.	
Gas 	The particles in a gas are far apart, and they move about very quickly. There are almost no forces holding them together. They collide with each other and bounce off in all directions.	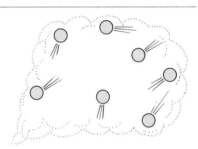

Changes of state

Melting When a solid is heated, its particles get more energy and vibrate more. This makes the solid **expand**. At the melting point, the particles vibrate so much that they break away from their positions. The solid becomes a liquid.

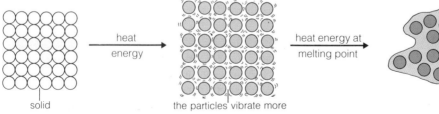 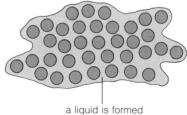

solid   heat energy   the particles vibrate more   heat energy at melting point   a liquid is formed

Boiling When a liquid is heated, its particles get more energy and move faster. They bump into each other more often and bounce further apart. This makes the liquid expand. At the boiling point, the particles get enough energy to overcome the forces holding them together. They break away from the liquid and form a gas.

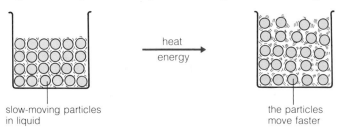

slow-moving particles in liquid

heat energy

the particles move faster

heat energy at boiling point

the particles get enough energy to escape

Evaporating Some particles in a liquid have more energy than others. Even when a liquid is well below boiling point, *some* particles have enough energy to escape and form a gas. This is called **evaporation**. It is why puddles of rain dry up in the sunshine.

Condensing and solidifying When a gas is cooled, the particles lose energy. They move more and more slowly. When they bump into each other, they do not have enough energy to bounce away again. They stay close together and a liquid forms. When the liquid is cooled, the particles slow down even more. Eventually they stop moving, except for tiny vibrations, and a solid forms.

Compressing a gas

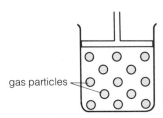

gas particles

plunger pushed in

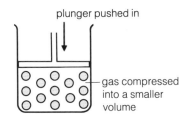

gas compressed into a smaller volume

There is a lot of space between the particles in a gas. You can force the particles closer by pushing in the plunger.

The gas gets squeezed or **compressed** into a smaller volume. We say gases are **compressible**.

If enough force is applied to the plunger, the particles get so close together that the gas turns into a liquid. But liquids and solids cannot be compressed because their particles are already close together.

These divers carry compressed air so that they can breathe underwater.

1 Using the idea of particles, explain why:
  a it is easy to pour a liquid;
  b a gas will completely fill any container;
  c a solid expands when it is heated.

2 Draw a diagram to show what happens to the particles in a liquid, when it boils.

3 Explain why a gas can be compressed into a smaller volume, but a solid can't.

1.4 A closer look at gases

When you blow up a balloon, you fill it with air particles moving at speed. The particles knock against the sides of the balloon and exert **pressure** on it. The pressure is what keeps the balloon inflated. In the same way, *all* gases exert pressure. The pressure depends on the **temperature** of the gas and the **volume** it fills, as you will see below.

When you blow air into a balloon, the gas particles exert pressure on the balloon and inflate it. The more you blow, the greater the pressure.

How gas pressure changes with temperature

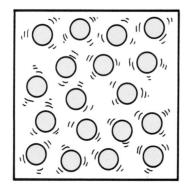

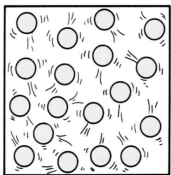

The particles in this gas are moving at speed. They hit the walls of the container and exert pressure on them.
If the gas is heated . . .

. . . the particles take in heat energy and move even faster. They hit the walls more often and with more force. So the gas pressure increases.

The same happens with all gases:
If the volume of a gas is kept constant, its pressure increases with temperature.

In a pressure cooker, water vapour is heated to well over 100°C. It is dangerous to open a pressure cooker without first letting it cool.

How gas pressure changes with volume

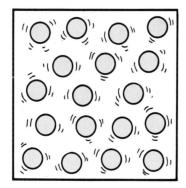

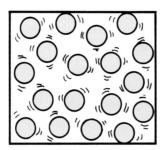

Here again is the gas from above. Its pressure is due to the particles colliding with the walls of the container.

This time the gas is squeezed into a smaller volume. So the particles hit the walls more often. The gas pressure increases.

The same thing is true for all gases:
When a gas is squeezed into a smaller volume, its pressure increases.

How gas volume changes with temperature

Now let's see what happens if the gas pressure is kept constant, but the temperature changes:

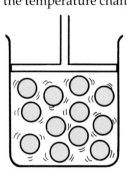 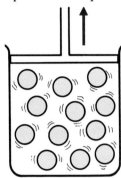

The plunger in this container can move in and out. When the gas is heated, the particles hit it more often . . .

. . . and with more energy, so it moves out. This means the pressure doesn't change. But now the gas fills a larger volume.

This shows that:
If the pressure of a gas is constant, its volume increases with temperature.

The diffusion of gases

On page 7 you saw that gases **diffuse**. A particle of ammonia gas has about half the mass of a particle of hydrogen chloride gas. So will it diffuse faster? Let's see:

1 Cotton wool soaked in ammonia solution is put into one end of a long tube. It gives off ammonia gas.
2 *At the same time*, cotton wool soaked in hydrochloric acid is put into the other end of the tube. It gives off hydrogen chloride gas.
3 The gases diffuse along the tube. White smoke forms where they meet.

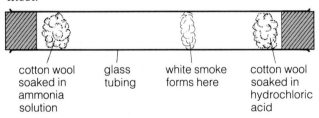

cotton wool soaked in ammonia solution | glass tubing | white smoke forms here | cotton wool soaked in hydrochloric acid

The white smoke forms closer to the right-hand end of the tube. So the ammonia particles have travelled further than the hydrogen chloride particles, in the same length of time.
The lighter the particles of a gas, the faster the gas will diffuse.

The sweet smell of freesias can fill a room. 'Smells' are caused by gas particles diffusing through the air. They dissolve in moisture in the lining of your nose. Then cells in the lining send a message to your brain.

1 What causes the pressure in a gas?
2 Why does a balloon burst if you keep on blowing?
3 A gas is in a sealed container. How do you think the pressure will change if the container is cooled? Explain your answer.
4 A gas flows from one container into a larger one. What happens to its pressure? Draw diagrams to explain.
5 Of all gases, hydrogen diffuses fastest. What can you tell from that?

1.5 Mixtures

A **mixture** contains more than one substance. So all the substances above are mixtures.

Solutions

A mixture of sugar in water is clear, and you cannot see the sugar. A mixture like this is called a **solution**. We say the sugar has **dissolved**. It is the **solute**, and water is the **solvent**:

> **solute + solvent = solution**

When sugar and water are mixed, the water particles get between the sugar particles and separate them. The separate particles are too small to be seen, which is why the solution looks clear:

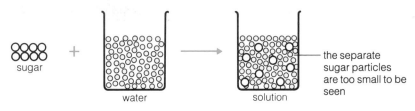

sugar + water → solution — the separate sugar particles are too small to be seen

A mixture of sugar and water

Suspensions

Look at the chalk in water on the right. There are white specks in the water because the chalk has not dissolved – it is **insoluble** in water. A mixture like this is called a **suspension**.
In a suspension, the particles of solid do not all separate. Instead they stay in clusters that are large enough to be seen.

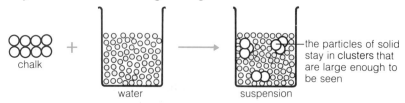

chalk + water → suspension — the particles of solid stay in clusters that are large enough to be seen

If the clusters of particles are heavy, they sink to the bottom and form a **sediment**.

A mixture of chalk and water

Other solvents

When water is the solvent, the solution is called an **aqueous solution** (from *aqua*, the Latin word for water).

Water is the most common solvent, but many others are used in industry and about the house. They are needed to dissolve substances that are insoluble in water. Some examples are:

Solvent	It dissolves
White spirit	Gloss paint
Propanone	Grease, nail polish
Dichloromethane	Grease. Used in dry cleaners
Trichloroethene	Grease and oil. Used in engineering works to 'degrease' metal parts
Ethanol	Glues, printing inks, the scented substances used in perfumes and aftershaves

All the solvents above evaporate easily at room temperature – they are **volatile**. This means that glues and paints dry easily. Aftershaves feel cool because ethanol cools the skin when it evaporates.

Other kinds of solution

Lots of solids dissolve in water to give solutions. So do liquids! For example:

a little washing-up liquid + water = a solution
a little ethanol + water = a solution

But if you add cooking oil to water you won't get a solution. The oil stays as a separate layer. Oil and water don't mix – they are **immiscible**.

Many gases dissolve in water. Carbon dioxide is dissolved in fizzy drinks and bottled drinking water to give them sparkle. The small amount of oxygen dissolved in river water is enough to keep fish alive.

Ethanol is used as a solvent for perfume. It is a volatile liquid. Why is that an advantage?

'Liquid' papers contain a volatile liquid.

1 Explain each term:
  solution   aqueous solution
  suspension   sediment
2 Name two solids other than chalk that are insoluble in water.
3 At home you make several solutions. Write down:
  a three solids you dissolve in water;
  b three solutions you make by dissolving a liquid in water.

4 What problems might there be if oxygen was *very* soluble in water?
5 Some gases *are* very soluble in water. Try to think of one you have come across in the lab.
6 Name two solvents other than water that are used in the home. What are they used for?
7 Is it a solution or suspension?
  a lemonade   **b** Milk of Magnesia
  c shampoo   **d** typists' correction fluid

1.6 More about solutes and solvents

Helping a solute dissolve

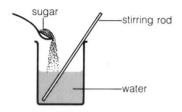

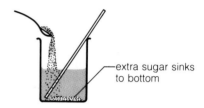

Sugar dissolves quite slowly in water at room temperature (20 °C). Stirring helps!

But if you keep on adding sugar to the water, even with plenty of stirring...

... eventually no more will dissolve. The solution is **saturated**.

A saturated solution is one that can dissolve no more solute at that temperature.

But look what happens when you heat the sugar solution:

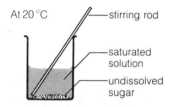

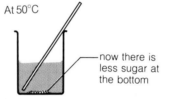

There is undissolved sugar at the bottom of the beaker.

Now some of it has dissolved – but there is still some left.

It has all dissolved. You might even be able to dissolve more!

So sugar is **more soluble** in hot water than in cold. In fact this is usually the case with solids that dissolve.
If a solid is soluble in a liquid, it usually gets more soluble as the temperature rises.

How soluble is it?

The solubility of a substance depends on the particles in it. So the solubility of each substance is different. Look at these examples:

Compound	Mass (g) dissolving in 100 g of water at 25 °C	
Silver nitrate	241.3	
Calcium nitrate	102.1	
Magnesium chloride	53.0	
Potassium nitrate	37.9	decreasing
Potassium sulphate	12.0	solubility
Calcium hydroxide	0.113	
Calcium carbonate	0.0013	
Silver chloride	0.0002	

As you can see, one compound of a metal may be highly soluble while another is almost insoluble. (Compare silver nitrate and silver chloride.) It depends on the particles.

The sodium halides

Compare the amounts that dissolve in 100 g of water at 25 °C:

sodium chloride	35.92 g
sodium bromide	94.56 g
sodium iodide	184.38 g

Now compare this with the order of reactivity of the halogens (page 36).
What do you notice? Can you explain?

Soluble and insoluble compounds

Look at the last three compounds in the table on the opposite page.
Although they are in fact **sparingly soluble**, they are usually
described as **insoluble** since so little dissolves. Compounds are
generally classed as either soluble or insoluble. They fall into
patterns, as you can see from this table:

Soluble compounds		Insoluble compounds
All sodium, potassium, and ammonium salts		
All nitrates		
Chlorides...	*except*	silver and lead chloride
Sulphates...	*except*	lead, barium, and calcium sulphates
Sodium, potassium, and ammonium carbonates...		but other common carbonates are insoluble
Sodium, potassium, and ammonium hydroxides...		but other common hydroxides are insoluble

Suppose you mix a solution of lead nitrate with a solution of sodium
chloride. What do you think will happen?

Comparing different solvents

As you saw opposite, the solubility of a substance depends on the
particles in it. But it also depends on the **solvent**:

Iodine is very slightly soluble in
water. Only 0.3 grams will
dissolve in 100 grams of water at
20 °C. So you get a very pale
solution.

It is much more soluble in
cyclohexane, a solvent which
smells like petrol. 2.8 grams of
iodine dissolve in 100 grams of
cyclohexane at 20 °C.

If you shake some cyclohexane
with a solution of iodine in
water, almost all the iodine
leaves the water and moves into
the cyclohexane layer.

So cyclohexane is much better than water at separating iodine
particles from each other. The iodine particles are more attracted to
cyclohexane than they are to water.

1 What is a *saturated* solution?

2 A solution of sugar in water is *just* saturated at 20 °C.
You heat it to 50 °C. Will it still be saturated? Explain.

3 A solution of sugar in water is saturated at 70 °C. What
will happen if you cool it to 40 °C? Why?

4 Why is the solubility of every solute different?

5 Calcium hydroxide is *sparingly* soluble. Explain.

6 Compounds are usually described as either *soluble* or
insoluble. Which term would you use for:
a calcium hydroxide?   b calcium nitrate?

7 Water won't dissolve grease. Name two solvents
which could be used for grease.

1.7 Solubility

The term *solubility* has a very precise meaning:
The solubility of a solute in water, at a given temperature, is the maximum amount of it that will dissolve in 100 grams of water at that temperature.

As you can see from the table on page 16, the solubility of calcium nitrate in water at 25 °C is 102.1 grams. When this amount is dissolved in 100 grams of water at that temperature, it gives a **saturated solution**.

Measuring the solubility of a solid in water

Let's take potassium sulphate as our example. This is what to do:

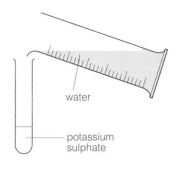

water

potassium
sulphate

heat

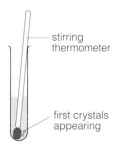

stirring
thermometer

first crystals
appearing

Put a weighed amount (say 2 g) of potassium sulphate in a test tube. Add a little water from a measuring cylinder.

Heat the test tube gently until the water is hot but not boiling. Add more water if necessary until the solid is *just* dissolved.

Let the solution cool, while stirring it with a thermometer. Note the temperature at which the first crystals appear.

Since you know the mass of solute and the volume of water you used, you can work out the solubility as shown in the calculation below.

Now look again at the last step. If you add a little more water, heat the solution again to make sure all the crystals have dissolved, and then let it cool, you'll be able to find the solubility at a lower temperature. You can repeat this for a range of temperatures.

Calculating solubility

2 grams of potassium sulphate were dissolved in 12.5 cm³ of water. On cooling, the first crystals appeared at 60 °C. What is the solubility of potassium sulphate in water at 60 °C?

12.5 cm³ of water weigh 12.5 g.
In 12.5 g of water 2 g of potassium sulphate dissolve, so
in 1 g of water $\dfrac{2}{12.5}$ g dissolve, and
in 100 g of water $\dfrac{2 \times 100}{12.5}$ g dissolve. $\qquad \dfrac{2 \times 100}{12.5} = 16.$
The solubility of potassium sulphate in water at 60 °C is **16 grams**.

Solubility curves

Solubility of copper(II) sulphate in water

Temperature/ °C	0	10	20	30	40	50	60	70
Solubility/g	14	17	21	24	29	34	40	47

Look at the table above. The results were obtained by experiment. You can use them to plot a graph called a **solubility curve**.

Look at the curve on the right. It shows clearly how the solubility of copper(II) sulphate increases with temperature. You can use it to find:

- the solubility of the salt at *any* temperature from 0 to 70 °C. For example, at 15 °C its solubility is 19 grams.
- the temperature at which crystals will first appear when you cool a given solution. If the solution contains 37 g of the salt in 100 g of water, crystals will appear just below 55 °C.
- the mass of crystals you'll obtain if you then cool this solution further. If you cool it to just below 15 °C, 18 g of crystals will form.

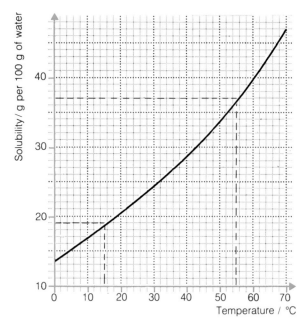

The solubility curve for copper(II) sulphate

The solubility of gases

Solid solutes usually get more soluble in water as the temperature rises. The *opposite* is true for gases. Look at this table:

Gas	Solubility (cm³ per 100 cm³ of water) at ...			
	0 °C	20 °C	40 °C	60 °C
Oxygen	4.8	3.3	2.5	1.9
Carbon dioxide	171	92.3	56.6	36.0
Sulphur dioxide	7980	4250	2170	
Hydrogen chloride	50 500	47 400	44 500	42 000

Look at carbon dioxide. It is quite soluble in water at room temperature (20 °C). But when it is pumped into soft drinks under pressure, a lot more dissolves. Then when you open the bottle it fizzes out of solution.

Look at hydrogen chloride. At room temperature it is over 14 000 times more soluble than oxygen. Its high solubility is the basis of the **fountain experiment** (page 217).

The solubility of gases

For gases, solubility changes with temperature *and* pressure. It:

- decreases with temperature
- increases with pressure.

---

1 What is: **a** solubility? **b** a solubility curve?
2 2.9 g of potassium sulphate were dissolved in 25 cm³ of water. The first crystals appeared at 31 °C. What is its solubility at 31 °C?
3 Use the solubility curve above to find the solubility of copper(II) sulphate in water at: **a** 36 °C **b** 66 °C.
4 How much copper(II) sulphate would you need to make a saturated solution in 100 g of water at 52 °C?
5 Some power stations use river water for cooling, and dump it back in the river still hot.
  a How will this affect the river's oxygen content?
  b How will it affect fish and other river life?

1.8 Separating mixtures (I)

How do you separate one substance from a mixture? It depends on the physical state of the substances in the mixture.

How to separate a solid from a liquid

By filtering Chalk can be separated from water by filtering the suspension through filter paper. The chalk gets trapped in the filter paper while the water passes through. The chalk is called the **residue**. The water is called the **filtrate**.

Other suspensions can be separated in the same way.

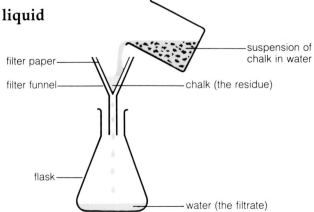

- suspension of chalk in water
- filter paper
- filter funnel
- chalk (the residue)
- flask
- water (the filtrate)

By centrifuging A centrifuge is used to separate *small* amounts of suspension. In a centrifuge, test-tubes of suspension are spun round very fast, so that the solid gets flung to the bottom:

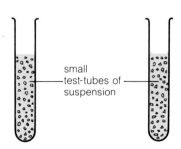

small test-tubes of suspension

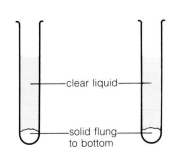

clear liquid

solid flung to bottom

A centrifuge

Before centrifuging, the solid is mixed all through the liquid.

After centrifuging, all the solid has collected at the bottom.

The liquid can be **decanted** (poured out) from the test-tubes, or removed with a small pipette. The solid is left behind.

By evaporating the solvent If the mixture is a *solution*, the solid cannot be separated by filtering or centrifuging. This is because it is spread all through the solvent in tiny particles. Instead, the solution is heated so that the solvent evaporates, leaving the solid behind. Salt is obtained from its solution by this method:

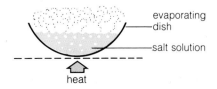

- evaporating dish
- salt solution
- heat

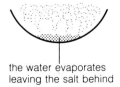

the water evaporates leaving the salt behind

Evaporating the water from a salt solution

By crystallizing   You can separate many solids from solution by letting them form crystals. Copper(II) sulphate is an example:

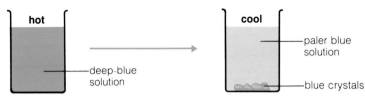

This is a saturated solution of copper(II) sulphate in water at 70 °C. If it is cooled to 20 °C ...

... crystals begin to appear, because the compound is *less soluble* at 20 °C than at 70 °C.

The process is called **crystallization**. It is carried out like this:

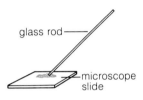

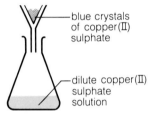

1 A solution of copper(II) sulphate is heated, to get rid of some water. As the water evaporates, the solution becomes more concentrated.

2 The solution can be checked to see if it is ready by placing one drop on a microscope slide. Crystals should form quickly on the cool glass.

3 Then the solution is left to cool and crystallize. The crystals are removed by filtering, rinsed with water and dried with filter paper.

How to separate a mixture of two solids

By dissolving one of them   A mixture of salt and sand can be separated like this:

1 Water is added to the mixture, and it is stirred. The salt dissolves.
2 The mixture is then filtered. The sand is trapped in the filter paper, but the salt solution passes through.
3 The sand is rinsed with water and dried in an oven.
4 The salt solution is evaporated to dryness.

This method works because salt is soluble in water but sand is not. Water could *not* be used to separate salt and sugar, because it dissolves both. You could use ethanol, because it dissolves sugar but not salt. Ethanol is inflammable, so it should be evaporated from the sugar solution over a water bath, as shown on the right.

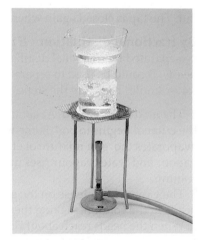

Evaporating ethanol from a sugar solution, over a water bath

1 What is: a filtrate?   a residue?
2 Describe two ways of separating the solid from the liquid in a suspension.
3 Sugar *cannot* be separated from sugar solution by filtering. Explain why.

4 What happens when a solution is evaporated?
5 Describe how you would crystallize potassium nitrate from its aqueous solution.
6 How would you separate salt and sugar? Mention any special safety precaution you would take.

1.9 Separating mixtures (II)

How to separate the solvent from a solution

By simple distillation   This is a way of getting pure solvent out of a solution. The apparatus is shown on the right. It could be used to obtain pure water from salt water, for example. This is what happens:

1 The solution is heated in the flask. It boils, and steam rises into the condenser. The salt is left behind.
2 The condenser is cold, so the steam condenses to water in it.
3 The water drips into the beaker. It is completely pure. It is called **distilled water**.
This method could also be used to obtain pure water from sea water, or from ink.

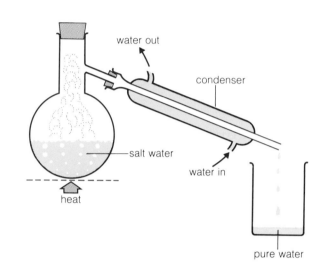

How to separate two liquids

Using a separating funnel   If two liquids are **immiscible**, they can be separated with a separating funnel. For example, when a mixture of oil and water is poured into the funnel, the oil floats to the top, as shown on the right. When the tap is opened, the water runs out. The tap is closed again when all the water has gone.

By fractional distillation   If two liquids are **miscible**, they must be separated by fractional distillation. The apparatus is shown below. It could be used to separate a mixture of ethanol and water, for example. These are the steps:

1 The mixture is heated. At about 78 °C, the ethanol begins to boil. Some water evaporates too, so a mixture of ethanol vapour and water vapour rises up the column.
2 The vapours condense on the glass beads in the column, making them hot.
3 When the beads reach about 78 °C, ethanol vapour no longer condenses on them. Only the water vapour does. The water drips back into the flask, while the ethanol vapour is forced into the condenser.
4 There it condenses. Liquid ethanol drips into the beaker.
5 Eventually, the thermometer reading rises above 78 °C. This is a sign that all the ethanol has been separated, so heating can be stopped.

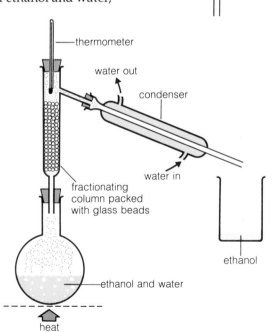

How to separate a mixture of coloured substances

Paper chromatography   This method can be used to separate a mixture of coloured substances. For example, it will separate the coloured substances in black ink:

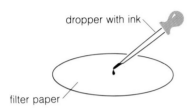

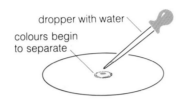

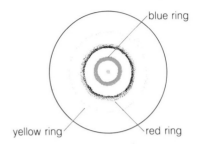

1 A small drop of black ink is placed at the centre of a piece of filter paper, and allowed to dry. Three or four more drops are added on the same spot.

2 Water is then dripped on to the ink spot, one drop at a time. The ink slowly spreads out and separates into rings of different colours.

3 Suppose there are three rings: yellow, red and blue. This shows that the ink contains three substances, coloured yellow, red and blue.

Each substance in the ink travels across the paper at a different rate. That's why they separate into rings. The filter paper showing the separate substances is called a **chromatogram**.

Paper chromotography is also used to **identify** the substances in a mixture. For example, mixture X is thought to contain the substances A, B, C, and D, which are all soluble in propanone. The mixture could be checked like this:

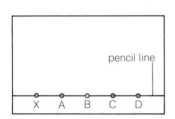

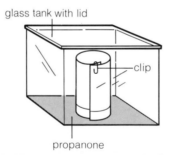

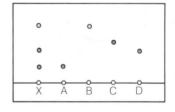

1 Concentrated solutions of X, A, B, C, and D are made up in propanone. A spot of each is placed on a line, on a sheet of filter paper, and labelled.

2 The paper is stood in a little propanone, in a covered glass tank. The solvent rises up the paper; when it's near the top, the paper is taken out again.

3 X has separated into three spots. Two are at the same height as A and B, so X must contain substances A and B. Does it also contain C and D?

The substances shown here are coloured. But you can also use paper chromatography for *colourless* substances. And identify a substance from tables, by *measuring* how far it travels. See Unit 20.8 for more.

1 How would you obtain pure water from ink? Draw the apparatus you would use, and explain how the method works.
2 Why are condensers so called? What is the reason for running cold water through them?
3 Water and turpentine are immiscible. How would you separate a mixture of the two?
4 Explain how fractional distillation works.
5 In the chromatogram above, how can you tell that X does not contain substance C?

Questions on Section 1

1 A large crystal of potassium manganate(VII) was placed in the bottom of a beaker of cold water and left for several hours.

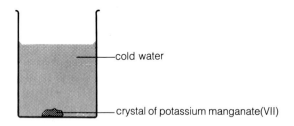

   —cold water

   —crystal of potassium manganate(VII)

  a Describe what would be seen after five minutes.
  b Describe what would be seen after several hours.
  c Explain your answers using the idea of particles.
  d Name the *two* processes which have taken place during the experiment.

2 Use the idea of particles to explain why:
  a solids have a definite shape;
  b solids cannot be poured;
  c liquids fill the bottom of a container;
  d you can't store gases in open containers;
  e you can't squeeze a sealed plastic syringe that's completely full of water;
  f a balloon expands as you blow into it.

3 Draw diagrams to show what happens to the particles when:
  a water freezes to ice;
  b steam condenses to water.

4 The graph below is a **heating curve** for a pure substance. It shows how the temperature rises with time, when the solid is heated until it melts, and then the liquid is heated until it boils.

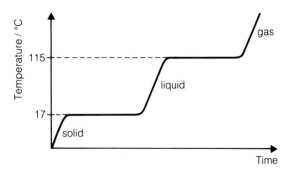

  a What is the melting point of the substance?
  b What is its boiling point?
  c What happens to the temperature while the substance changes state?
  d How can you tell that the substance is not water?

5 Sketch the heating curve for pure water, between −10 °C and 110 °C. Mark in the temperatures at which water changes state, and its state for each sloping part of the graph.

6 A **cooling curve** shows how the temperature of a substance changes with time, as it is cooled from a gas to a solid. Below is the cooling curve for one substance:

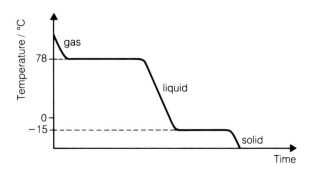

  a What is the state of the substance at room temperature (20 °C)?
  b Use the list of melting and boiling points on page 9 to identify the substance.

7 Using the idea of particles explain why:
  a the smell of burnt toast travels all over the house;
  b when two solids are placed on top of each other, they don't mix;
  c a liquid is used in a car's breaking system;
  d if you pump up your bike tyres quite hard you get a smooth ride;
  e halving the volume of a gas will double its pressure;
  f pollution from just one factory can affect a large area of land.

8 A test-tube of ammonia gas is placed above a test-tube of air. Ammonia is an alkaline gas that turns litmus blue. It is lighter than air.

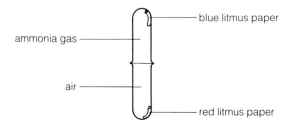

  blue litmus paper

ammonia gas

air

  red litmus paper

  a After a short time the red litmus paper turns blue. Explain why.
  b Would it make any difference if you reversed the test-tubes? Explain your answer.
  c What would you *see* if the test-tube of air was replaced by one containing hydrogen chloride?

9 The graph shows the solubility curves for copper(II) sulphate (A) and sodium chloride (B), both plotted on the same axes.

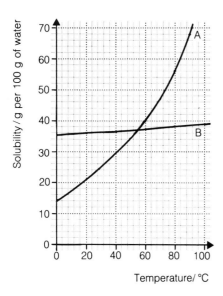

Temperature/ °C

a What is the solubility of each substance at:
  i 40 °C? **ii** 60 °C?
b Which substance is more soluble at:
  i 30 °C? **ii** 70 °C?
c At what temperature do both substances have the same solubility?
d At what temperature would 45 g of copper(II) sulphate just dissolve in 100 g of water?
e Would it be possible to dissolve 50 g of sodium chloride in 100 g of water?

10 Look again at the solubility curves in question 9.
a If a saturated solution of copper(II) sulphate in 100 g of water is cooled from 80 °C to 20 °C, how much copper(II) sulphate will crystallize out of solution?
b If a saturated solution of sodium chloride in 100 g of water is cooled from 80 °C to 20 °C, how much sodium chloride will crystallize out of solution?
c To obtain sodium chloride from its solution in water, the solution must be evaporated to dryness, rather than left to crystallize. Why is this?

11 At 20 °C (room temperature) 75 000 cm^3 of ammonia will dissolve in 100 cm^3 of water.
a Look at the table on page 19. How does the solubility of ammonia compare with that of:
  i oxygen? **ii** hydrogen chloride?
b What can you say about the solubility of ammonia at: **i** 0 °C? **ii** 40 °C?
c Explain how the fountain experiment works.
d Would you expect the fountain experiment to work with ammonia? Why?
e Would you expect ammonia to have the same solubility in other solvents? Why?

12 Describe the relationship that exists between:
a gas volume and pressure;
b gas volume and temperature;
c gas pressure and temperature.

13

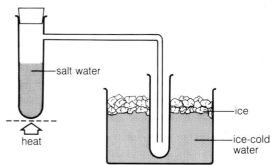

The apparatus above can be used to obtain pure water from salt water.
a What is the purpose of the ice-cold water?
b Why must the glass arm from the first tube reach far down into the second tube?
c Explain how the method works.
d What is this separation method called?

14 A mixture of salt and sugar has to be separated, using the solvent ethanol.
a Which of the two substances is soluble in ethanol?
b Draw a diagram to show how you would separate the salt.
c How could you obtain sugar crystals from the sugar solution, *without* losing the ethanol in the process?
d Draw a diagram of the apparatus for **c**.

15 Eight coloured substances were spotted on to a piece of filter paper, which was then stood in a covered glass tank containing a little propanone. Three of the substances were the basic colours red, blue, and yellow. The others were dyes, labelled A, B, C, D, E. The resulting chromatogram is below:

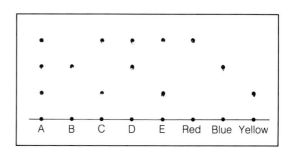

a Which dye contains only one basic colour?
b Which dye contains all three basic colours?
c Which basic colour is most soluble in propanone?

2.1 Atoms, elements, and compounds

Atoms

The pieces of sodium in this photograph are made of billions of tiny particles called **sodium atoms**.

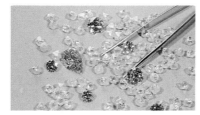

Diamond is a form of carbon. These diamonds are made of billions of **carbon atoms**, which are different from sodium atoms.

Mercury is made of **mercury atoms**, which are different from both sodium atoms and carbon atoms.

It is almost impossible for us to imagine how small atoms are. For example, about four billion sodium atoms would fit side by side on the full stop at the end of this sentence. However, in spite of the small size of atoms, scientists have managed to find out a great deal about them. They have found that every atom consists of a **nucleus**, and a cloud of particles called **electrons** that whizz non-stop round the nucleus.

The drawing on the right shows what a sodium atom might look like, greatly magnified.

(You can find out more about the nucleus and electrons on page 28.)

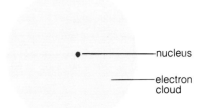

nucleus

electron cloud

A typical atom

Elements

Sodium is made of sodium atoms only, so it is an **element**.
An element is a substance that is made of only one kind of atom.
Diamond (carbon) and mercury are also elements.
Altogether, 105 different elements are known. Of these, 90 have been obtained from the Earth's crust and atmosphere, and 15 have been artificially made by scientists.
Every element has a name and a symbol. Here are some of them:

Element	Symbol	Element	Symbol
Aluminium	Al	Bromine	Br
Copper	Cu	Carbon	C
Iron	Fe	Chlorine	Cl
Lead	Pb	Hydrogen	H
Magnesium	Mg	Nitrogen	N
Mercury	Hg	Oxygen	O
Potassium	K	Phosphorus	P
Silver	Ag	Sulphur	S
Sodium	Na	Silicon	Si

It is easy to remember that the symbol for aluminium is Al, and for carbon is C. But some symbols are harder to remember, because they are taken from the Latin names for the elements. For example: potassium has the symbol K, from its Latin name kalium. Sodium has the symbol Na, from its Latin name natrium.

A few elements, such as copper, have been known for thousands of years. But most were discovered in the last 400 years. This painting shows Henry Brand, the German alchemist, on his discovery of phosphorus in 1659. He extracted it from urine, by accident, during his search for the elixir of life. To his amazement it glowed in the dark.

The metals   The elements in the list on page 26 are in two columns for a good reason. The ones on the left are **metals**, while those on the right are **nonmetals**. Over 80 of the elements are metals. Although they all look different, they have many properties in common. Here are a few:

1 They allow electricity and heat to pass through them easily – they are all good **conductors** of electricity and heat.
2 They are all solids at room temperature, except mercury, and most of them have high melting points.
3 Most of them can be hammered into different shapes (they are **malleable**) and drawn into wires (they are **ductile**).

The nonmetals   Only about one-fifth of the elements are nonmetals. They are quite different from metals:

1 They are poor conductors of electricity and heat. (Carbon is an exception.)
2 They usually have low melting points (eleven of them are gases and one is a liquid at room temperature).
3 When solid nonmetals are hammered, they break up – they are **brittle**.

Nerves of steel? 16 elements (including iron, but mainly oxygen, carbon, nitrogen, calcium, and phosphorus) combine to make the hundreds of compounds in the human body. There's enough iron in your body to make a large nail.

Compounds

Elements can combine with each other to form **compounds**. **A compound contains atoms of different elements joined together**. Although there are only 105 elements, there are millions of compounds. This table shows three common ones:

Name of compound	Elements in it	How the atoms are joined up
Water	Hydrogen and oxygen	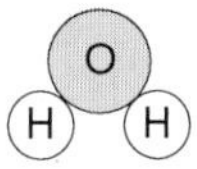
Carbon dioxide	Carbon and oxygen	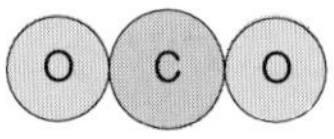
Ethanol	Carbon, hydrogen and oxygen	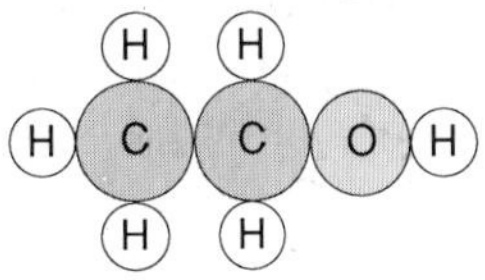

Symbols for compounds   The symbol for a compound is called its **formula**. It is made up from the symbols of the elements. The formula for water is H_2O and for ethanol is C_2H_5OH. Can you see why? Can you guess the formula for carbon dioxide? (Check your answer on page 68.) Note that the plural of **formula** is **formulae**.

O_2C

CO_2

1 What is an atom?
2 What is the centre part of an atom called?
3 Explain what an element is.
4 Explain what these words mean:
   malleable   ductile   brittle
5 Write down three properties of nonmetals.
6 What is: **a** a compound?   **b** a formula?
7 What is H_2O? What does the $_2$ in it show?
8 A certain compound has the formula NaOH. What elements does it contain?

2.2 More about the atom

Protons, neutrons, and electrons

On page 26 you saw that all atoms consist of a **nucleus** and a cloud of **electrons** that move round the nucleus. The nucleus is itself a cluster of two sorts of particle, **protons** and **neutrons**.

All the particles in an atom are very light. Their mass is measured in **atomic mass units**, rather than grams. Protons and electrons also have an **electric charge**:

Particle in atom	Mass	Charge
Proton	1 unit	Positive charge (+1)
Neutron	1 unit	None
Electron	Almost nothing	Negative charge (−1)

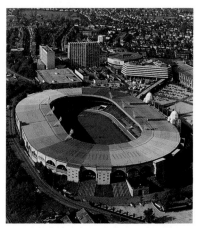

The nucleus is very tiny compared with the rest of the atom. If the atom was the size of a football stadium, the nucleus (sitting on the centre spot) would be the size of a pea!

How the particles are arranged

The sodium atom is a good one to start with. It has **11** protons, **11** electrons and **12** neutrons. They are arranged like this:

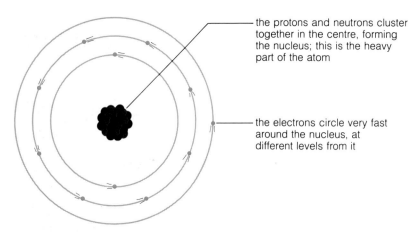

the protons and neutrons cluster together in the centre, forming the nucleus; this is the heavy part of the atom

the electrons circle very fast around the nucleus, at different levels from it

The different energy levels for the electrons are called **electron shells**. Each shell can hold only a limited number of electrons:

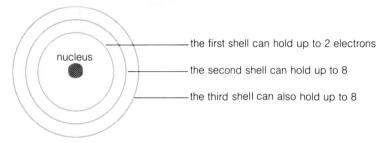

nucleus

the first shell can hold up to 2 electrons

the second shell can hold up to 8

the third shell can also hold up to 8

Notice how the electrons are arranged in the sodium atom:
    2 in the first shell (it is full);
    8 in the second shell (it is also full);
    1 in the third shell (it is not full).
The atom is often written as **Na(2,8,1)**. Can you see why?
The (2,8,1) is its electron arrangement or **electronic configuration**.

Niels Bohr, a Danish scientist, was the first person to put forward the idea of electron shells. He died in 1962.

Proton number and mass number

Proton number   Look again at the sodium atom on the opposite page. It has **11** protons. This fact could be used to identify it, because *only* a sodium atom has 11 protons. Every other atom has a different number of protons.
You can identify an atom by the number of protons in it.
The number of protons in an atom is called its **proton number** or **atomic number**.
The proton number of sodium is 11.

The sodium atom also has **11** electrons. So it has an equal number of protons and electrons. The same is true for every sort of atom:
Every atom has an equal number of protons and electrons.
Because of this, atoms have no overall charge. The charge on the electrons cancels the charge on the protons. You can check this for a sodium atom, on the right.

Mass number   The electrons in an atom have almost no mass. So the mass of an atom is nearly all due to its protons and neutrons. For this reason, the number of protons and neutrons in an atom is called its **mass number** or **nucleon number**.
The mass number = the number of protons + neutrons in an atom.
A sodium atom has 11 protons and 12 neutrons, so the mass number of sodium is 23.

Since the proton number is the number of protons only, then:
    mass number – proton number = number of neutrons.
So, for a sodium atom, the number of neutrons = (23 –11) = 12.

Shorthand for an atom

The sodium atom can be described in a short way, using:
    the symbol for sodium (Na)
    its proton number (11)
    its mass number (23)

The information is written as $^{23}_{11}\mathbf{Na}$.

From it you can tell that the sodium atom has 11 protons, 11 electrons and 12 neutrons (23 –11 = 12). Chemists often describe atoms in this way. The information is always put in the same order:
    $\overset{\textbf{mass number}}{\underset{\textbf{proton number}}{}}$ **symbol**

The charge on a sodium atom:

●●●● 11 protons
●●●● Each has a charge of +1
●●● Total charge +11

• • • • 11 electrons
• • • • Each has a charge of −1
• • • Total charge −11

Adding the charges: +11
                    −11
                    ———
                      0

The answer is zero.
The atom has no overall charge.

A hundred years ago, hardly anything was known about the atom. For example, the neutron was discovered only in 1932, by this British scientist, Sir James Chadwick.

---

1 Name the particles that make up the atom.
2 Which particle has:
   a a positive charge?   b no charge?
   c almost no mass?
3 Draw a sketch of the sodium atom.
4 What does *electronic configuration* mean?

5 What does the term mean?   a proton number
   b mass number   c nucleon number
6 Name each of these atoms, and say how many protons, electrons and neutrons it has:

   $^{12}_{6}C$ $^{16}_{8}O$ $^{24}_{12}Mg$ $^{27}_{13}Al$ $^{64}_{29}Cu$

2.3 Some different atoms

What makes two atoms different?

On page 28 you saw that sodium atoms have 11 protons. This is what makes them different from all other atoms. *Only* sodium atoms have 11 protons, and any atom with 11 protons *must* be a sodium atom.

In the same way, an atom with 6 protons must be a carbon atom, and an atom with 7 protons must be a nitrogen atom:
You can identify an atom by the number of protons in it.

The first twenty elements   There are 105 elements altogether. Of these, hydrogen has the smallest atoms, with only 1 proton each. Helium atoms have 2 protons each, lithium atoms have 3 protons each, and so on up to hahnium atoms, which have 105 protons each. Below are the first twenty elements, arranged in order according to the number of protons they have:

Element	Symbol	Number of protons (proton number)	Number of electrons	Number of neutrons	Number of protons + neutrons (mass number)
Hydrogen	H	1	1	0	1
Helium	He	2	2	2	4
Lithium	Li	3	3	4	7
Beryllium	Be	4	4	5	9
Boron	B	5	5	6	11
Carbon	C	6	6	6	12
Nitrogen	N	7	7	7	14
Oxygen	O	8	8	8	16
Fluorine	F	9	9	10	19
Neon	Ne	10	10	10	20
Sodium	Na	11	11	12	23
Magnesium	Mg	12	12	12	24
Aluminium	Al	13	13	14	27
Silicon	Si	14	14	14	28
Phosphorus	P	15	15	16	31
Sulphur	S	16	16	16	32
Chlorine	Cl	17	17	18	35
Argon	Ar	18	18	22	40
Potassium	K	19	19	20	39
Calcium	Ca	20	20	20	40

Drawing the different atoms

It is easy to draw the different atoms, if you remember these rules:
1 The protons and neutrons form the nucleus at the centre.
2 The electrons are in electron shells around the nucleus.
3 The first electron shell can hold a maximum of 2 electrons, the second can hold 8, and the third can also hold 8.

In the drawings below, **p** = proton, **e** = electron and **n** = neutron.

A hydrogen atom 1p, 1e	A lithium atom 3p, 3e, 4n	A magnesium atom 12p, 12e, 12n
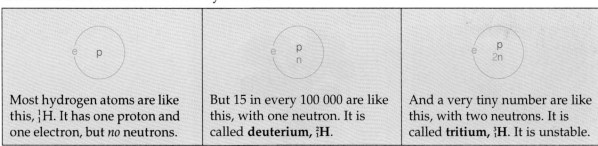		
Note that it has no neutrons. So its proton number is 1 *and* its mass number is 1. It is described as $^1_1$H.	It has three electrons. The first shell can hold only two electrons, so a second shell is used. The atom is described as $^7_3$Li. Can you explain why?	The first two shells together hold only ten electrons. The remaining two electrons must go in a third shell. The atom is described as $^{24}_{12}$Mg. Can you explain why?

Isotopes

The atoms of an element are not always identical:

Most hydrogen atoms are like this, $^1_1$H. It has one proton and one electron, but *no* neutrons.	But 15 in every 100 000 are like this, with one neutron. It is called **deuterium**, $^2_1$H.	And a very tiny number are like this, with two neutrons. It is called **tritium**, $^3_1$H. It is unstable.

All three atoms belong to the element hydrogen, *because they all have 1 proton*. They are called **isotopes** of hydrogen.
Isotopes are atoms of the same element, with the same number of protons but different numbers of neutrons.

Most elements have more than one isotope. For example carbon has three: $^{12}_6$C, $^{13}_6$C and $^{14}_6$C. Chlorine has two: $^{35}_{17}$Cl and $^{37}_{17}$Cl.

Unstable isotopes
- Many isotopes (like tritium) are unstable. They are called **radioisotopes**.
- Sooner or later the nucleus breaks up, giving out **radiation** (see page 294).

---

1 Which element has atoms with:
   a 5 protons? **b** 15 protons? **c** 20 protons?
2 A hydrogen atom can be described as $^1_1$H. Use the same method to describe the atoms of all the elements listed on page 30.

3 Draw a sketch of: **a** a lithium atom;
   b a carbon atom; **c** a neon atom.
   Give the electronic configuration for each.
4 What is an isotope? Name the isotopes of hydrogen and write symbols for them.

2.4 The periodic table

Arranging the elements in groups

Two groups of elements are shown below. For each element, the electronic configuration of its atoms is given.
The three elements in each group have something in common:

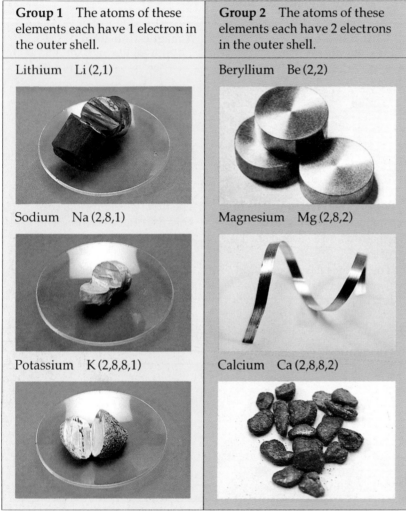

Group 1   The atoms of these elements each have 1 electron in the outer shell.

Lithium   Li (2,1)

Sodium   Na (2,8,1)

Potassium   K (2,8,8,1)

Group 2   The atoms of these elements each have 2 electrons in the outer shell.

Beryllium   Be (2,2)

Magnesium   Mg (2,8,2)

Calcium   Ca (2,8,8,2)

Using the same idea, scientists have put all the elements into groups.

1 First, the elements are listed in order of increasing proton number. Hydrogen comes first, since its proton number is 1.
2 Next, the list is sub-divided like this:
   The elements whose atoms have 1 outer-shell electron are picked out of the list, in order, and called **Group 1**.
   The elements whose atoms have 2 outer-shell electrons are picked out and called **Group 2**.
   Group 3, Group 4 and so on are formed in the same way.
3 Then the groups are arranged side by side to give the **periodic table**. This is shown opposite and again on page 298.

The Russian chemist Mendeleev, who drew up the first version of the periodic table in 1869. You can find out more about him and his table in Unit 20.1.

The periodic table

Group 1	2													3	4	5	6	7	0
1 $^1_1$H hydrogen																			$^4_2$He helium
2 $^7_3$Li lithium	$^9_4$Be beryllium													$^{11}_5$B boron	$^{12}_6$C carbon	$^{14}_7$N nitrogen	$^{16}_8$O oxygen	$^{19}_9$F fluorine	$^{20}_{10}$Ne neon
3 $^{23}_{11}$Na sodium	$^{24}_{12}$Mg magnesium	The transition metals												$^{27}_{13}$Al aluminium	$^{28}_{14}$Si silicon	$^{31}_{15}$P phosphorus	$^{32}_{16}$S sulphur	$^{35\cdot5}_{17}$Cl chlorine	$^{40}_{18}$Ar argon
4 $^{39}_{19}$K potassium	$^{40}_{20}$Ca calcium	$^{45}_{21}$Sc scandium	$^{48}_{22}$Ti titanium	$^{51}_{23}$V vanadium	$^{52}_{24}$Cr chromium	$^{55}_{25}$Mn manganese	$^{56}_{26}$Fe iron	$^{59}_{27}$Co cobalt	$^{59}_{28}$Ni nickel	$^{64}_{29}$Cu copper	$^{65}_{30}$Zn zinc			$^{70}_{31}$Ga gallium	$^{73}_{32}$Ge germanium	$^{75}_{33}$As arsenic	$^{79}_{34}$Se selenium	$^{80}_{35}$Br bromine	$^{84}_{36}$Kr krypton
5 $^{85}_{37}$Rb rubidium	$^{88}_{38}$Sr strontium	$^{89}_{39}$Y yttrium	$^{91}_{40}$Zr zirconium	$^{93}_{41}$Nb niobium	$^{96}_{42}$Mo molybdenum	$^{98}_{43}$Tc technetium	$^{101}_{44}$Ru ruthenium	$^{103}_{45}$Rh rhodium	$^{106}_{46}$Pd palladium	$^{108}_{47}$Ag silver	$^{112}_{48}$Cd cadmium			$^{115}_{49}$In indium	$^{119}_{50}$Sn tin	$^{122}_{51}$Sb antimony	$^{128}_{52}$Te tellurium	$^{127}_{53}$I iodine	$^{131}_{54}$Xe xenon
6 $^{133}_{55}$Cs caesium	$^{137}_{56}$Ba barium	$^{139}_{57}$La lanthanum	$^{178\cdot5}_{72}$Hf hafnium	$^{181}_{73}$Ta tantalum	$^{184}_{74}$W tungsten	$^{186}_{75}$Re rhenium	$^{190}_{76}$Os osmium	$^{192}_{77}$Ir iridium	$^{195}_{78}$Pt platinum	$^{197}_{79}$Au gold	$^{201}_{80}$Hg mercury			$^{204}_{81}$Tl thallium	$^{207}_{82}$Pb lead	$^{209}_{83}$Bi bismuth	$^{210}_{84}$Po polonium	$^{210}_{85}$At astatine	$^{222}_{86}$Rn radon
7 $^{223}_{87}$Fr francium	$^{226}_{88}$Ra radium	$^{227}_{89}$Ac actinium																	

$^{140}_{58}$Ce cerium	$^{141}_{59}$Pr praesodium	$^{144}_{60}$Nd neodimium	$^{147}_{61}$Pm promethium	$^{150}_{62}$Sm samarium	$^{152}_{63}$Eu europium	$^{157}_{64}$Gd gadolinium	$^{159}_{65}$Tb terbium	$^{162}_{66}$Dy dysprosium	$^{165}_{67}$Ho holmium	$^{167}_{68}$Er erbium	$^{169}_{69}$Tm thulium	$^{173}_{70}$Yb ytterbium	$^{175}_{71}$Lu lutetium
$^{232}_{90}$Th thorium	$^{231}_{91}$Pa protactinium	$^{238}_{92}$U uranium	$^{237}_{93}$Np neptunium	$^{242}_{94}$Pu plutonium	$^{243}_{95}$Am americium	$^{247}_{96}$Cm curium	$^{247}_{97}$Bk berkelium	$^{251}_{98}$Cf californium	$^{254}_{99}$Es einsteinium	$^{256}_{100}$Fm fermium	$^{256}_{101}$Md mendelevium	$^{254}_{102}$No nobelium	$^{257}_{103}$Lw lawrencium

The groups The table has eight groups of elements, plus a block of **transition metals**. The eight groups are numbered. Group 4 contains the elements carbon (C), silicon (Si), germanium (Ge), tin (Sn), and lead (Pb). Their atoms each have 4 electrons in the outer shell. The atoms of Group 5 elements each have 5 electrons in the outer shell, and so on. Now look at the last group, Group 0. Their atoms all have *full outer shells*.

Some of the groups have special names:
  Group 1 is often called **the alkali metals.**
  Group 2 is **the alkaline earth metals.**
  Group 7 is **the halogens.**
  Group 0 is **the noble gases**.
Look at the zigzag line through the groups. It separates the **metals** from the **nonmetals**. The metals are on the left.

The periods The horizontal rows in the table are called **periods**. Period 2 contains lithium (Li), beryllium (Be), boron (B), carbon (C), nitrogen (N), oxygen (O), fluorine (F), and neon (Ne).

The transition metals The atoms of these have more complicated electron arrangements. Note that the group contains many common metals, such as iron (Fe), nickel (Ni), and copper (Cu).

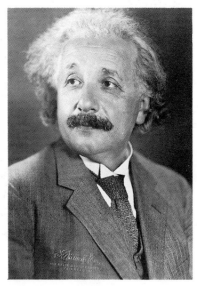

One of the elements in the periodic table is named after this famous scientist. Which one?

1 Explain why beryllium, magnesium and calcium are all in Group 2 of the periodic table.
2 Copy and complete:
  a In Group 4, the atoms have 4 ...
  b In Group ..., the atoms have 6 ...
  c In Group 0, the atoms have ...
3 What are the rows in the table called?
4 Use the larger table on page 298 to help you name the elements in: a Group 5 b Period 1 c Period 3
5 What is the special name for the elements in:
  a Group 1? b Group 7? c Group 0?
6 Draw a large outline of the periodic table and mark in the names and symbols for the first twenty elements. Use the table on page 298 to help you.

2.5 Groups 0 and 7 of the periodic table

On the last two pages you saw that the periodic table shows all the elements arranged in groups.
A group of elements is sometimes called **a family**, because its elements resemble each other. Sometimes they look alike, and usually they behave alike. As you'll see, their behaviour depends mainly on the number of electrons in their atoms' outer shells.

Group 0 – the noble gases

This group contains the elements helium, neon, argon, krypton and xenon. These elements are all:
- nonmetals;
- colourless gases (they occur naturally in air);
- **monatomic** – they exist as single atoms.

The striking thing about these elements is how unreactive they are. They will not normally react with anything.

Why they have similar properties These elements have similar properties because their atoms have full outer electron shells.
A full outer electron shell makes an atom unreactive.

Atoms that don't have a full outer shell react with other atoms in order to obtain one.
The noble gases are unreactive, and monatomic, because their atoms already have full outer electron shells.

Looking for trends The noble gases are not *identical*, however. For example if you filled five balloons with the same volume of each gas at the same temperature and let them go, this is what you'd find:

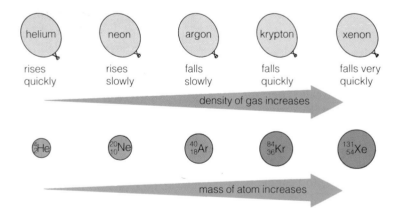

helium — rises quickly
neon — rises slowly
argon — falls slowly
krypton — falls quickly
xenon — falls very quickly

density of gas increases

$^4_2$He $^{20}_{10}$Ne $^{40}_{18}$Ar $^{84}_{36}$Kr $^{131}_{54}$Xe

mass of atom increases

The balloon experiment shows that the 'heaviness' or **density** of the gases increases from helium to xenon. This is an example of a gradual change or **trend** in the periodic table.
As you go down Group 0, the density of the gases increases.

The density increases because the mass of the atoms increases. Each balloon contains exactly the same number of atoms. (More about this when you meet Avogadro!)

All from the same family. Can you see the resemblance? Like humans, the elements in a family or **group** resemble each other.

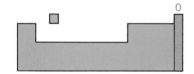

Helium is lighter than air, and unreactive. A good choice for balloons!

Group 7 – the halogens

Chlorine, bromine and iodine are the most common halogens.
- All three are poisonous nonmetals.
- They have coloured vapours – see the diagram below.
- They exist as **diatomic** molecules (each with two atoms).

They also react in similar ways. For example with iron:

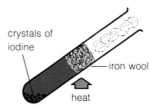

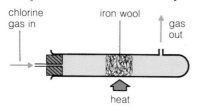

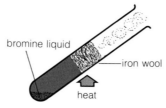

Hot iron wool glows brightly when chlorine passes over it. Brown smoke forms, and a brown solid is left behind.

It's the same with bromine – but this time the iron glows less brightly because bromine is less reactive.

And the same with iodine – but the iron glows even less brightly since iodine is the least reactive of the three.

Why they have similar properties The halogens have similar properties because their atoms all have 7 electrons in the outer shell.

Why they are reactive The halogens are reactive because their atoms are just one electron short of a full outer shell. They can gain this electron by reacting with atoms of other elements.

Why they are diatomic They are diatomic because two atoms can gain full shells by sharing electrons with each other.

Trends within the group Look at these trends:

Atom	Element		Melts at ... (°C)	Boils at ... (°C)	
$^{35}_{17}Cl$	chlorine (green gas)	reactivity decreases as atoms get larger	−101	−35	melting and boiling points increase
$^{80}_{35}Br$	bromine (red liquid)		−7	59	
$^{127}_{53}I$	iodine (black solid)		114	184	

Melting and boiling points *increase* down the group because the attraction between molecules increases. More energy is needed to help them escape from the solid to form liquid, and from the liquid to form gas.

At the same time, reactivity *decreases* because of size. A halogen atom is able to attract an extra electron into its outer shell because of the positive charge on the nucleus. (Opposite charges attract.) But as the atoms get bigger, their outer shells get further from the nucleus. The force of attraction gets less. So the element gets less reactive.

1 Members of a group have similar properties. Why?
2 Explain why: **a** the noble gases are unreactive;
   b their density increases as we go down Group 0.
3 Explain why the halogens are:
   a reactive; **b** *not* monatomic.

4 The first element in Group 7 is fluorine. Would you expect it to be:
   a a gas, a liquid or a solid? Why?
   b coloured or colourless? **c** harmful or harmless?
   d more reactive or less reactive than chlorine?

2.6 Groups 1 and 2 of the periodic table

Group 1 – the alkali metals

The first three elements of Group 1 are lithium, sodium, and potassium. All three:

- are soft metals that can be cut with a knife.
- are so light they float on water.
- are silvery and shiny when freshly cut, but quickly tarnish.
- have low melting and boiling points compared with other metals.

They also react in a similar way. For example:

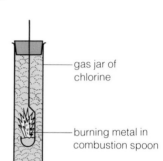

gas jar of chlorine

burning metal in combustion spoon

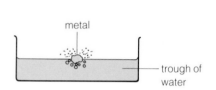

metal

trough of water

They react violently with water to give hydrogen and an alkaline solution. Lithium floats, and sodium shoots across the water, while potassium melts and the hydrogen catches fire.

They burn quickly in chlorine with a bright flame. Potassium burns fastest and sodium next. White solids are left behind. These solids are potassium, sodium and lithium chloride.

Lithium is the least reactive of the three metals – it reacts the most slowly. Potassium is the most reactive of the three.

Why they have similar properties Alkali metals have similar properties because their atoms all have 1 electron in the outer shell.

Trends within the group Look at these trends:

Atom	Metal		Melts at ... (°C)	Boils at ... (°C)	
$^{7}_{3}$Li	lithium		181	1342	
$^{23}_{11}$Na	sodium	reactivity increases	98	883	melting and boiling points decrease
$^{39}_{19}$K	potassium	as atoms get larger	63	760	
$^{85}_{37}$Rb	rubidium		39	686	

As we go down Group 1, reactivity increases while melting and boiling points decrease. (This is the opposite of Group 7.)
Alkali metals react to *lose* an outer electron and obtain a full outer shell. The further the electron is from the nucleus, the easier this is. So the bigger the atom, the more reactive the metal will be.
Meanwhile melting and boiling points *decrease* because the attraction between the atoms gets less strong as the atoms get larger.

Group 2 – the alkaline earth metals

The elements in Group 2 are also silvery metals. But if you compare them with Group 1 metals in the same period, you'll find differences:

Period	Group 1	Group 2		
2			Magnesium is harder than sodium. It reacts only very slowly with water. It has a much higher melting and boiling point.	reactivity increases down both groups
	sodium	magnesium		
	melts at 98 °C	649 °C		
	boils at 883 °C	1107 °C		
3			Calcium is much harder than potassium. It reacts much less vigorously with water. It has a very much higher melting and boiling point.	
	potassium	calcium	But it is more reactive than magnesium, just as potassium is more reactive than sodium.	
	melts at 63 °C	839 °C		
	boils at 760 °C	1484 °C		

From Group 1 to Group 2:

hardness increases

melting and boiling points increase

reactivity decreases

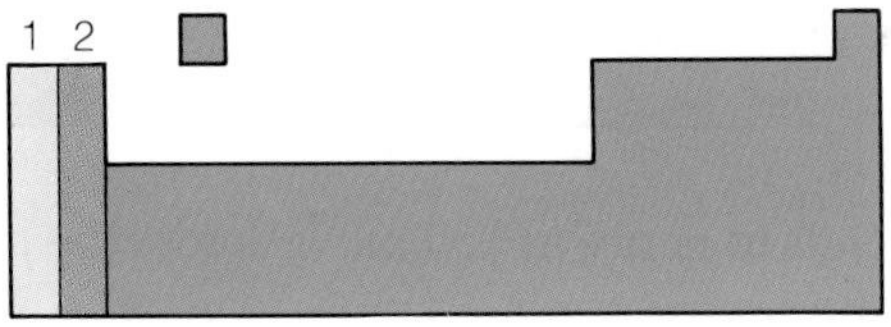

Why alkaline earth metals are less reactive When Group 2 metals react they have to give up *two* outer electrons to obtain a full outer shell. This is more difficult than losing just one electron, so they are less reactive than Group 1 metals.

But as the atoms get bigger it gets easier to lose the two electrons, so the metals get more reactive as you go down the group. This was also the trend in Group 1.

Now look at the melting and boiling points for Group 2, shown on the right. Some of the values seem out of place. But the *overall* trend is a decrease as you go down the group.

Groups 1 and 2 share two trends: increasing reactivity down the group, and an overall decrease in melting and boiling points.

Group 2	Melting point in °C	Boiling point in °C
Beryllium	1278	2970
Magnesium	649	1107
Calcium	839	1484
Strontium	769	1384
Barium	725	1640

The trend is not so smooth. (Magnesium seems out of place, for example.) But the overall trend is a decrease from top to bottom.

1 Sodium is more reactive than lithium. Why?
2 Caesium comes below rubidium in Group 1.
   Compared with rubidium, would you expect it:
   a to have a higher or a lower melting point?
   b to be more reactive or less reactive?

3 Why do the Group 2 metals have similar properties?
4 Magnesium is less reactive than sodium. Why?
5 Lithium is the first element in Group 1 and beryllium the first in Group 2. Which do you think:
   a is more reactive?   b has a higher melting point?

2.7 Across the periodic table

Trends across a period

Look at the elements from Period 3 of the periodic table:

Group	1	2	3	4	5	6	7	0
Element	sodium	magnesium	aluminium	silicon	phosphorus	sulphur	chlorine	argon
Outer electrons	1	2	3	4	5	6	7	8
Element is a ...	metal	metal	metal	metalloid	nonmetal	nonmetal	nonmetal	nonmetal
Reactivity	high ⟶			low ⟶			high	unreactive
Melting point (°C)	98	649	660	1410	590	119	−101	−189
Boiling point (°C)	883	1107	2467	2355	(ignites)	445	−35	−186
Oxide formed ...	Na_2O	MgO	Al_2O_3	SiO_2	P_2O_5	SO_3	Cl_2O_7	none
Oxide is ...	basic		amphoteric				acidic	−

There are several trends to notice as you move across the period:

1 The number of outer-shell electrons increases by 1 each time – like the group number. By argon (Group 0) the shell is full.
2 The elements go from metals to nonmetals. Silicon is in between, like a metal in some ways and a nonmetal in others. It is called a **metalloid**.
3 The metal atoms have few outer shell electrons, and *lose* them during reactions to achieve full shells. So sodium atoms lose 1 electron and aluminium atoms lose 3 (like their group numbers). But the nonmetal atoms react to *gain* or *share* electrons. Group 7 atoms need to share just 1 electron, Group 6 atoms need to share 2, and so on.
4 The melting and boiling points increase to the middle of the period, then decrease again. They are lowest on the right. (Only chlorine and argon are gases at room temperature.)
5 All the elements except argon react with oxygen to form oxides. (Why doesn't argon?) The ratio of oxygen atoms to metal atoms in the oxide increases steadily across the period.
6 The oxides on the left are **basic**, which means they react with acids to form salts. Those on the right are **acidic** – they react with alkalis to form salts. Aluminium oxide is in between – it reacts with both acids and alkalis to form salts. It is called an **amphoteric oxide**. (There is more about oxides on page 226.)

The elements in Period 2 show similar trends.

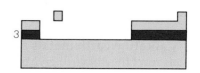

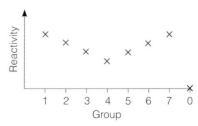

The trend in reactivity across a period

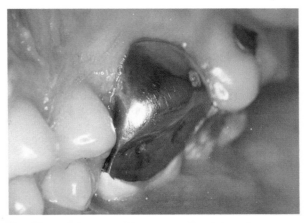

Transition metals are used for, among other things, the 'colour' for glass and gold for teeth.

The transition metals

The long block in the middle of the periodic table contains only metals – the *transition metals*. They have these general properties:

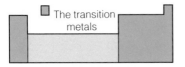
The transition metals

- They are hard and dense, with high melting points.
- They are not very reactive.
- They form coloured compounds. (Compare with Group 1 and 2 metals which form white compounds.)
- They have **variable valency**. That means their atoms can combine in different ratios with atoms of other elements. For example iron forms iron(II) chloride, $FeCl_2$, and iron(III) chloride, $FeCl_3$. The number in brackets shows how many electrons their atoms have lost.
- Many are used in making alloys. (See page 205.)
- Many transition metals and their compounds act as catalysts. (See pages 125 and 129.)

Hydrogen

Hydrogen stands on its own in the periodic table. This is because it has one outer electron, like the Group 1 metals, but unlike them it is a gas, and it usually reacts like a nonmetal.

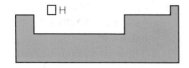
H

Some artificial elements

Not all elements occur naturally. At least 15 are artificial, created by scientists during nuclear reactions. Most of these are in the bottom block of the periodic table, and include the elements neptunium (Np) to lawrencium (Lr) in the very bottom row.

All the isotopes of the artificial elements are **radioactive**. Their atoms have an unstable nucleus which breaks down, giving out radiation.

1 A challenge! Make a table like the one opposite, but for *Period 2*. Show the group numbers, names of elements and numbers of outer electrons. Now try to predict:
   a the melting and boiling points for the elements;
   b the formulae for their oxides.

2 Name ten transition metals. (Page 33 will help!) Now write down one use for each of the ten.

3 Most paints contain compounds of transition elements. Why do you think this is?

4 One transition metal is a liquid at room temperature. Which one?

Questions on Section 2

1 Turn to page 30 and learn the names and symbols for the first twenty elements, in order. Then close the book and write them out.

2 a Give one difference between an element and a compound.
   b The formulae of some compounds are given below. Write down the names of the elements they contain.
   CO_2 $CaCl_2$ H_2S $PbCO_3$ KOH   HgO

3 Hydrogen, deuterium and tritium are isotopes. Their structures are shown below.

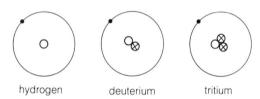

   hydrogen   deuterium   tritium

   a Copy and complete the following key:
   ● represents ....
   ○ represents ....
   ⊗ represents ....
   b What are the mass numbers of hydrogen, deuterium and tritium?
   c Copy and complete this statement:
   Isotopes of an element always contain the same number of .... and .... but different numbers of ....
   d The average mass number of naturally-occurring hydrogen is 1.008. Which isotope is present in the highest proportion, in naturally-occurring hydrogen?

4 Copy and complete the following table for isotopes of some common elements:

Isotope	Name of element	Proton number	Mass number	Number of p	e	n
$^{16}_{8}O$	oxygen	8	16	8	8	8
$^{18}_{8}O$						
$^{12}_{6}C$						
$^{13}_{6}C$						
$^{25}_{12}Mg$						
$^{26}_{12}Mg$						

5 For each of the six elements aluminium (Al), boron (B), nitrogen (N), oxygen (O), phosphorus (P), sulphur (S), write down:
   a the period of the periodic table to which it belongs;
   b its group number in the periodic table;
   c its proton number (atomic number);
   d the number of electrons in one atom;
   e its electronic configuration;
   f the number of outer-shell electrons in one atom.
   Which of the above elements would you expect to have similar properties? Why?

6 The statements below are about metals and non-metals. Say whether each is true or false. (If false, give a reason.)
   a All metals conduct electricity.
   b All metals are solid at room temperature.
   c Nonmetals are good conductors of heat but poor conductors of electricity.
   d Many nonmetals are gases at room temperature.
   e Most metals are brittle and break when hammered.
   f Most nonmetals are ductile.
   g There are about four times as many metals as nonmetals.

7 a Make a larger copy of this outline of the periodic table:

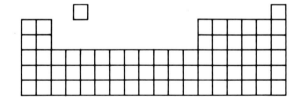

   b Write in the group and period numbers.
   c Draw a zigzag line to show how the metals are separated from the nonmetals in the table.
   d Now put the letters A to L in the correct places in the table, to fit these descriptions:
   A The lightest element.
   B Any noble gas.
   C The element with proton number 5.
   D The element with 6 electrons in its atoms.
   E Any element with 6 outer-shell electrons.
   F The most reactive alkali metal.
   G The least reactive alkali metal.
   H The most reactive halogen.
   I   A Group 3 metal.
   J   An alkaline earth metal.
   K A transition metal.
   L A Group 5 nonmetal.

8 This table gives data for some elements:

Name	Symbol	Melting point (°C)	Boiling point (°C)	Electrical conductivity
Aluminium	Al	660	2450	good
Bromine	Br	−7	58	poor
Calcium	Ca	850	1490	good
Chlorine	Cl	−101	−35	poor
Copper	Cu	1083	2600	good
Helium	He	−270	−269	poor
Iron	Fe	1540	2900	good
Lead	Pb	327	1750	good
Magnesium	Mg	650	1110	good
Mercury	Hg	−39	357	good
Nitrogen	N	−210	−196	poor
Oxygen	O	−219	−183	poor
Phosphorus	P	44	280	poor
Potassium	K	64	760	good
Sodium	Na	98	890	good
Sulphur	S	119	445	poor
Tin	Sn	230	2600	good
Zinc	Zn	419	906	good

Use the table to answer these questions.
a What is the melting point of iron?
b Which element melts at −7 °C?
c Which element boils at 280 °C?
d Which element has a boiling point only 1 °C higher than its melting point?
e Over what temperature range is sulphur a liquid?
f Which element has:
  i the highest melting point?
  ii the lowest melting point?
g Which element remains liquid over the widest temperature range?
h Divide the elements into two groups, one of metals and the other of nonmetals.
i List the *metals* in order of increasing melting point.
j Which metal is a liquid at room temperature (20 °C)?
k Which nonmetal is a liquid at room temperature?
l List the elements which are gases at room temperature. What do you notice about the elements in this list?

9 Before the development of the periodic table, a scientist called Döbereiner discovered sets of three elements with similar properties, which he called **triads**. Some of these triads are shown as members of the same group, in the modern periodic table.
Complete the following triads by inserting the missing middle element:
    chlorine (Cl), ..., iodine (I);
    lithium (Li), ..., potassium (K)
    calcium (Ca), ..., barium (Ba).

10 Many scientists contributed to the development of the modern periodic table. One of them was the Russian chemist Mendeleev. In 1869 he arranged the elements that were then known, in a table very similar to the one in use today. He realized that gaps should be left for elements that had not yet been discovered, and even went so far as to predict the properties of several of these elements.
Rubidium is an alkali metal that lies below potassium in Group 1. Here is some data for Group 1:

Element	Proton number	Melting point (°C)	Boiling point (°C)	Chemical reactivity
Lithium	3	180	1330	quite reactive
Sodium	11	98	890	reactive
Potassium	19	64	760	very reactive
Rubidium	37	?	?	?
Caesium	55	29	690	violently reactive

a Using your knowledge of the periodic table, predict the missing data for rubidium.
b In a rubidium atom:
  i how many electron shells are there?
  ii how many electrons are there?
  iii how many outer-shell electrons are there?

11 This question is about elements from the families called: alkali metals, alkaline earth metals, transition metals, halogens, noble gases.

Element A is a soft, silvery metal which reacts violently in water.
Element B is a gas at room temperature. It reacts violently with other elements, without heating.
Element C is a gas that sinks in air. It does not react readily with any other element.
Element D is a hard solid at room temperature and forms coloured compounds.
Element E conducts electricity and reacts slowly with water. During the reaction its atoms each give up two electrons.

a Place the elements in their correct families. Give further information about the position of the element within the family.
b Describe the outer shell of electrons for each element described above.
c How does the arrangement of electrons in their atoms make some elements very reactive and others unreactive?
d Name elements which fit descriptions A to E.

3.1 Why compounds are formed

Most elements form compounds

On page 36 you saw that sodium reacts with chlorine:

When sodium is heated and placed in a jar of chlorine, it burns with a bright flame.

The result is a white solid, which has to be scraped from the sides of the jar.

The white solid is called **sodium chloride**. It is formed by atoms of sodium and chlorine joining together, so it is a **compound**. The reaction can be described like this:

sodium + chlorine ⟶ sodium chloride

The + means *reacts with*, and the ⟶ means *to form*. **Like sodium and chlorine, most elements react to form compounds.**

The noble gases do not usually form compounds

The noble gases, however, do *not* form compounds, as you saw on page 34. For this reason their atoms are described as **unreactive** or **stable**. They are stable because their outer electron shells are *full*: **A full outer shell makes an atom stable.**

Helium atom, full outer shell: *stable*

Neon atom, full outer shell: *stable*

Argon atom, full outer shell: *stable*

Only the noble gas atoms have full outer shells. The atoms of all other elements have incomplete outer shells. That is why they react. **By reacting with each other, atoms can obtain full outer shells and so become stable.**

When atoms react, they *lose* or *gain* or *share* electrons to form full shells. These electrons are called **valency electrons**. The **valency** of an element tells you the number of electrons its atoms lose or gain or share.

Several of the noble gases are used in lighting. Xenon is used in lighthouse lamps, like this one. It gives a beautiful blue light.

Losing or gaining electrons

The atoms of some elements can obtain full shells by *losing* or *gaining* electrons, when they react with other atoms:

Losing electrons The sodium atom is a good example. It has just 1 electron in its outer shell. It can obtain a full outer shell by losing this electron to another atom. The result is a **sodium ion**:

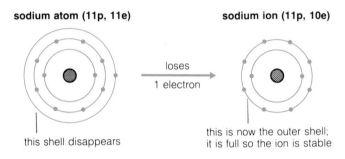

sodium atom (11p, 11e) sodium ion (11p, 10e)

loses 1 electron

this shell disappears

this is now the outer shell; it is full so the ion is stable

The charge on the sodium ion:	
the charge on 11 protons is	$+11$
the charge on 10 electrons is	-10
total charge	$+1$

The sodium ion has 11 protons but only 10 electrons, so it has a charge of $+1$, as you can see from the box on the right above.

The symbol for sodium is Na, so the sodium ion is called **Na$^+$**. The $^+$ means *1 positive charge*. Na$^+$ is a **positive ion**.

Gaining electrons A chlorine atom has 7 electrons in its outer shell. It can reach a full shell by accepting just 1 electron from another atom. It becomes a **chloride ion**:

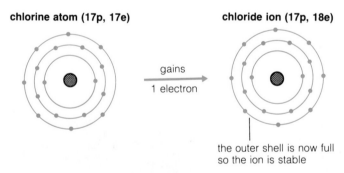

chlorine atom (17p, 17e) chloride ion (17p, 18e)

gains 1 electron

the outer shell is now full so the ion is stable

The charge on the chloride ion:	
the charge on 17 protons is	$+17$
the charge on 18 electrons is	-18
total charge	-1

The chloride ion has a charge of -1, so it is a **negative ion**. Its symbol is **Cl$^-$**.

Ions Any atom becomes an ion if it loses or gains electrons. **An ion is a charged particle. It is charged because it contains an unequal number of protons and electrons.**

1 What is another word for *unreactive*?
2 Why are the noble gas atoms unreactive?
3 Explain why all other atoms are reactive.
4 Draw a diagram to show how a sodium atom obtains a full outer shell.
5 Explain why the sodium ion has a charge of +1.
6 What is the valency of sodium?

7 Draw a diagram to show how a chlorine atom can obtain a full outer shell.
8 Write down the symbol for a chlorine ion. Why does this ion have a charge of –1?
9 What is the valency of chlorine?
10 Explain what an ion is, in your own words.
11 Why do noble gas atoms *not* form ions?

3.2 The ionic bond

The reaction between sodium and chlorine

As you saw on the last page, a sodium atom can lose one electron, and a chlorine atom can gain one, to obtain full outer shells. So when a sodium atom and a chlorine atom react together, the sodium atom loses its electron *to the chlorine atom*, and two ions are formed. Here, sodium electrons are shown as • and chlorine electrons as ×, but remember that all electrons are exactly the same:

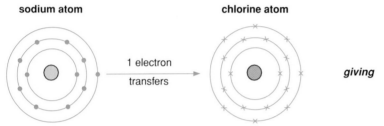

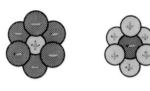

stable ions with full shells

The two ions have opposite charges, so they attract each other. The force of attraction between them is strong. It is called an **ionic bond**, or sometimes an **electrovalent bond**.

How solid sodium chloride is formed

When sodium reacts with chlorine, billions of sodium and chloride ions form, and are attracted to each other. But the ions do not stay in pairs. Instead, they cluster together, so that each ion is surrounded by six ions of opposite charge. They are held together by strong ionic bonds. (Each ion forms six bonds.)
The pattern grows until a giant structure of ions is formed. It contains equal numbers of sodium and chloride ions.
This giant structure is the compound **sodium chloride**, or **salt**.

Because sodium chloride is made of ions, it is called an **ionic compound**. It contains one Na^+ ion for each Cl^- ion, so its formula is **NaCl**.
The charges in the structure add up to zero:

the charge on each sodium ion is	+1
the charge on each chloride ion is	−1
total charge	0

The compound therefore has no overall charge.

These polystyrene spheres have been given opposite charges, so they attract each other. The same happens with ions of opposite charge.

Other ionic compounds

Sodium is a **metal**, and chlorine is a **nonmetal**. They react together to form an **ionic compound**. Other metals can also react with non-metals to form ionic compounds. Below are two more examples.

Magnesium and oxygen   A magnesium atom has 2 outer electrons and an oxygen atom has 6. Magnesium burns fiercely in oxygen. During the reaction, each magnesium atom loses its 2 outer electrons to an oxygen atom. Magnesium ions and oxide ions are formed:

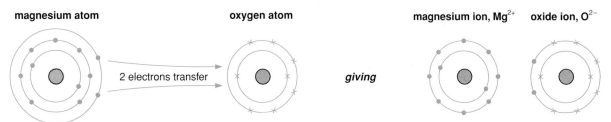

The ions attract each other because of their opposite charges. Like the ions on the last page, they group together into a giant ionic structure. The resulting compound is called **magnesium oxide**. Magnesium oxide contains one magnesium ion for each oxide ion, so its formula is **MgO**. The compound has no overall charge:

the charge on each magnesium ion is	2+
the charge on each oxide ion is	2−
total charge	0

Magnesium and chlorine   To obtain full outer shells, a magnesium atom must lose 2 electrons, and a chlorine atom must gain 1 electron. So when magnesium burns in chlorine, each magnesium atom reacts with *two* chlorine atoms, to form **magnesium chloride**:

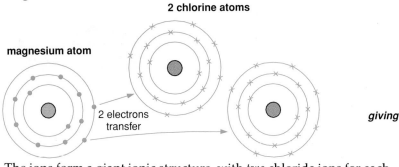

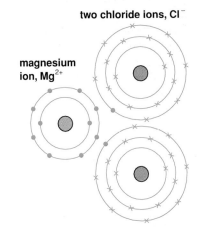

The ions form a giant ionic structure, with *two* chloride ions for each magnesium ion. The formula of magnesium chloride is therefore **MgCl₂**. The compound has no overall charge. Can you explain why?

1 Draw a diagram to show what happens when a sodium atom reacts with a chlorine atom.
2 What is an ionic bond? What is the other name for it?
3 Sketch the structure of sodium chloride, and explain why its formula is NaCl.

4 Explain why:
  a a magnesium ion has a charge of 2+;
  b the ions in magnesium oxide stay together;
  c magnesium chloride has no overall charge;
  d the formula of magnesium chloride is MgCl₂.

3.3 Some other ions

Ions of the first twenty elements

Not every element forms ions during reactions. In fact, out of the first twenty elements in the periodic table, only twelve easily form ions. These ions are given below, with their names.

Group 1	2	H^+ hydrogen		3	4	5	6	7	0 none
Li^+ lithium	Be^{2+} beryllium			none	none	N^{3-} nitride	O^{2-} oxide	F^- fluoride	none
Na^+ sodium	Mg^{2+} magnesium			Al^{3+} aluminium	none	none	S^{2-} sulphide	Cl^- chloride	none
K^+ potassium	Ca^{2+} calcium	transition metals							

Note that hydrogen and the metals form **positive ions**, which have the same names as the atoms. The nonmetals form **negative ions** and their names end in *-ide*.

The elements in Group 4 do not form ions, because their atoms would have to gain or lose four electrons, and that takes too much energy. The elements in Group 0 do not form ions because their atoms already have full shells.

The names and formulae of their compounds

The names   To name an ionic compound, you just put the names of the ions together, with the positive one first:

Ions in compound	Name of compound
K^+ and F^-	Potassium fluoride
Ca^{2+} and S^{2-}	Calcium sulphide

The formulae   The formulae of ionic compounds can be worked out by the following steps. Two examples are given below.

1 Write down the name of the ionic compound.
2 Write down the symbols for its ions.
3 The compound must have no overall charge, so **balance** the ions, until the positive and negative charges add up to zero.
4 Write down the formula without the charges.

Bath salts contain Na^+ and CO_3^{2-} ions. Epsom salts contain Mg^{2+} and SO_4^{2-} ions. Can you give the names and formulae of the three main compounds contained in these cartons?

Example 1
1 Lithium fluoride.
2 The ions are Li^+ and F^-.
3 One Li^+ is needed for every F^-, to make the total charge zero.
4 The formula is LiF.

Example 2
1 Sodium sulphide.
3 The ions are Na^+ and S^{2-}.
3 Two Na^+ ions are needed for every S^{2-} ion, to make the total charge zero: $Na^+ Na^+ S^{2-}$.
4 The formula is Na_2S. (What does the $_2$ show?)

Transition metal ions

Some transition metals form only one type of ion:
- silver forms only Ag^+ ions
- zinc forms only Zn^{2+} ions

but most of them can form more than one type. For example, copper and iron can each form two:

Ion	Name	Example of compound
Cu^+	copper(I) ion	copper(I) oxide, Cu_2O
Cu^{2+}	copper(II) ion	copper(II) oxide, CuO
Fe^{2+}	iron(II) ion	iron(II) chloride, $FeCl_2$
Fe^{3+}	iron(III) ion	iron(III) chloride, $FeCl_3$

The (II) in a name shows that the ion has a charge of $2+$. What do the (I) and (III) show?

Compound ions

So far, all the ions have been formed from single atoms. But ions can also be formed from groups of joined atoms. These are called **compound ions**. The most common ones are shown on the right.
Remember, each is just one ion, even though it contains more than one atom.
The formulae for their compounds can be worked out as before. Some examples are shown below.

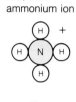

NH_4^+, the ammonium ion

OH^-, the hydroxide ion

NO_3^-, the nitrate ion

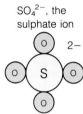

SO_4^{2-}, the sulphate ion

CO_3^{2-}, the carbonate ion

HCO_3^-, the hydrogen carbonate ion

Example 3
1 Sodium carbonate.
2 The ions are Na^+ and CO_3^{2-}.
3 Two Na^+ are needed to balance the charge on one CO_3^{2-}.
4 The formula is Na_2CO_3.

Example 4
1 Calcium nitrate.
2 The ions are Ca^{2+} and NO_3^-.
3 Two NO_3^- are needed to balance the charge on one Ca^{2+}.
4 The formula is $Ca(NO_3)_2$. Note that a bracket is put round the NO_3, before the $_2$ is put in.

1 Explain why a calcium ion has a charge of $2+$.
2 Why is the charge on an aluminium ion $3+$?
3 Write down the symbols for the ions in:
  a potassium chloride b calcium sulphide
  c lithium sulphide d magnesium fluoride
4 Now work out the formula for each compound in question 3.
5 Work out the formula for each compound:
  a copper(II) chloride b iron(III) oxide
6 Write a name for each compound: $CuCl$, FeS, Na_2SO_4, $Mg(NO_3)_2$, NH_4NO_3, $Ca(HCO_3)_2$
7 Work out the formula for:
  a sodium sulphate b potassium hydroxide
  c silver nitrate d ammonium nitrate

3.4 The covalent bond

Sharing electrons

When two nonmetal atoms react together, *both of them need to gain electrons* to reach full shells. They can manage this only by sharing electrons between them. Atoms can share only their outer electrons, so just the outer electrons are shown in the diagrams below.

A model of the hydrogen molecule. The molecule can be shown as H—H. The line represents a single bond.

Hydrogen   A hydrogen atom has only one electron. Its shell can hold two electrons, so is not full. When two hydrogen atoms get close enough, their shells overlap and then they can share electrons:

two hydrogen atoms   **a hydrogen molecule, H₂**

a shared pair of electrons

Because the atoms share electrons, there is a strong force of attraction between them, holding them together. This force is called a **covalent bond**. The bonded atoms form a **molecule**.
A molecule is a small group of atoms which are held together by covalent bonds.
Hydrogen gas is made up of hydrogen molecules, and for this reason it is called a **molecular** substance. Its formula is **H₂**.
Several other nonmetals are also molecular. For example:

chlorine, Cl_2	iodine, I_2	oxygen, O_2
nitrogen, N_2	sulphur, S_8	phosphorus, P_4

In H_2, the $_2$ tells you the number of hydrogen atoms in each molecule.
Because it has two atoms in each molecule, hydrogen is described as **diatomic**.

Chlorine   A chlorine atom needs a share in one more electron, to obtain a full shell. So two chlorine atoms bond covalently like this:

two chlorine atoms   **a chlorine molecule, Cl₂**

A model of the chlorine molecule

Oxygen   The formula for oxygen is O_2, so each molecule must contain two atoms. Each oxygen atom has only six outer electrons; it needs a share in two more to reach a full shell:

two oxygen atoms   **an oxygen molecule, O₂**

two shared pairs of electrons

Since the oxygen atoms share two pairs of electrons, the bond between them is called a **double covalent bond**, or just a **double bond**.

A model of the oxygen molecule. The molecule can be shown as O=O. The lines represent a double bond.

Covalent compounds

On the opposite page you saw that several nonmetal *elements* exist as molecules. A huge number of *compounds* also exist as molecules. In a molecular compound, atoms of *different* elements share electrons with each other. These compounds are often called **covalent compounds** because of the covalent bonds in them. Water, ammonia and methane are examples.

Water   Its formula is H_2O. Each oxygen atom shares electrons with two hydrogen atoms, and they all reach full shells.

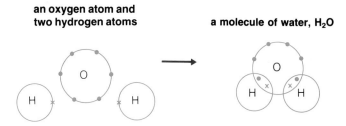

A model of the water molecule

Ammonia   Its formula is NH_3. Each nitrogen atom shares electrons with three hydrogen atoms, and they all reach full shells:

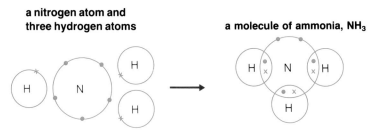

A model of the ammonia molecule

Methane   Its formula is CH_4. Each carbon atom shares electrons with four hydrogen atoms, and they all obtain full shells:

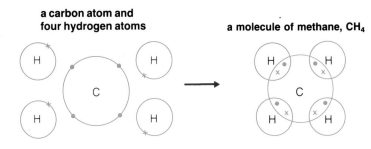

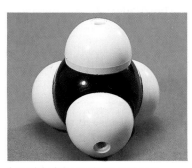

A model of the methane molecule

1 Name the bond formed when atoms share electrons.
2 What is a molecule?
3 Give: **a** five examples of molecular elements
   b three examples of diatomic gases
4 Draw a diagram to show the bonding in:
   a chlorine   **b** oxygen
5 The bond between two oxygen atoms is called a double bond. Why?

6 **a** How many electrons must nitrogen share, to gain a full outer shell?
   b Draw a diagram to show the bonding in a nitrogen molecule. (Then check page 215.)
   c The bond is called a **triple bond**. Why?
7 Show the bonding in a molecule of water.
8 Hydrogen chloride (HCl) is molecular. Draw a diagram to show its bonding. (Then check page 234.)

3.5 Ionic and molecular solids

On page 10 you saw that solids are made of particles packed closely together in a regular pattern. This regular arrangement is called a **lattice**. If the particles are **ions**, the solids are called **ionic solids**. If they are **molecules**, the solids are **molecular solids**. The two types of solid have quite different properties, as you will see below.

> Some ionic solids you can find in the laboratory:
>
> sodium chloride
> sodium hydroxide
> copper(II) sulphate
> iron(II) chloride
> silver nitrate

Ionic solids

One of the most common ionic solids is sodium chloride – it is ordinary table salt. You already know quite a lot about its structure:

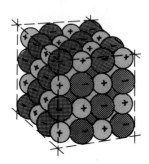

In sodium chloride, the sodium and chloride ions are packed in a lattice like this, and held by strong **ionic bonds**.

The result is a **crystal** with straight edges and flat faces. This is a crystal of sodium chloride, magnified 35 times.

The crystals look white and shiny. A box of table salt contains millions of them.

Sodium chloride is typical of ionic solids. In *all* ionic solids, the ions are packed in a lattice, and held together by strong ionic bonds. This means that all ionic solids are **crystalline**.

Their properties   Ionic solids have these properties:

1 They have high melting and boiling points. For example:

Substance	Melting point/°C	Boiling point/°C
sodium chloride	801	1413
magnesium oxide	2852	3600
aluminium oxide	2072	2980

This is because ionic bonds are very strong, so it takes a lot of heat energy to break up the lattice and form a liquid. In fact all ionic substances are solid at room temperature.

2 They shatter when hit with a hammer: they are **brittle**.
3 They are usually soluble in water, but insoluble in other solvents such as tetrachloromethane and petrol.
4 They do not conduct electricity when solid. Electricity is a stream of moving charges. Although the ions *are* charged, they cannot move, because the ionic bonds hold them firmly in position. (You will learn more about this in Section 7.)
5 They *do* conduct electricity when they are melted or dissolved. This is because the ions are then free to move.

> **Making use of melting points**
>
> - Their high melting points mean some ionic solids are useful as **refractory** (heat resistant) materials.
>
> - For example aluminium oxide is used to make laser tubes, and bricks for lining furnaces.

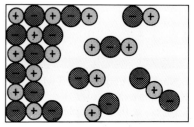

An ionic solid melting – eventually!

Molecular solids

Iodine is a good example of a molecular solid. Each iodine molecule contains two atoms, held together by a strong covalent bond.

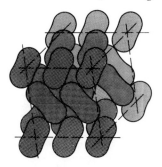  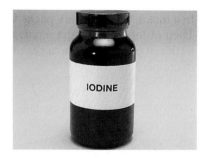

Here ▭ represents an iodine molecule. The molecules are packed in a lattice, and held together by weak forces.

The pattern repeats millions of times, and the result is a crystal. Above is a single iodine crystal, magnified 15 times.

Iodine crystals are grey-black, and shiny, and brittle. This jar contains hundreds of thousands of them.

Iodine is typical of molecular solids. In *all* molecular solids, the molecules are held in a lattice. So the solids are crystalline. The forces that hold the molecules together are weak – much weaker than the forces *within* the molecules.

Their properties Molecular solids have these properties:

1 They have low melting points and boiling points – much lower than ionic solids do. This is because it doesn't take much heat energy to overcome the weak forces between molecules. In fact many molecular substances melt, and even boil, *below* room temperature, so are liquids or gases at room temperature. Here are some examples:

Substance	Melting point/°C	Boiling point/°C
Oxygen	−219	−183
Chlorine	−101	−35
Water	0	100
Naphthalene	80	218

Some molecular substances and their state at room temperature:

Solids iodine, sulphur, naphthalene

Liquids bromine, water, ethanol

Gases oxygen, nitrogen, carbon dioxide

2 They shatter when hit with a hammer: they are brittle.
3 Unlike ionic solids, molecular solids are usually insoluble in water, but soluble in solvents such as tetrachloromethane and petrol.
4 They do not conduct electricity. Molecules are not charged, so molecular substances cannot conduct, even when melted.

1 What is:
 a an ionic solid? b a molecular solid?
 Give three examples of each.
2 What is another name for molecular solids?
3 Explain how a crystal of sodium chloride is formed. Can you think of a reason why its faces are flat?
4 Why do ionic solids have high melting points?
5 List four properties of covalent solids.
6 Explain why many molecular substances are gases or liquids at room temperature, and give four examples.
7 You can buy solid air-fresheners in shops. Do you think these substances are ionic or covalent? Why?

3.6 Metals

The metallic bond

In a metal, the atoms are packed tightly together in a regular pattern. Their outer electrons get separated from the atoms. The result is a lattice of positive ions in a sea of electrons. Look at the arrangement in copper:

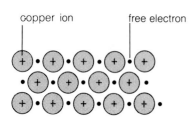

copper ion   free electron

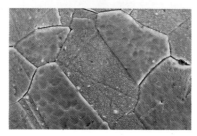

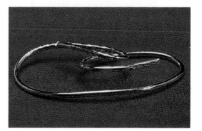

The copper ions are held together by their attraction to the electrons between them. The strong forces of attraction are called **metallic bonds**.

The regular arrangement of atoms results in **crystals** of copper. This shows the crystals in a piece of copper magnified 1000 times.

Metal crystals are called **grains**. A piece of copper wire like this contains millions of tiny grains. You'd need a microscope to see them.

The way the atoms are packed together depends on the metal. There are several possible arrangements. But the arrangements are always regular so they always form grains.

The properties of metals

All metals are different. But they all share these general properties to some extent:

1   They are usually hard. They won't cut or dent easily. (But think of sodium.)
2   They are usually **tough**. When you hammer them they won't break up easily. *Tough* is the opposite of *brittle*.
3   They usually have high **compressive strength**. You have to compress (squash) them very hard before they deform.
4   They usually have high **tensile strength**. You have to put them under a lot of tension before they stretch and break.
5   But with enough force, most metals can be bent or hammered into shape – they are **malleable** – or drawn into wires – they are **ductile**. This is because the layers of atoms can slide over each other without the metallic bonds breaking. The bonds just rearrange.
6   They are good conductors of electricity, because the free electrons can move through the lattice carrying charge.
7   They are good conductors of heat. The free electrons take in heat energy, which makes them move faster. They spread it through the lattice.
8   They usually have high melting points because it takes a lot of heat energy to break the strong metallic bonds. For example, copper melts at 1083 °C. (But think of mercury.)

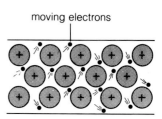

moving electrons

Metals contain electrons that are free to move, so they are good conductors of heat and electricity.

9 They are usually 'heavy' or **dense**, because the atoms are packed together tightly. But compare lead and aluminium:

Lead atoms have almost eight times the mass of aluminium atoms. They are tightly packed which makes lead dense.

Although much lighter than lead atoms, aluminium atoms are not that much smaller. So aluminium is much less dense.

Calculating density

$$\text{Density} = \frac{\text{mass (g)}}{\text{volume (cm}^3)}$$

The density of aluminium is $2.7\,\text{g/cm}^3$. The density of lead is $11.34\,\text{g/cm}^3$.

More about metal grains

The grains in metals are not perfect. If you could slice through a metal and look at the atoms in the grains, you'd see something like this.

Grain boundary   This is the edge of the grain, where it meets other grains. Here the atoms are disorderly. Note that atoms *inside* grains are all packed in the same way, but the grains are at different angles.

Vacant site   Here an atom is missing. A neighbouring atom can move into the space. In effect a space can 'diffuse' along a row. This happens more easily when the metal is warmed. The process helps to remove strains within the metal, and this makes the metal less brittle.

Dislocation   Here a whole row of atoms is missing. When you apply force to a metal, the layers of atoms slide over each other at dislocations, as shown below.

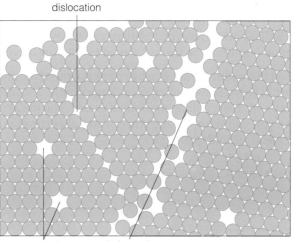

dislocation

vacant sites   grain boundary

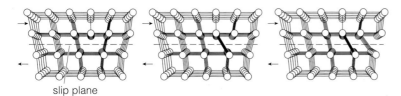

slip plane

Manufacturers want metals to be malleable and ductile so that they can be shaped easily. But they also want them to be hard and strong, so that they are useful. You can find out how they make metals stronger in the next Unit.

---

1 Describe the bonding in metals.
2 Name two properties of metals that depend on the free electrons.
3 What do they mean?   **a** high compressive strength   **b** high tensile strength   **c** malleable   **d** ductile.

4 Why is lead so dense?
5 What are metal crystals usually called?
6 What is a *dislocation*?
7 Dislocations help to make metals malleable and ductile. Explain why.

3.7 Making metals stronger

If the layers in a metal slip over each other easily, the metal will be easily bent and stretched. No good for planes, cars, or bridges! But there are several things manufacturers can do to make a metal harder and stronger. They can:

1 control the **grain size**;
2 add other substances to it to form **alloys**;
3 **work harden** it;
4 put it through **heat treatment**.

We will look at each of these in turn.

1 Grain size

When you apply force to a metal, the layers of atoms slide over each other at dislocations. A dislocation may move to the edge of the grain (page 53). It can even cross into the next grain, but that takes much more force. In effect, the grain boundary acts as a barrier.

Smaller grains mean dislocations can't move so far before they hit a grain boundary. So the metal won't stretch or bend so easily.

You can control grain size by controlling the rate at which a molten metal cools. Slow cooling gives large grains. The metal is **coarse-grained**. Fast cooling gives small grains. The metal is **fine-grained**. As you will see opposite, you can also alter the grain size by heating a metal *without* melting it.

Metals used to build rockets must be light but very strong – and that means alloys. The DC rocket, designed to put satellites into orbit, is almost 36 metres long.

Slow cooling: coarse grains

Fast cooling: fine grains

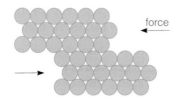

Pure metal: layers slide over each other

2 Making alloys

An alloy is formed by mixing at least one other element with the molten metal, and letting the mixture solidify. The added atoms spread through the crystal structure. Because they are a different size they prevent layers from moving, as you can see on the right.

Alloys are used instead of pure metal for most engineering purposes. For example, commercially pure aluminium is very light, which is an advantage. But it is too soft and ductile to stand up to much stress. It is usually alloyed with small amounts of other metals like these:

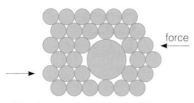

Alloy: layers can no longer slide

Name of alloy	Made of aluminium plus...	Properties	Used for
3103 H4	1.25% manganese	Strong, tough	Building, packaging
7075 TF	6% zinc 2.5% magnesium 1.25% copper	Very high strength/weight ratio	Aircraft

By alloying, aluminium can be made up to four times stronger than mild steel. You can find out more about alloys on pages 204 and 209.

3 Work hardening

Molten metals and alloys are usually allowed to harden as big blocks or **ingots**. These are then worked into the required shape. For example, they are **rolled** between heavy rollers into slabs, plates, and sheets. Plates and sheets are pressed or **forged** into different shapes in huge presses.

When a metal is worked at room temperature, its grains get deformed and more dislocations are introduced. But this doesn't mean the layers slide over each other more easily. The dislocations all try to move in different directions, so they get into a jam and can't move at all. The metal has been **work hardened**.

There comes a point when the metal can't be worked any more. It is too brittle. If more force is applied it will crack.

A process called **annealing** is used to solve this problem. The worked metal is heated in a furnace. Heat causes metal atoms to diffuse within the grains and the grains to recrystallize. The result is that strains are removed, so the metal can be worked again.

By carefully controlling the annealing process, grains can be made to recrystallize smaller than before. This makes the metal stronger. By repeating the cycle of work hardening and annealing, a metal can be brought to exactly the strength and thickness that's required.

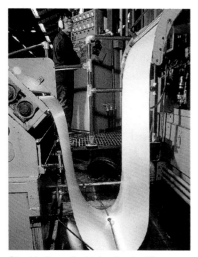

Steel being rolled into sheets. The process puts the grains under a lot of pressure.

4 Heat treatment

Heat treatment is used to harden **alloys**. It involves:
- heating the alloy to a high temperature;
- cooling it very quickly, for example in oil or cold water. This is called **quenching**;
- reheating the material to a lower temperature.

How does it work? It uses the fact that at high temperatures, metal grains can hold *more* of the alloying substances in their structures.

If the grains are cooled very fast, they may continue to hold more of the substances than usual. That's what happens when carbon steel is quenched. But the resulting steel is very brittle. Reheating makes it less brittle. This is called **tempering** the steel.

When aluminium alloys are quenched they behave differently. The extra material precipitates out as very fine particles inside the grains or at their boundaries. This helps to harden and strengthen the alloy. Reheating increases the precipitation. The reheating process for aluminium is called **artificial ageing**.

The heat treatment of carbon steel

- Carbon steel is an alloy of iron and carbon.

- At high temperatures iron crystals are in a form called **austenite**. These can hold around 2% carbon.

- If they are cooled slowly they turn into a form called **ferrite** which can hold only 0.03% carbon.

- But if they are cooled very fast they turn into **martensite** which can hold 0.6% carbon, making the alloy harder than the ferrite form.

1 List four ways to make a metal harder.
2 How would you produce a metal with large grains?
3 A metal with large grains is softer than the same metal with fine grains. Why?
4 Why is an alloy stronger than the pure metal?

5 When it is rolled or forged at room temperature a metal gets *work hardened*. What does that mean?
6 What is *annealing*? Why is it needed?
7 What are the stages in heat treatment?
8 Explain how heat treatment works for carbon steel.

3.8 Clues from diamond and graphite

Where does a solid get its properties? You already have some clues. A study of diamond and graphite will give you more. These two substances look different and have different properties. But both are made *only* of carbon atoms. So how do the differences arise?

> **Allotropes**
>
> Some elements can exist in different forms in the same state. These are called **allotropes**. Diamond and graphite are allotropes of carbon.

Diamond   In diamond, the carbon atoms form a lattice:

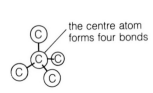

the centre atom forms four bonds

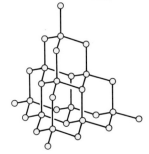

A carbon atom forms covalent bonds to *four* others, as shown above. Each outer atom then bonds to three more, and so on.

Eventually millions of carbon atoms are bonded together, in a giant covalent structure. This is part of it.

The result is a single crystal of diamond like this one. It has been cut, shaped and polished to make it sparkle.

Graphite   In graphite, the carbon atoms form flat sheets:

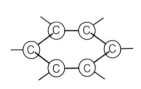

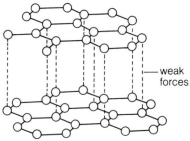

weak forces

Each carbon atom forms covalent bonds to *three* others. This gives rings of six atoms.

These join to make flat sheets that lie on top of each other, held together by weak forces.

Even at low magnification the layered structure of graphite shows up clearly.

Now compare their properties:

Property	Because ...	So it is used ...
Diamond		
the hardest substance known, with a very high melting point (3550 °C)	the bonds in it are very strong	in drilling and cutting tools
does not conduct electricity	there are no ions or free electrons to carry charge	
sparkles when cut	it has a regular structure	for jewellery
Graphite		
soft and slippery	the layers can slide over each other	as a lubricant for locks (and in some car engines) and as 'lead' in pencils
conducts electricity (the only nonmetal that does!)	each atom has four outer electrons, but forms only three bonds. The fourth electron can move through the graphite, carrying charge.	for electrodes, and as connecting 'brushes' in dynamos and motors

The properties of solids

You saw opposite that the different properties of diamond and graphite are a result of the different ways their atoms are bonded and arranged. We can extend that idea to *all* solids.

The properties of a solid depend on three things:
1 what particles it is made of;
2 how the particles are bonded together;
3 how they are arranged.

For example:

Lead is dense because its atoms are heavy and tightly packed. It conducts electricity because of its free electrons. It can be rolled into sheets because the layers of atoms slide over each other.

Sand is impure silicon(IV) oxide. Each silicon atom forms covalent bonds to four oxygen atoms, making a giant structure. So sand is hard. Embedded in paper, it is ideal for sanding.

Rubber is made of long molecules, mostly carbon atoms. It is elastic because the molecules are coiled but they stretch when you pull. There is no rigid lattice.

Drawing conclusions from properties

If you know the properties of a solid, you can tell quite a lot about the atoms in it, and how they are bonded and arranged. For example, if a solid . . .

- conducts electricity, it must be a metal (unlesss it's graphite!);
- is hard and rigid and strong, and doesn't conduct electricity even when melted, it must have a giant structure of atoms held together by covalent bonds;
- is flexible (easily bent) and doesn't conduct electricity, it must consist of separate molecules held together by the forces of attraction between them;
- is elastic, its molecules must be able to coil and uncoil.

In the next Unit, see how glass and plastics fit these rules.

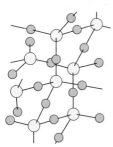

The structure of silicon(IV) oxide, the main compound in sand. It melts at 1723 °C. (Not all the atoms are shown.)

---

1 Draw a diagram to show the structure of diamond. Use it to explain why diamond:
   a is very hard   b does not conduct electricity
2 Draw a diagram to show the structure of graphite. Use it to explain why graphite:
   a conducts electricity   b can be used as a lubricant.
3 State three things that affect the properties of a solid.

4 This is about silicon(IV) oxide.
   a In what ways is its structure like diamond's?
   b i Its melting point is high. Why?
     ii But it is lower than diamond's. Suggest a reason.
   c What can you say about its conductivity?
5 Predict, with reasons, the type of structure you'd find in:
   a a kitchen tile   b chewing gum   c chalk

3.9 Comparing glass and plastics

Glass and plastics have very different properties. That suggests they have very different structures too. Let's see!

Glass

There are several kinds of glass, but the main ingredient of all of them is **sand**. The most common glass around the house is **soda-lime** glass. It is made by heating sand, limestone (calcium carbonate), sodium carbonate, and recycled glass in a furnace to around 1300 °C.

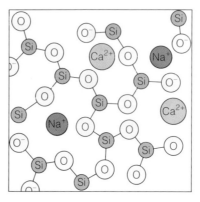

In the furnace the carbonates break down to give a molten mixture of oxides. On cooling, this forms a giant irregular structure made of silicon and oxygen atoms . . .

. . . with sodium and calcium ions trapped inside. This represents a cross-section through it. Each silicon atom is covalently bonded to *four* oxygen atoms, but one is hidden from view.

While still soft, the glass can be moulded and blown into shape to make items such as bottles, jars, and drinking glasses. Or it can be cooled as flat sheets, for making windows.

The properties of glass

1 It is hard and will not bend, because of the strong covalent bonds in the giant structure.
2 Because the bonding is covalent, it does not conduct electricity. (But *molten* glass can conduct. Why is this?)
3 It is a good conductor of heat. Think of glass casseroles!
4 It can withstand being squashed without crumbling: it is strong under compression.
5 But it can't withstand being stretched. It is weak under tension and will crack.
6 It will shatter when dropped: it is brittle. Its structure is irregular (unlike diamond) and weaker at some points than others. Energy from an impact spreads through the structure causing it to break at the weak points.
7 For the same reason a sudden temperature change may shatter it.
8 It is unreactive because of its stable covalent bonds. It is not affected by water, oxygen, or most chemicals, which is useful. Why?
9 It has a high melting point because of its strong covalent bonds and giant structure. It melts over a range of temperature because it is not a pure substance.

Plastics

There are lots of different plastics. Polythene is one example.

Ethene is a gas. Its molecules have double bonds between carbon atoms. These break and the molecules join to form long carbon chains...

...in a **polymerization** reaction. The result is polythene, a solid. The chains can be thousands of atoms long. They are all tangled together like this.

Polythene is used to make plastic bags, bottles, dustbins, and cling film. You can make it 'heavy' or 'light' by changing the reaction conditions.

The properties of plastics

1 They are all carbon compounds.
2 Their molecules are long chains of atoms, made by joining lots of small molecules together. Carbon atoms form the spine.
3 Because they are molecular compounds, they do not conduct electricity. They are electrical and thermal insulators.
4 Plastics like polythene are flexible because they don't contain a rigid structure. For the same reason they are not brittle: they don't shatter when you drop them.
5 They are light compared with glass. But they are also strong because their molecules are so long. The longer the molecules, the larger the force of attraction between them. (Some plastics are used to make climbing ropes.)
6 If you pull them hard enough they stretch: the chains get pulled apart from each other. With enough force they will tear because the chains separate completely. (But as you'll see in Unit 16.6, the chains in some plastics are cross-linked, which makes them strong and rigid, more like glass. Those plastics, called **thermosets**, don't bend or stretch.)
7 Plastics are quite unreactive because of their stable covalent bonds. Most are not affected by water, oxygen, or other chemicals, which is useful. But some catch fire easily and may give out poisonous gases when they burn. (PVC gives out hydrogen chloride.)
8 Because they don't have a giant structure, plastics like polythene soften and melt at quite low temperatures. They melt over a range of temperatures because they contain chains of different lengths.

You can find out how plastics are made in Units 6.2 and 16.6.

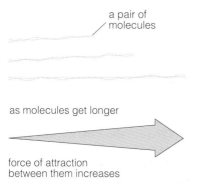

a pair of molecules

as molecules get longer

force of attraction between them increases

Pollution from plastics

- Most plastics are unreactive, so they do not rot away.
- That means plastic cartons and wrappers fill up more and more space in rubbish dumps.
- They also pollute beaches and other places where people throw litter.
- **Biodegradeable plastics** help to solve this problem. They are designed to break down in water and rubbish dumps. Some can disappear in weeks.

---

1 Explain why:
  a a glass jug will shatter if you drop it but a polythene one won't;
  b polythene is flexible but glass is not;
  c both melt over a range of temperature;
  d neither conducts when solid.

2 Suggest why plastic bags:
  a are strong enough to carry shopping;
  b but stretch a bit if it's a heavy load.
3 The melting point for glass is much higher than that for plastic. Why?
4 Plastic may burn but glass doesn't. Suggest a reason.

Questions on Section 3

1 The table below shows the structure of several particles:

Particle	Electrons	Protons	Neutrons
A	12	12	12
B	12	12	14
C	10	12	12
D	10	8	8
E	9	9	10

a Which three particles are neutral atoms?
b Which particle is a negative ion? What is the charge on this ion?
c Which particle is a positive ion? What is the charge on this ion?
d Which two particles are isotopes?
e Use the table on page 30 to identify the particles A to E.

2 This question is about the ionic bond formed between the metal lithium (atomic number 3) and the non-metal fluorine (atomic number 9).
a How many electrons are there in a lithium atom? Draw a diagram to show its electron structure. (You can show the nucleus as a dark circle at the centre.)
b How does a metal atom obtain a full outer shell of electrons?
c Draw the structure of a lithium ion, and write a symbol for the ion.
d How many electrons are there in a fluorine atom? Draw a diagram to show its electron structure.
e How does a nonmetal atom become a negative ion?
f Draw the structure of a fluoride ion, and write a symbol for the ion.
g Draw a diagram to show what happens when a lithium atom reacts with a fluorine atom.
h Draw the arrangement of ions in the compound that forms when lithium and fluorine react together.
i Write a name and a formula for the compound in part **h**.
j Write names and formulae for *two* other compounds which have a similar structure to the compound in **h**.

3 Na^+ O_2 Al CH_4 N I^-
a From the list above, select
  i two atoms
  ii two molecules
  iii two ions
b What do the following symbols represent?
  i Na^+
  ii I^-
c Name the compound made up from Na^+ and I^- ions, and write a formula for it.

4 a The electronic configuration of a neon atom is (2,8). What is special about the outer shell of a neon atom?
b The electronic configuration of a calcium atom is (2,8,8,2). What must happen to a calcium atom for it to achieve noble gas structure?
c Draw a diagram of an oxygen atom, showing its eight protons (p), eight neutrons (n), and eight electrons (e).
d What happens to the outer shell electrons of a calcium atom, when it reacts with an oxygen atom?
e Name the compound that is formed when calcium and oxygen react together. What type of bonding does it contain?
f Write a formula for the compound in **e**.

5 a Write down the formula for each of the following:
  i a nitrate ion
  ii a sulphate ion
  iii a carbonate ion
  iv a hydroxide ion
  v a nitride ion
  vi a hydrogen carbonate ion
b The metal strontium forms ions with the symbol Sr^{2+}. Write down the formula for each of the following:
  i strontium oxide
  ii strontium chloride
  iii strontium nitrate
  iv strontium sulphate
  v strontium hydrogen carbonate

6 A molecule of a certain gas can be represented by the diagram on the right.
a What is the gas? What is its formula?
b What type of bonding holds the atoms together?
c Name another compound with this type of bonding.
d What do the symbols ● and × represent?

7 Draw diagrams to show how the electrons are shared in the following molecules:
a fluorine, F_2
b water, H_2O
c methane, CH_4
d trichloromethane, $CHCl_3$
e oxygen, O_2
f hydrogen sulphide, H_2S
g hydrogen chloride, HCl
Draw the shapes of molecules **a**, **b**, and **e**.

8 a An oxygen molecule is represented as O=O. What does the double line mean? How many electrons from each atom take part in bonding?
b A molecule of carbon dioxide (CO_2) can be drawn as O=C=O. Draw a diagram to show how the electrons are shared in the molecule.

9 Nitrogen is in Group 5 of the periodic table, and its atomic number is 7. It exists as molecules, each containing two atoms.
  a Write the formula for nitrogen.
  b What type of bonding would you expect between the two atoms?
  c How many electrons does a nitrogen atom have in its outer shell?
  d How many electrons must each atom obtain a share of, in order to gain a full shell of 8 electrons?
  e Draw a diagram to show the bonding in a nitrogen molecule. You need show only the outer-shell electrons. (It may help you to look back at the bonding in an oxygen molecule, on page 48.)
  f The bond in a nitrogen molecule is called a **triple bond**. Can you explain why?
  g Nitrogen (N_2), oxygen (O_2), chlorine (Cl_2), and hydrogen (H_2) all exist as diatomic molecules. What does *diatomic* mean?
  h One common gaseous element can form *triatomic* molecules. Which one?

10 Bricks, tiles, mugs, and earthenware pots are all **ceramics**. Ceramics are made of clay, which is made of silicon, oxygen, aluminium, and other elements. When clay is heated in a kiln, a series of reactions occurs. The result is a mass of tiny mineral crystals bonded together by glass.
  The atoms on the surface of a ceramic form a giant lattice where all the bonds are strong. But below the hard surface layer the ceramic contains tiny holes or pores.
  a What is meant by a *giant lattice*?
  b Why are ceramics able to withstand temperatures of over 1500 °C without melting?
  c Name a use of ceramics which uses this property.
  d What is the main disadvantage of ceramics?
  e Explain how this disadvantage is related to the structure of the ceramic.

11 This table gives information about some properties of substances A to G.

Substance	M.pt. °C	Electrical conductivity		Solubility in water
		solid	liquid	
A	−112	poor	poor	insoluble
B	680	poor	good	soluble
C	−70	poor	poor	insoluble
D	1495	good	good	insoluble
E	610	poor	good	soluble
F	1610	poor	poor	insoluble
G	660	good	good	insoluble

  a Which of the substances are metals? Explain.
  b Which of the substances are ionic compounds? Give reasons for your choice.
  c Two of the substances have very low melting points, compared with the rest. Explain why these could *not* be ionic compounds.
  d Two of the substances are molecular. Which ones?
  e Which substance is a giant covalent structure?
  f Which substance do you expect to be very hard?
  g Which substances do you expect to be soluble in tetrachloromethane?

12 Silicon lies directly below carbon in Group 4 of the periodic table. Here are the melting and boiling points for silicon, carbon (diamond), and their oxides.

Substance	Symbol or formula	M.pt. °C	B.pt. °C
Carbon	C	3730	4530
Silicon	Si	1410	2400
Carbon dioxide	CO_2	sublimes at −78 °C	
Silicon dioxide	SiO_2	1610	2230

  a In which state are the two *elements*, at room temperature (20 °C)?
  b Is the structure of carbon (diamond) giant covalent or molecular?
  c What type of structure would you expect silicon to have? Give reasons.
  d In which state are the two oxides, at room temperature?
  e What type of structure does carbon dioxide have?
  f Does silicon dioxide have the same structure as carbon dioxide? What is the evidence for your answer?

13 Hydrogen bromide is a compound of the two elements hydrogen and bromine. It has a melting point of −87 °C and a boiling point of −67 °C. Bromine is one of the halogens (Group 7 of the periodic table).
  a Is hydrogen bromide a solid, a liquid, or a gas at room temperature (20 °C)?
  b Is it made of molecules, or does it have a giant structure? How can you tell?
  c What type of bond is formed between the hydrogen and bromine atoms in hydrogen bromide? Show this on a diagram.
  d Write a formula for hydrogen bromide.
  e Name two other compounds that would have bonding similar to hydrogen bromide.
  f Write formulae for these two compounds.

14 a Use their structures to explain why:
    i graphite is used for the 'lead' in pencils
    ii diamonds are used in cutting tools.
  b Give two reasons why:
    i copper is used in electrical wiring
    ii steel is used for domestic radiators.
  c Ethanol is used as the solvent in perfume and aftershave, because it evaporates easily. What does that tell you about the bonding in it?

4.1 The masses of atoms

Relative atomic mass

A single atom weighs hardly anything. For example, the mass of a single hydrogen atom is only about 0.000 000 000 000 000 000 000 002 grams. Very small numbers like that are awkward to use, so scientists had to find a simpler way to express the mass of an atom. Here is what they did.

First, they chose a carbon atom to be the standard atom.

Next, they fixed its mass as exactly 12 units. (It has 6 protons and 6 neutrons. They ignored the electrons.)

Then they compared all the other atoms with this standard atom, using a machine called a mass spectrometer, and found values for their masses, like this:

The mass spectrometer was invented in 1919 by a British scientist called Aston.

This is the standard atom, $^{12}_{6}C$. Its mass is exactly 12.	This magnesium atom is twice as heavy as the standard atom, so its mass must be 24.	This hydrogen atom is $\frac{1}{12}$ as heavy as the standard atom, so its mass must be 1.

The mass of an atom found by comparing it with the $^{12}_{6}C$ atom is called its **relative atomic mass** or **RAM**. (It is also shown as $\mathbf{A_r}$.) So the RAM of hydrogen is 1 and the RAM of magnesium is 24.

RAMs and isotopes   Not all atoms of an element are exactly the same. For example, when scientists examined chlorine in the mass spectrometer, they found there were two types of chlorine atom:

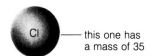 this one has a mass of 35

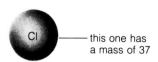 this one has a mass of 37

These atoms are the **isotopes** of chlorine. (They have different masses because one has two neutrons more than the other.)

It was found that out of every four chlorine atoms, three have a mass of 35 and one has a mass of 37. Using this information, the *average* mass of a chlorine atom was calculated to be 35.5.

Most elements have more than one isotope, and these have to be taken into account when finding RAMs:

The RAM of an element is the average mass of its isotopes compared with an atom of $^{12}_{6}C$.

For most elements, the RAMs work out very close to whole numbers. They are usually rounded off to whole numbers, to make calculations easier.

Calculating the RAM of chlorine

In every 4 chlorine atoms, 3 have a mass of 35 and 1 has a mass of 37.

$$3 \times 35 = 105$$
$$1 \times 37 = 37$$

So total mass of 4 atoms = 142

Average mass of an atom = $\frac{142}{4}$

$$= 35.5$$

So the RAM of chlorine is 35.5.

You can work out the RAM of any element in this way, if you know the masses and abundance of its isotopes.

The RAMs of some common elements   Here is a list of them:

Element	Symbol	RAM	Element	Symbol	RAM
Hydrogen	H	1	Chlorine	Cl	35.5
Carbon	C	12	Potassium	K	39
Nitrogen	N	14	Calcium	Ca	40
Oxygen	O	16	Iron	Fe	56
Sodium	Na	23	Copper	Cu	64
Magnesium	Mg	24	Zinc	Zn	65
Sulphur	S	32	Iodine	I	127

Finding the mass of an ion:

Mass of sodium atom = 23, so mass of sodium ion = 23, since a sodium ion is just a sodium atom minus an electron, and an electron has hardly any mass. **An ion has the same mass as the atom from which it is made.**

Formula mass

Using a list of RAMs, it is easy to work out the mass of any molecule or group of ions. Check these examples using the information above:

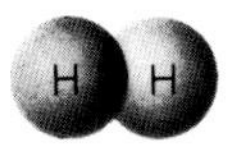

Hydrogen gas is made of molecules. Each molecule contains 2 hydrogen atoms, so its mass is 2. $(2 \times 1 = 2)$

The formula of water is H_2O. A water molecule contains 2 hydrogen atoms and 1 oxygen atom, so its mass is 18. $(2 \times 1 + 16 = 18)$

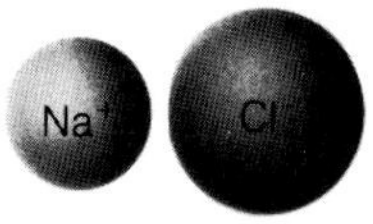

Sodium chloride (NaCl) forms a giant structure with 1 sodium ion for every chloride ion. The mass of a 'unit' of sodium chloride is 58.5 $(23 + 35.5 = 58.5)$.

The mass of a substance found in this way is called its **formula mass**, because is obtained by adding up the masses of the atoms in the formula. If the substance is made of *molecules*, its mass is known as the **relative molecular mass**, or **RMM**. (It is also shown as M_r.) So the RMM of hydrogen is 2 and the RMM of water is 18. Here are some examples:

Substance	Formula	Atoms in formula	RAM of atoms	Formula mass
Nitrogen	N_2	2N	N = 14	$2 \times 14 = \mathbf{28}$
Ammonia	NH_3	1N 3H	N = 14 H = 1	$1 \times 14 = 14$ $3 \times 1 = \underline{3}$ Total $= \underline{\mathbf{17}}$
Magnesium nitrate	$Mg(NO_3)_2$	1Mg 2N 6O	Mg = 24 N = 14 O = 16	$1 \times 24 = 24$ $2 \times 14 = 28$ $6 \times 16 = \underline{96}$ Total $= \underline{\mathbf{148}}$

1 What is the relative atomic mass of an element?
2 What is the RAM of the iodide ion, I^-?
3 Show that the formula mass for chlorine (Cl_2) is 71.
4 What is the RMM of butane, C_4H_{10}?

5 Work out the formula mass of:
 a oxygen, O_2 b iodine, I_2
 c methane, CH_4 d ethanol, C_2H_5OH
 e ammonium sulphate $(NH_4)_2SO_4$

4.2 The mole

What is a mole?

On the last two pages you read about relative atomic masses and formula masses. These are not just boring numbers – they are very important for a chemist to know.

If you work out the RAM or formula mass of a substance, and then weigh out that number of grams of the substance, you can say how many atoms or molecules it contains.

This is very useful, since single atoms and molecules are far too small to be seen or counted.

For example, the RAM of carbon is 12. The photograph on the right shows 12 grams of carbon. The heap contains 602 000 000 000 000 000 000 000 carbon atoms.

This is called a **mole** of atoms.
The number is called **Avogadro's number** or **Avogadro's constant** after the Italian scientist who proposed it.
It is usually written in a shorter way as **6.02×10^{23}**.
(The 10^{23} shows that you must move the decimal point 23 places to the right to get the full number.)
Now look at these:

Sodium is made up of single sodium atoms. Its RAM is **23**.	Iodine is made up of iodine molecules. Its formula is I_2. Its formula mass is **254**.	Water is made up of water molecules. Its formula is H_2O. Its formula mass is **18**.

This is **23 grams** of sodium. It contains 6.02×10^{23} sodium atoms, or **1 mole** of sodium atoms.	Above is **254 grams** of iodine. It contains 6.02×10^{23} iodine molecules, or **1 mole** of iodine molecules.	The beaker contains **18 grams** of water or 6.02×10^{23} water molecules or **1 mole** of water molecules.

From these examples you should see that:
One mole of a substance is 6.02×10^{23} particles of the substance. It is obtained by weighing out the RAM or formula mass, in grams.

Finding the mass of a mole

You can find the mass of 1 mole of any substance, by these steps:

1 Write down the symbol or formula of the substance.
2 Find out its RAM or formula mass.
3 Express that mass in grams.

This table shows more examples:

Substance	Symbol or formula	RAMs	Formula mass	Mass of 1 mole
Helium	He	He = 4	4	4 grams
Oxygen	O_2	O = 16	$2 \times 16 = 32$	32 grams
Ethanol	C_2H_5OH	C = 12 H = 1 O = 16	$2 \times 12 = 24$ $6 \times 1 = 6$ $1 \times 16 = \underline{16}$ $\underline{46}$	46 grams

Some calculations on the mole

Example 1 Calculate the mass of:
a 0.5 moles of bromine atoms.
b 0.5 moles of bromine molecules.

a The RAM of bromine is 80, so 1 mole of bromine *atoms* has a mass of 80 grams. Therefore 0.5 moles of bromine *atoms* has a mass of 0.5×80 grams, or 40 grams.
b A bromine *molecule* contains 2 atoms, so its formula mass is 160. Therefore 0.5 moles of bromine *molecules* has a mass of 0.5×160 grams, or 80 grams.

Formula mass or RMM = 160

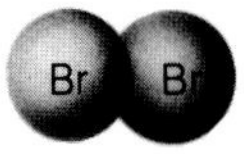

So, to find the mass of a given number of moles:
mass = mass of 1 mole × number of moles

Example 2 How many moles of oxygen molecules are there in 64 grams of oxygen, O_2?

The formula mass of oxygen gas is 32, so 32 grams of it is 1 mole.

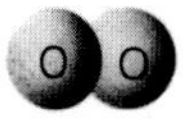

Therefore 64 grams is $\frac{64}{32}$ moles, or 2 moles.

Formula mass or RMM = 32

So, to find the number of moles in a given mass:

$$\text{number of moles} = \frac{\text{mass}}{\text{mass of 1 mole}}$$

---

1 How many atoms are in 1 mole of atoms?
2 How many molecules are in 1 mole of molecules?
3 What name is given to the number 6.02×10^{23}?
4 Find the mass of 1 mole of:
  a hydrogen atoms   b iodine atoms
  c chlorine atoms   d chlorine molecules

5 Find the mass of 2 moles of:
  a oxygen atoms   b oxygen molecules
6 Find the mass of 3 moles of ethanol, C_2H_5OH.
7 How many moles of molecules are there in:
  a 18 grams of hydrogen, H_2?
  b 54 grams of water?

4.3 Percentage composition of a compound

Percentage composition

The percentage composition of a compound tells you *which elements* are in the compound and *how much* of each there is, as a percentage of the total mass.

Calculating the percentage of an element by mass

RAMs:
$C = 12, H = 1$
Formula mass $= 16$

Methane is a compound of carbon and hydrogen, with the formula CH_4. The mass of a carbon atom is 12, and the mass of each hydrogen atom is 1, so the formula mass of methane is 16.

You can find what fraction of the total mass is carbon, and what fraction is hydrogen, like this:

$$\text{Mass of carbon as fraction of total mass} = \frac{\text{mass of carbon}}{\text{total mass}}$$
$$= \frac{12}{16} \text{ or } \frac{3}{4}$$

$$\text{Mass of hydrogen as fraction of total mass} = \frac{\text{mass of hydrogen}}{\text{total mass}}$$
$$= \frac{4}{16} \text{ or } \frac{1}{4}$$

These fractions are usually written as percentages. To change a fraction to a percentage, you just multiply it by 100:

$$\frac{3}{4} \times 100 = \frac{300}{4} = 75 \text{ per cent or } 75\% \qquad \frac{1}{4} \times 100 = \frac{100}{4} = 25\%$$

So 75% of the mass of methane is carbon, and 25% is hydrogen. We say that the **percentage composition** of methane is 75% carbon, 25% hydrogen.

The famous scientist John Dalton and a pupil, collecting methane from a pond.

The Law of Constant Composition

Methane is formed by the action of bacteria on organic material such as animal remains, dead plants, and human sewage. It occurs as natural gas (e.g. North Sea gas) formed by the decay of dead sea animals. You also find it in ponds and marshes, in sewage works, and in the compost heap at the bottom of the garden.

But no matter where you find it, a methane molecule *always* contains one carbon atom and four hydrogen atoms. The percentage composition of methane is *always* 75% carbon, 25% hydrogen. This follows one of the basic laws of chemistry:

The Law of Constant Composition states that every pure sample of a given compound has exactly the same composition.

The methane you cook with, in North Sea gas, has exactly the same composition as John Dalton's.

Calculating the percentage composition of a compound

Here is how to calculate the percentage of an element in a compound:
1 Write down the formula of the compound.
2 Using a list of RAMs, work out its formula mass.
3 Write the mass of the element you want, as a fraction of the total.
4 Multiply the fraction by 100, to give a percentage.

Example 1   The gas sulphur dioxide is a pollutant formed during the burning of fossil fuels. It dissolves in rain water to make acid rain. It has the formula SO_2. Calculate the percentage of oxygen in the compound.

The formula mass of the compound is 64, as shown on the right.
Mass of oxygen in the formula $= 32$

Mass of oxygen as a fraction of the total $= \dfrac{32}{64}$

Mass of oxygen as a percentage of the total $= \dfrac{32}{64} \times 100 = 50\%$

So the compound is **50% oxygen**.
This means it is also 50% sulphur ($100\% - 50\% = 50\%$).

Finding the formula mass for sulphur dioxide:

The RAMs are: $S = 32$, $O = 16$.
The formula contains 1 S and 2 O, so the formula mass is:

$1\,S \quad\quad = 32$
$2\,O = 2 \times 16 = \underline{32}$
$\quad\quad\text{Total} = \underline{64}$

Example 2   Fertilizers contain nitrogen which plants need to make them grow. One important fertilizer which is rich in nitrogen is ammonium nitrate, which has the formula NH_4NO_3. Calculate:
i the percentage of nitrogen in ammonium nitrate;
ii the mass of nitrogen in a 20 kg bag of the fertilizer.

i The formula mass of the compound is 80, as shown on the right. The element we are interested in is nitrogen.

Mass of nitrogen in the formula $\quad\quad = 28$

Mass of nitrogen as a fraction of the total $= \dfrac{28}{80}$

Mass of nitrogen as a percentage of the total $= \dfrac{28}{80} \times 100 = \mathbf{35\%}$

So the fertilizer is **35% nitrogen**.

ii The fertilizer is 35% nitrogen.

So the mass of nitrogen in a 20 kg bag $= \dfrac{35}{100} \times 20\,kg$
$= 7\,kg$

The bag contains **7 kg** of nitrogen.

Now see if you can work out how many *atoms* of nitrogen the bag contains!

Finding the formula mass for ammonium nitrate:

The RAMs are: $N = 14$, $H = 1$, $O = 16$.
The formula contains 2 N, 4 H and 3 O, so the formula mass is:

$2\,N = 2 \times 14 = 28$
$4\,H = 4 \times 1 = 4$
$3\,O = 3 \times 16 = \underline{48}$
$\quad\quad\text{Total} = \underline{80}$

1 A compound contains just oxygen and sulphur. It is 40% sulphur. What percentage is oxygen?
2 Find the percentage of:
  i hydrogen ii oxygen
  in ammonium nitrate, NH_4NO_3.
3 Calculate the percentage of copper in copper(II) oxide, CuO. (The RAMs are: $Cu = 64$, $O = 16$.)
4 12 g of copper is completely converted into copper(II) oxide. What mass of copper(II) oxide is formed?

4.4 The formula of a compound (I)

What a formula tells you

The formula of carbon dioxide is CO_2. Some molecules of it are shown on the right. You can see that:

| 1 carbon atom | combines with | 2 oxygen atoms |

. It follows that

RAMs: C = 12, O = 16

| 1 mole of carbon atoms | combines with | 2 moles of oxygen atoms |

.

Moles can be changed to grams, using RAMs. So we can write:

| 12 g of carbon | combines with | 32 g of oxygen |

.

In the same way:
6 g of carbon combines with 16 g of oxygen
24 kg of carbon combines with 64 kg of oxygen, and so on.
The masses of each substance taking part in the reaction are *always in the same ratio*.

Therefore, from the formula of a compound, you can tell:
- **how many moles of the different atoms combine;**
- **how many grams of the different elements combine.**

Finding the empirical formula, starting from masses

From the formula of a compound you can tell what masses of the elements combine. But you can also do things the other way round. If you know what masses combine, you can work out the formula. These are the steps:

| Find the masses that combine (in **grams**) by experiment | → | Change grams to **moles of atoms** | → | This tells you the **ratio** in which atoms combine | → | So you can write a **formula** |

A formula found by this method is called the **empirical formula**.
The empirical formula shows the simplest ratio in which atoms combine.

Below are some examples of how to work it out.

Example 1 An experiment shows that 32 g of sulphur combine with 32 g of oxygen to form the compound sulphur dioxide. What is its empirical formula?

First, change the masses to moles of atoms.
The RAM of sulphur is 32, and the RAM of oxygen is 16, so:

$\frac{32}{32}$ moles of sulphur atoms combines with $\frac{32}{16}$ moles of oxygen atoms

or 1 mole of sulphur atoms combines with 2 moles of oxygen atoms.
The atoms therefore combine in a ratio of 1 : 2.
So the empirical formula of sulphur dioxide is SO_2.

> Remember, to change masses to moles:
>
> no. of moles = $\dfrac{\text{mass}}{\text{mass of 1 mole}}$

Example 2 Compound X is a **hydrocarbon**: it contains only carbon and hydrogen atoms. 0.84 g of X was completely burned in air. This produced 2.64 g of carbon dioxide (CO_2) and 1.08 g of water (H_2O). Find the empirical formula of X.

RAM = 16

RAM = 12

The formula mass of carbon dioxide is $12 + 16 + 16 = 44$.

1 In carbon dioxide, $\frac{12}{44}$ of the mass is carbon, so

2.64 g of carbon dioxide contains $\frac{12}{44} \times 2.64$ g or 0.72 g of carbon.

2 All this carbon came from X. So X contained **0.72 g** of carbon and therefore **0.12 g** of hydrogen. $(0.84 - 0.72 = 0.12.)$

3 Converting mass to moles of atoms: X contained

$\frac{0.72}{12}$ or 0.06 moles of carbon atoms and $\frac{0.12}{1}$ or 0.12 moles of hydrogen atoms.

So the ratio of carbon to hydrogen atoms in X is 0.06 : 0.12 or 6 : 12 or, at its simplest, 1 : 2. The empirical formula of X is **CH₂.**

An experiment to find the empirical formula

To work out the empirical formula, you need to know the masses of elements that combine. *The only way to do this is by experiment.*

For example, magnesium combines with oxygen to form magnesium oxide. The masses that combine can be found like this:

1 Weigh a crucible and lid, empty. Then add a coil of magnesium ribbon and weigh it again.
2 Heat the crucible. Raise the lid carefully at intervals to let oxygen in. The magnesium burns brightly.
3 When burning is complete, let the crucible cool (still with its lid on). Then weigh it again. The increase in mass is due to oxygen.

tongs to raise lid

the magnesium burns

heat

The results Here are some sample results, and the calculation:

Mass of crucible + lid = 25.2 g
Mass of crucible + lid + magnesium = 27.6 g
Mass of crucible + lid + magnesium oxide = 29.2 g

Mass of magnesium = 27.6 g − 25.2 g = 2.4 g
Mass of magnesium oxide = 29.2 g − 25.2 g = 4.0 g
Mass of oxygen therefore = 4.0 g − 2.4 g = 1.6 g

So 2.4 g of magnesium combines with 1.6 g of oxygen.
The RAMs are: Mg = 24, O = 16. Changing masses to moles:

$\frac{2.4}{24}$ moles of magnesium atoms combine with $\frac{1.6}{16}$ moles of oxygen atoms

0.1 moles of magnesium atoms combine with 0.1 moles of oxygen atoms
So the atoms combine in a ratio of 0.1 : 0.1, or, more simply, 1 : 1.
The empirical formula of magnesium oxide is **MgO**.

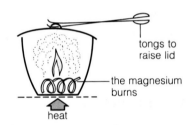

RAM = 16

RAM = 24

---

1 a How many moles of carbon atoms combine with 4 moles of hydrogen atoms, to form methane, CH_4?
  b How many grams of hydrogen combine with 12 grams of carbon, to form methane?
2 How would you change *grams* to *moles of atoms*?

3 To form iron(II) sulphide, 56 g of iron combines with 32 g of sulphur. Find its empirical formula. (The RAMs are: Fe = 56, S = 32.)
4 1.84 g of hydrocarbon Y burns to give 6.16 g of carbon dioxide and 1.44 g of water. Find its empirical formula.

4.5 The formula of a compound (II)

The formulae of ionic compounds

You saw in the last Unit that the empirical formula shows the *simplest ratio* in which atoms combine.

The diagram on the right shows the structure of sodium chloride. You can see that sodium and chlorine atoms combine in a ratio of 1:1 to form this compound. So its empirical formula is NaCl.
The formula of an ionic compound is always the same as its empirical formula.

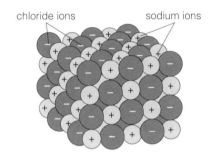

chloride ions   sodium ions

The structure of sodium chloride

The formulae of molecular compounds

The gas ethane is one of the **alkane** family of compounds. An ethane molecule is drawn on the right. Ethane contains only hydrogen and carbon atoms so it is a **hydrocarbon**.

As you can see, the ratio of carbon to hydrogen atoms in ethane is 2:6. The simplest ratio is therefore 1:3. So the *empirical* formula of ethane is CH_3. But the *molecular* formula is C_2H_6.
The molecular formula shows the *actual numbers* of atoms that combine to form a molecule.

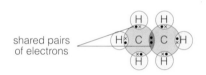

shared pairs of electrons

An ethane molecule

The molecular formula is more useful than the empirical formula because it gives you more information. For some molecular compounds both formulae are the same. For others they are different. Compare the members of the alkane family in the table on the right. What do you notice?

Name of alkane	Molecular formula	Empirical formula
Methane	CH_4	CH_4
Ethane	C_2H_6	CH_3
Propane	C_3H_8	C_3H_8
Butane	C_4H_{10}	C_2H_5
Pentane	C_5H_{12}	C_5H_{12}
Hexane	C_6H_{14}	C_3H_7

Working out the molecular formula

To find the molecular formula for an unknown compound, you need to know:
- the **empirical formula**. This is found by experiment.
- the **formula mass** of the compound. This is also found by experiment. For example, you could turn the compound into a gas and find its volume, then use Avogadro's Law. (See next page.) Or you could use a mass spectrometer.

Once you have these two pieces of information you can work out the molecular formula. Two examples follow.

Example 1   Octane is a member of the alkane family. Its percentage composition is 84.2% carbon and 15.8% hydrogen. Its formula mass is 114. What is its molecular formula?

The mass spectrometer

The best way to find the formula mass of an unknown molecular compound is to use a **mass spectrometer** (page 62). In this machine, the mass of the molecule is compared with the mass of an atom of carbon-12.

1 Find the empirical formula for the compound
   84.2 g of carbon combines with 15.8 g of hydrogen.
   Changing masses to moles:
   $\frac{84.2}{12}$ moles of carbon atoms combine with $\frac{15.8}{1}$ moles of hydrogen atoms
   which means that...

7.02 moles of carbon atoms combines with 15.8 moles of hydrogen atoms

1 mole of carbon atoms combines with $\dfrac{15.8}{7.02}$ moles of hydrogen atoms

1 mole of carbon atoms combines with 2.25 moles of hydrogen atoms.
So atoms combine in the ratio of 1 : 2.25 or **4 : 9**.
(You must change the ratio into whole numbers since only whole atoms can combine.)
The empirical formula of octane is therefore **C_4H_9**.

2 Use the formula mass to find the molecular formula
The formula mass based on the empirical formula (C_4H_9) is 57.
The actual formula mass $= 114 = 2 \times 57$
So the molecular formula $= 2 \times C_4H_9 = $ **C_8H_{18}**.

In general, to find the molecular formula:
i calculate $\dfrac{\text{actual formula mass}}{\text{empirical formula mass}}$ for the substance.
ii multiply the numbers in the empirical formula by this number.

Example 2 An unknown organic liquid has the empirical formula CH_2. 0.21 g of the liquid was injected into a gas syringe. This was heated in an oven until all the liquid had turned to gas. Its volume was measured, and the reading converted to the corresponding value at room temperature and pressure. The result was 60 cm³. What is the molecular formula for the compound?

Here you are given the empirical formula but not the formula mass. So you need to calculate that. You can use Avogadro's Law.

1 Find the formula mass for the compound
A volume of 60 cm³ is occupied by 0.21 g of the compound at rtp, so

1 cm³ is occupied by $\dfrac{0.21}{60}$ g at rtp, and

24 000 cm³ is occupied by $0.21 \times \dfrac{24\,000}{60}$ g at rtp

$0.21 \times \dfrac{24\,000}{60} = 84$

So the formula mass is **84**.

2 Compare this with the empirical formula mass
The formula mass based on the empirical formula $= 14$.
So $\dfrac{\text{actual formula mass}}{\text{empirical formula mass}} = \dfrac{84}{14} = 6$.
Therefore the molecular formula $= 6 \times CH_2$ or **C_6H_{12}**.

Avogadro's Law

1 mole of any gas occupies a volume of 24 dm³ at 20 °C and 1 atmosphere pressure.

20 °C and 1 atmosphere pressure are known as room temperature and pressure (rtp).

24 dm³ = 24 000 cm³

See page 82 for more on Avogadro's Law.

---

1 In the ionic compound magnesium chloride, magnesium and oxygen atoms combine in the ratio 1 : 2. What is the formula of magnesium chloride?
2 What is the difference between an empirical formula and a molecular formula?
3 For some compounds the empirical and molecular formulae are the same. Give two *new* examples.

4 A compound has the empirical formula CH_2 and a formula mass of 28. What is its molecular formula?
5 An oxide of phosphorus has a formula mass of 220. It is 56.4% phosphorus. Find its molecular formula.
6 A gas has the empirical formula CH_4. 0.16 g of the gas occupies a volume of 240 cm³ at rtp. What is its molecular formula?

4.6 The concentration of a solution

A

This solution contains 2.5 grams of copper(II) sulphate in 1 dm³ of water. Its concentration is **2.5 grams / dm³**.

B

This one contains 25 grams of the salt in 1 dm³ of water. Its concentration is **25 grams / dm³**.

C

This one contains 125 grams of the salt in 0.5 dm³ of water. Its concentration is **250 grams / dm³**.

The concentration of a solution is the amount of solute, in grams or moles, that is dissolved in 1 dm³ of solution.

Finding the concentration in moles

Example   Find the concentrations of solutions A and C above in moles per dm³.

You must first change the mass of the solute to moles. The formula mass of copper(II) sulphate is 250, as shown on the right below. So 1 mole of the compound has a mass of 250 grams.

Solution A   It has 2.5 grams of the compound in 1 dm³ of solution.

$$2.5 \text{ grams} = \frac{2.5}{250} \text{ moles} = 0.01 \text{ moles}$$

so its concentration is **0.01 moles per dm³**.
Mole per dm³ is often shortened just to M, so the concentration of solution A can be written as **0.01 M**.

Solution C   It has 250 grams of the compound in 1 dm³ of solution.
  250 grams = 1 mole
  so its concentration is **1 mole per dm³**, or **1 M** for short.
A solution that contains 1 mole of solute per dm³ is often called a **molar solution**, so C is a molar solution.

> In general, to find the concentration of a solution in moles per dm³:
> $$\text{concentration (mol/dm}^3\text{)} = \frac{\text{amount of solute (mol)}}{\text{volume of solution (dm}^3\text{)}}$$

Use the equation to check that the last column in this table is correct:

Amount of solute (mol)	Volume of solution (dm³)	Concentration of solution (mol/dm³)
1.0	1.0	1.0
0.2	0.1	2.0
0.5	0.2	2.5
1.5	0.3	5.0

> **Remember**
> - 1 dm³ = 1 litre
>   = 1000 cm³
>   = 1000 ml
>
> - All these mean the same thing:
>   moles per dm³
>   mol / dm³
>   mol dm⁻³
>   moles per litre

> The formula of copper(II) sulphate is $CuSO_4.5H_2O$. The formula contains 1 Cu, 1 S, 9 O, and 10 H, so the formula mass is:
>
> | | | |
> |---|---|---|
> | 1 Cu = | 1 × 64 = | 64 |
> | 1 S = | 1 × 32 = | 32 |
> | 9 O = | 9 × 16 = | 144 |
> | 10 H = | 10 × 1 = | 10 |
> | | Total = | 250 |

Finding the amount of solute in a solution

If you know the concentration of a solution, and its volume:
- You can work out how much solute it contains, in moles. Just rearrange the equation from the opposite page:
  amount of solute (mol) = concentration (mol/dm³) × volume (dm³)
- You can then convert moles to grams by multiplying the number of moles by the formula mass.

Sample calculations

This table shows some examples. Check that you understand the calculations. Are the results correct?

Use the calculation triangle

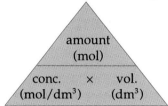

Cover the one you want to find – and you'll see how to calculate it.

	Solution A sodium hydroxide (NaOH)	Solution B sodium thiosulphate $Na_2S_2O_3$	Solution C lead nitrate $Pb(NO_3)_2$	Solution D silver nitrate $AgNO_3$
	2 dm³	250 cm³	100 cm³	25 cm³
concentration (mol/dm³)	1	2	0.1	0.05
amount of solute (moles)	$1 \times 2 = 2$	$2 \times \frac{250}{1000} = 0.5$	$0.1 \times \frac{100}{1000} = 0.01$	$0.05 \times \frac{25}{1000} = 0.00125$
formula mass (g/mol)	40	158	331	170
mass of solute (grams)	80	79	3.31	0.2125

1 What does *concentration of a solution* mean?
2 Find the concentration of solution B on the opposite page, in mol/dm³.
3 Think of a way to find the concentration of a 1 litre solution of copper(II) sulphate by experiment. (You can use up some of the solution.)
4 How many moles of solute are in:
  a 500 cm³ of solution, concentration 2 mol/dm³?
  b 2 litres of solution, concentration 0.5 mol/dm³?
  c 20 cm³ of solution, concentration 0.4 mol/dm³?
5 What is the concentration of a solution containing:
  a 4 moles in 2 dm³ of solution?
  b 0.5 moles in 0.1 dm³ of solution?
  c 3 moles in 200 cm³ of solution?

6 What volume of :
  a a 4 mol/dm³ solution contains 2 moles?
  b a 6 mol/dm³ solution contains 0.03 moles?
7 The formula mass of sodium hydroxide is 40. How many grams of sodium hydroxide are there in:
  a 500 cm³ of a 1 M solution?
  b 25 cm³ of a 0.5 M solution?
8 What is the concentration of:
  a a sodium carbonate solution containing 53 g of the salt (Na_2CO_3) in 1 litre?
  b a copper(II) sulphate solution containing 62.5 g of the salt ($CuSO_4.5H_2O$) in 1 litre?
9 What mass of silver nitrate ($AgNO_3$) would you need to make 50 cm³ of a 0.2 M solution.

Questions on Section 4

Relative atomic masses are given on page 298. Use the approximate values given in the table.

1 How many grams are there in:
 a 1 mole of copper atoms?
 b 1.5 moles of sulphur atoms?
 c 2 moles of magnesium atoms?
 d 5 moles of carbon atoms?
 e 10 moles of chlorine atoms?
 f 0.1 moles of nitrogen atoms?
 g 0.2 moles of neon atoms?
 h 0.6 moles of hydrogen atoms?
 i 1.5 moles of oxygen atoms?

2 How many grams are there in:
 a 1 mole of hydrogen molecules, H_2?
 b 2 moles of hydrogen molecules, H_2?
 c 1 mole of oxygen molecules, O_2?
 d 0.5 moles of chlorine molecules, Cl_2?
 e 2 moles of phosphorus molecules, P_4?
 f 4 moles of sulphur molecules, S_8?
 g 3 moles of ozone molecules, O_3?

3 Find how many moles of atoms there are, in:
 a 32 g of sulphur
 b 48 g of magnesium
 c 23 g of sodium
 d 14 g of lithium
 e 1.4 g of lithium
 f 3.1 g of phosphorus
 g 6.4 g of oxygen
 h 5.4 g of aluminium
 i 2 g of hydrogen
 j 0.6 g of carbon
 k 12 kg of carbon

4 For each pair, decide which of the two substances contains the greater number of atoms.
 a 80 g of sulphur, 80 g of calcium
 b 80 g of sulphur, 80 g of oxygen
 c 1 mole of sulphur atoms, 8 moles of chlorine atoms
 d 1 mole of sulphur atoms, 1 mole of oxygen molecules
 e 4 moles of sulphur atoms, $\frac{1}{8}$ mole of sulphur molecules (S_8)

5 How many grams are there in:
 a 1 mole of water, H_2O?
 b 5 moles of water?
 c 1 mole of anhydrous copper(II) sulphate, $CuSO_4$?
 d 1 mole of hydrated copper(II) sulphate, $CuSO_4.5H_2O$?
 e 2 moles of ammonia, NH_3?
 f 0.5 moles of ammonium carbonate, $(NH_4)_2CO_3$?
 g 0.3 moles of calcium carbonate, $CaCO_3$?
 h $\frac{1}{5}$ mole of magnesium oxide, MgO?
 i 0.1 moles of sodium thiosulphate, $Na_2S_2O_3$?
 j 2 moles of iron(III) chloride, $FeCl_3$?

6 1 mole of sodium carbonate (Na_2CO_3) contains 2 moles of sodium atoms, 1 mole of carbon atoms, and 3 moles of oxygen atoms.
In the same way, write down the number of moles of each atom present in 1 mole of:
 a lead oxide, Pb_3O_4
 b ammonium nitrate, NH_4NO_3
 c calcium hydroxide, $Ca(OH)_2$
 d dinitrogen tetroxide, N_2O_4
 e ethanol, C_2H_5OH
 f ethanoic acid, CH_3COOH
 g hydrated iron(II) sulphate, $FeSO_4.7H_2O$
 h iron(III) ammonium sulphate, $NH_4Fe(SO_4)_2$
 i calcium carbide, CaC
 j nitroglycerine, $C_3H_5(NO_3)_3$

7 The formula of calcium oxide is CaO. The RAMs are: Ca $=40$, O $=16$.
Complete the following statements:
 a 1 mole of Ca (......g) and 1 mole of O (......g) combine to form ...... mole of CaO (......g).
 b 4.0 g of calcium and ......g of oxygen combine to form ......g of calcium oxide.
 c When 0.4 g of calcium reacts with oxygen, the increase in mass is ......g.
 d If 6 moles of CaO were decomposed to calcium and oxygen, ... moles of Ca and ...... moles of O_2 would be obtained.
 e The percentage by mass of calcium in calcium oxide is ......%.

8 In a reaction to make manganese from manganese oxide, the following results were obtained:
174 g of manganese oxide produced 110 g of manganese. (RAMs: Mn $=55$, O $=16$)
 a What mass of oxygen is there in 174 g of manganese oxide?
 b How many moles of oxygen atoms is this?
 c How many moles of manganese atoms are there in 110 g of manganese?
 d What is the empirical formula of manganese oxide?
 e What mass of manganese would be obtained from 1000 g of manganese oxide?

9 27 g of aluminium burns in a stream of chlorine to form 133.5 g of aluminium chloride.
(RAMs: Al $=27$, Cl $=35.5$)
 a What mass of chlorine is present in 133.5 g of aluminium chloride?
 b How many moles of chlorine atoms is this?
 c How many moles of aluminium atoms are present in 27 g of aluminium?
 d Use your answers for parts **b** and **c** to find the simplest formula of aluminium chloride.
 e 1 dm^3 of an aqueous solution is made using 13.35 g of aluminium chloride. What is its concentration in moles per dm^3?

10 Copy and complete the following table.
(The RAMs are: H = 1, C = 12, N = 14, O = 16.)

Compound	Molar mass g/mol	Empirical formula	Molecular formula
hydrazine	32	NH_2	
cyanogen	52	CN	
nitrogen oxide	92	NO_2	
glucose	180	CH_2O	

11 Hydrocarbons A and B both contain 85.7% carbon. The molar masses of A and B are 42 and 84 g/mol.
a What two elements are present in a hydrocarbon?
b Calculate the empirical formulae of A and B. (The RAMs are: H = 1, C = 12.)
c Calculate the molecular formulae of A and B.

12 Zinc phosphide is made by heating zinc and phosphorus together. It is found that 8.4 g of zinc combines with 3.1 g of phosphorus.
a Find the empirical formula for the compound. (The RAMs are: Zn = 56, P = 31.)
b Calculate the percentage of phosphorus in it.

13 Phosphorus forms two oxides which have the empirical formulae P_2O_3 and P_2O_5.
a Which oxide contains the higher percentage of phosphorus? (RAMs: P = 31, O = 16.)
b What mass of phosphorus will combine with 1 mole of oxygen molecules (O_2) to form P_2O_3?
c What is the molecular formula of the oxide which has a formula mass of 284?
d Assuming that the other oxide is similar, suggest what its molecular formula might be.

14 For each pair, decide which of the two solutions contains the greater number of moles of solute.
a 1 dm^3 of 1 M sodium chloride (NaCl), 1 dm^3 of 2 M sodium chloride
b 500 cm^3 of 1 M sodium chloride, 1 dm^3 of 1 M sodium chloride
c 1 dm^3 of 0.1 M sodium chloride, 100 cm^3 of 2 M sodium chloride
d 250 cm^3 of 2 M sodium chloride, 1 dm^3 of 1 M sodium hydroxide (NaOH)
e 20 cm^3 of 0.5 M sodium chloride, 40 cm^3 of 1 M sodium chloride

15 Using 10 g of this solute, what volume of solution will you prepare so that its concentration is 2 M? (RAMs: H = 1, Li = 7, N = 14, O = 16, Mg = 24, S = 32, Ca = 40)
a lithium sulphate, Li_2SO_4
b magnesium sulphate, $MgSO_4$
c ammonium nitrate, NH_4NO_3
d calcium nitrate, $Ca(NO_3)_2.2H_2O$

16 An oxide of copper can be converted to copper by heating it in a stream of hydrogen in the apparatus shown below:

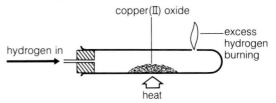

The hydrogen supply was turned on, and hydrogen was allowed to pass through the test-tube for some time before the excess gas was lit. The test-tube was heated until all the copper oxide was converted to copper. The apparatus was allowed to cool, with hydrogen still passing through, before it was dismantled.
a Copy and complete the word equation for the reaction:
copper oxide + hydrogen → copper + ...
b Why was the hydrogen allowed to pass through the apparatus for some time, before the excess was lit? Why was the excess hydrogen burned?

The experiment was repeated several times, by different groups in the class. Each group used a different mass of oxide. The results are shown below.

Group	Mass of copper oxide/g	Mass of copper produced/g	Mass of oxygen lost/g
1	0.62	0.55	0.07
2	0.90	0.80	0.10
3	1.12	1.00	0.12
4	1.69	1.50	
5	1.80	1.60	

c Work out the missing figures for the table.
d On graph paper, plot the mass of copper against the mass of oxygen. (Show the mass of copper along the x-axis and the mass of oxygen along the y-axis.) Then draw the best straight line through the origin and the set of points.
e From the graph, find the mass of oxygen which would combine with 1.28 g of copper.
f Calculate the mass of oxygen which would combine with 128 g of copper. (Remember, 1.28 × 100 = 128)
g How many moles of copper atoms are there in 128 g of copper? (Cu = 64)
h How many moles of oxygen atoms combine with 128 g of copper? (O = 16)
i What is the simplest formula of this oxide of copper?
j What is another name for the *simplest formula* found in this way?

5.1 Physical and chemical change

A substance can be changed by heating it, adding water to it, mixing another substance with it, and so on. The change that takes place will be either a **chemical** change or a **physical** one.

Chemical change

mixture of iron filings and sulphur

iron filings cling to the magnet
magnet

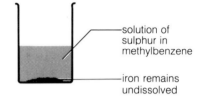
solution of sulphur in methylbenzene
iron remains undissolved

1 Some yellow sulphur and black iron filings are mixed together.

2 The mixture is easily separated again, by using a magnet to attract the iron . . .

3 . . . or by dissolving the sulphur in methylbenzene (a solvent).

the contents glow even after the bunsen is removed

black solid
magnet

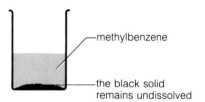
methylbenzene
the black solid remains undissolved

4 But when the mixture is *heated*, it glows brightly. The yellow specks of sulphur disappear. A black solid forms.

5 This black solid is not at all like the mixture. It is not affected by a magnet . . .

6 . . . and none of it dissolves in methylbenzene.

The black solid is obviously a new chemical substance.
When it produces a new chemical substance, a change is called a chemical change.
So in step 4, a chemical change has taken place. The iron and sulphur have **reacted** together, to form the compound iron sulphide.

The difference between the mixture and the compound In the mixture above, iron and sulphur particles are mixed closely together, but they are not bonded to each other – the iron particles still behave like iron, and the sulphur particles like sulphur. During the reaction, however, iron and sulphur atoms form ions which bond to each other. The magnet and solvent now have no effect.

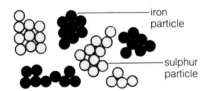
iron particle
sulphur particle

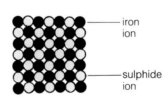
iron ion
sulphide ion

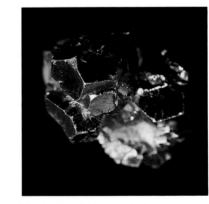
The compound iron sulphide occurs in the earth as **pyrite**.

Mixture Iron and sulphur particles mixed together. Each particle contains many atoms.

Compound Iron ions and sulphide ions bonded together to form iron sulphide.

The signs of a chemical change

A chemical change is usually called a **chemical reaction**. You can tell when a chemical reaction has taken place, by these signs:

1 One or more new chemical substances are formed.
The new substances usually look quite different from the starting substances. For example:

$$\text{iron} \quad + \quad \text{sulphur} \quad \longrightarrow \quad \text{iron sulphide}$$
(black filings)   (yellow powder)   (black solid)

2 Energy is taken in or given out, during the reaction.
In step 4, a little energy in the form of heat from a Bunsen is needed to start off the reaction between iron and sulphur. But the reaction gives out heat once it begins.
A reaction that gives out heat energy is **exothermic**.
A reaction that takes in heat energy is **endothermic**.
So the reaction between iron and sulphur is exothermic. The reactions that take place when you fry an egg are endothermic.
Energy can be given out in the form of light or sound, as well as heat. For example magnesium burns in air with a bright white light and a hiss.

3 The change is usually difficult to reverse.
You would need to carry out several other reactions, to get back the iron and sulphur from iron sulphide.

Fireworks contain magnesium and other substances. When they burn, the reactions give out energy in the form of heat, light, and sound.

The reactions that take place when you fry an egg are endothermic.

Physical change

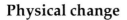
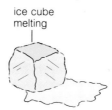

ice cube melting

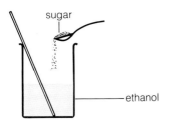

sugar

ethanol

Ice turns to water at 0 °C. It is easy to change the water back to ice again, by cooling it.

Sugar dissolves in ethanol. You can separate the two again by distilling the solution.

No new chemical substances are formed in these changes.
For example, although ice and water *look* different, they are both made of water molecules, and have the formula H_2O.
When no new chemical substance is formed, a change is called a physical change.
So the changes above are both physical changes. Physical changes are usually easy to reverse.

1 Explain the difference between a *mixture* of iron and sulphur and the *compound* iron sulphide.
2 What are the signs of a chemical change?
3 What is: **a** an exothermic reaction?
   b an endothermic reaction?

4 Is the change chemical or physical? Give reasons.
  a Glass bottle breaking.
  b Butter and sugar being made into toffee.
  c Wool being knitted into a sweater.
  d Coal burning in air.

5.2 Equations for chemical reactions

The reaction between carbon and oxygen   When carbon is heated in oxygen, they react together and carbon dioxide is formed. The carbon and oxygen are called **reactants**, because they react together. Carbon dioxide is the **product** of the reaction. You could show the reaction by a diagram, like this:

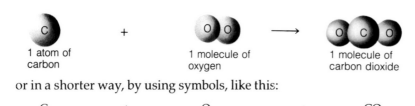

or in a shorter way, by using symbols, like this:

C   +   O₂   ⟶   CO₂

This short way to describe the reaction is called a **chemical equation**.

The reaction between hydrogen and oxygen   When hydrogen and oxygen react together, the product is water. The diagram is:

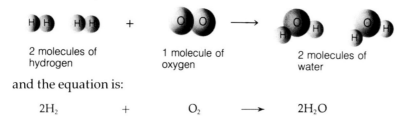

and the equation is:

2H₂   +   O₂   ⟶   2H₂O

Can you see why there is a 2 in front of H_2 and H_2O, in the equation? Now look at the number of atoms on each side of the equation:

On the left:	On the right:
4 hydrogen atoms	4 hydrogen atoms
2 oxygen atoms	2 oxygen atoms

The numbers of hydrogen and oxygen atoms are the same on both sides of the equation. This is because atoms do not *disappear* during a reaction – they are just *rearranged*, as shown in the diagram.
When the numbers of different atoms are the same on both sides, an equation is said to be **balanced**. An equation which is not balanced is not correct. Check the equation for the reaction between carbon and oxygen above. Is it balanced?

Adding more information to equations   Reactants and products may be solids, liquids, gases or solutions. You can show their states by adding **state symbols** to the equations. The state symbols are:

   (*s*) for solid   (*l*)   for liquid
   (*g*) for gas   (*aq*) for aqueous solution (solution in water)

For the two reactions above, the equations with state symbols are:
   C (*s*) + O₂ (*g*)   ⟶   CO₂ (*g*)
   2H₂ (*g*) + O₂ (*g*) ⟶   2H₂O (*l*)

The main reaction in a coal fire is $C + O_2 \rightarrow CO_2$.

The reaction between hydrogen and oxygen gives out so much energy that it is used to power rockets. The hydrogen and oxygen are carried as liquids in the fuel tanks. See page 216 for more.

How to write the equation for a reaction

These are the steps to follow, when writing an equation:

1 Write the equation in words.
2 Now write the equation using symbols. Make sure all the formulae are correct.
3 Check that the equation is balanced, for each type of atom in turn. *Make sure you do not change any formulae.*
4 Add the state symbols.

Example 1 Calcium burns in chlorine to form calcium chloride, a solid. Write an equation for the reaction, using the steps above.

1 Calcium + chlorine $\longrightarrow$ calcium chloride
2 Ca + Cl_2 $\longrightarrow$ $CaCl_2$
3 Ca: 1 atom on the left and 1 atom on the right.
   Cl: 2 atoms on the left and 2 atoms on the right.
   The equation is balanced.
4 $Ca(s)$ + $Cl_2(g)$ $\longrightarrow$ $CaCl_2(s)$

Example 2 In industry, hydrogen chloride is formed by burning hydrogen in chlorine. Write an equation for the reaction.

1 Hydrogen + chlorine $\longrightarrow$ hydrogen chloride
2 H_2 + Cl_2 $\longrightarrow$ HCl
3 H: 2 atoms on the left and 1 atom on the right.
   Cl: 2 atoms on the left and 1 atom on the right.
   The equation is *not* balanced. It needs another molecule of hydrogen chloride on the right. So a 2 is put *in front of* the HCl.
   $H_2 + Cl_2 \longrightarrow 2HCl$
   The equation is now balanced. Do you agree?
4 $H_2(g)$ + $Cl_2(g)$ $\longrightarrow$ $2HCl(g)$

Example 3 Magnesium burns in oxygen to form magnesium oxide, a white solid. Write an equation for the reaction.
1 Magnesium + oxygen $\longrightarrow$ magnesium oxide
2 Mg + O_2 $\longrightarrow$ MgO
3 Mg: 1 atom on the left and 1 atom on the right.
   O: 2 atoms on the left and 1 atom on the right.
   The equation is *not* balanced. Try this:
   $Mg + O_2 \longrightarrow 2MgO$ (Note, the 2 goes *in front of* the MgO.)
   Another magnesium atom is now needed on the left:
   $2Mg + O_2 \longrightarrow 2MgO$
   The equation is balanced.
4 $2Mg(s)$ + $O_2(g)$ $\longrightarrow$ $2MgO(s)$

Magnesium burning in oxygen

---

1 What do + and $\longrightarrow$ mean, in an equation?
2 Balance the following equations:
   a $Na(s) + Cl_2(g) \longrightarrow NaCl(s)$
   b $H_2(g) + I_2(g) \longrightarrow HI(g)$
   c $Na(s) + H_2O(l) \longrightarrow NaOH(aq) + H_2(g)$

d $NH_3(g) \longrightarrow N_2(g) + H_2(g)$
e $C(s) + CO_2(g) \longrightarrow CO(g)$
f $Al(s) + O_2(g) \longrightarrow Al_2O_3(s)$

3 Aluminium burns in chlorine to form aluminium chloride, $AlCl_3(s)$. Write an equation for the reaction.

5.3 Calculations from equations

What an equation tells you

When carbon burns in oxygen, the reaction can be shown as:

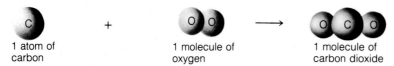

| 1 atom of carbon | + | 1 molecule of oxygen | → | 1 molecule of carbon dioxide |

or in shorthand using an equation:

$$C(s) + O_2(g) \longrightarrow CO_2(g)$$

This equation tells you that:

| 1 carbon atom | reacts with | 1 molecule of oxygen | to give | 1 molecule of carbon dioxide |

Now suppose there was 1 *mole* of carbon atoms. These would react with 1 *mole* of oxygen molecules:

| 1 mole of carbon atoms | reacts with | 1 mole of oxygen molecules | to give | 1 mole of carbon dioxide molecules |

Moles can be changed to grams, using RAMs and formula masses. The RAMs are: C = 12, O = 16. So the formula mass of CO_2 is: (12 + 16 + 16) = 44, and we can write:

| 12 g of carbon | reacts with | 32 g of oxygen | to give | 44 g of carbon dioxide |

This also means that:

| 6 g of carbon | reacts with | 16 g of oxygen | to give | 22 g of carbon dioxide |

and so on. The masses of each substance taking part in the reaction *are always in the same ratio.*

You can find out the same kind of information from any equation.
From the equation for a reaction you can tell:
- **how many moles of each substance take part;**
- **how many grams of each substance take part.**

Does the mass change during a reaction?

Now look what happens to the total mass, during the above reaction:

Mass of carbon and oxygen at the start: 12 g + 32 g = **44 g**
Mass of carbon dioxide at the end: **44 g**

The total mass has not changed, during the reaction. This is because no atoms have disappeared. They have just been rearranged. This is one of the basic laws of chemistry.

The Law of Conservation of Mass states that the total mass remains unchanged during a chemical reaction.

Calculations from equations

Example 1 Hydrogen burns in oxygen to form water. The equation for the reaction is: $2H_2(g) + O_2(g) \longrightarrow 2H_2O(l)$
How much oxygen is needed to burn 1 gram of hydrogen?

1 The RAMs are: $H = 1$, $O = 16$. So $H_2 = 2$ and $O_2 = 32$.
2 $2H_2(g) + O_2(g) \longrightarrow 2H_2O(l)$
   2 moles of hydrogen molecules need 1 mole of oxygen molecules
   4 g of hydrogen needs 32 g of oxygen (moles changed to grams)
   1 g of hydrogen needs 8 g of oxygen
3 The reaction needs **8 g** of oxygen.

Example 2 The equation for the reaction between iron and sulphur is: $Fe(s) + S(s) \longrightarrow FeS(s)$. When 7 g of iron is heated with excess sulphur, how much iron(II) sulphide is formed? (*Excess* sulphur means *more than enough* sulphur for the reaction.)

1 The RAMs are: $Fe = 56$, $S = 32$. So $FeS = 56 + 32 = 88$.
2 $Fe(s) + S(s) \longrightarrow FeS(s)$
   1 mole of iron atoms gives 1 mole of iron sulphide units so
   56 g of iron gives 88 g of iron sulphide

   1 g of iron gives $\dfrac{88}{56}$ g of iron sulphide

   7 g of iron gives $7 \times \dfrac{88}{56}$ g of iron sulphide

3 $7 \times \dfrac{88}{56} = 11$ so **11 g** of iron(II) sulphide is produced.

Example 3 The main ore of iron is haematite, Fe_2O_3. How much iron will be extracted by the reduction of 100 tonnes of haematite?

1 The RAMs are: $Fe = 56$, $O = 16$. So Fe_2O_3 is 160.
2 The reduction reaction is:
   $Fe_2O_3(s) + 3CO(g) \longrightarrow 2Fe(l) + 3CO_2(g)$
   1 mole of haematite yields 2 moles of iron so
   160 g of haematite yields 112 g of iron and

   100 g of haematite yields $\dfrac{112}{160} \times 100$ g of iron.

3 $\dfrac{112}{160} \times 100 = 70$ so 100 g of haematite yields 70 g of iron.

Since the amounts are always in the same ratio, this means that 100 tonnes of haematite will yield **70 tonnes of iron**.

These models show how the atoms are rearranged during the reaction between hydrogen and oxygen. The equation is $2H_2 + O_2 \rightarrow 2H_2O$.

The reaction between iron and sulphur. It is exothermic, but heat is needed to start it off.

---

1 The total mass does not change during a reaction. Why?
2 The reaction between magnesium and oxygen is:
   $2Mg(s) + O_2(g) \longrightarrow 2MgO(s)$
   a Write a word equation for the reaction.
   b How many moles of magnesium atoms react with 1 mole of oxygen molecules?
   c The RAMs are: $Mg = 24$, $O = 16$.
   How many grams of oxygen react with:
      i 48 g of magnesium?   **ii** 12 g of magnesium?

3 Copper(II) carbonate breaks down on heating:
   $CuCO_3(s) \xrightarrow{\text{heat}} CuO(s) + CO_2(g)$
   a Write a word equation for the reaction.
   b Find the mass of 1 mole of each substance in the reaction. ($Cu = 64$, $C = 12$, $O = 16$)
   c When 31 g of copper(II) carbonate is used:
      i how many grams of carbon dioxide form?
      ii what mass of solid remains after heating?
   d What is the name of the solid produced?

5.4 Reactions involving gases

Molar gas volume

The volume of a gas depends on its temperature and pressure. But what about the size of its molecules? For example, carbon dioxide molecules are larger than oxygen or hydrogen molecules. Does that make a difference to the volume of these gases?

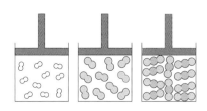

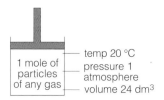

temp 20 °C
1 mole of particles of any gas — pressure 1 atmosphere
— volume 24 dm³

You might expect a mole of carbon dioxide molecules to take up more space than a mole of hydrogen or oxygen molecules.

But this doesn't happen. At a given temperature and pressure, a mole of particles of *any* gas occupies the same volume.

At room temperature and pressure (rtp) the volume of 1 mole of a gas is 24 dm³. (rtp is 20 °C and 1 atmosphere.)

The volume of one mole of a gas is called the **molar gas volume**. **The molar gas volume of any gas at rtp is 24 dm³.**
 This is called **Avogadro's Law** after the Italian scientist who proposed it in 1811.

Remember:
24 dm³ = 24 litres
 = 24 000 cm³

24 dm³ of gas would just fit into 8 of the biggest size of plastic Coca Cola bottles you can buy in the shops.

Calculations involving molar gas volume

You can find the volume of any gas at rtp using these steps:
1 Find how many moles of the gas are present.
2 Multiply this by 24 dm³, the molar gas volume at rtp.

Example 1   What volume does 0.25 moles of a gas occupy at rtp?

Volume = number of moles × 24 dm³
 = 0.25 × 24 dm³
 = 6 dm³
so 0.25 moles of any gas occupies **6 dm³** (or **6000 cm³**) at rtp.

Example 2   What volume does 22 g of carbon dioxide occupy at rtp?

The molar mass of carbon dioxide is 44 g, so
44 g = 1 mole
22 g = 0.5 mole
therefore the volume occupied = 0.5 × 24 dm³ = **12 dm³**.

Example 3   A reaction produces 100 cm³ of hydrogen at rtp. How many moles of H_2 is this?

1 mole of any gas occupies 24 dm³ or 24 000 cm³ at rtp.
24 000 cm³ = 1 mole

$$100 \text{ cm}^3 = \frac{1}{24\,000} \times 100 \text{ moles} = 0.00417 \text{ moles}.$$

The reaction produces **0.00417 moles of hydrogen, H_2**.

Use the calculation triangle

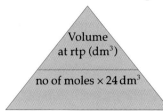

Volume at rtp (dm³)

no of moles × 24 dm³

Cover the one you want to find – and you'll see how to calculate it.

Calculations based on equations

From the equation for a reaction, you can tell how many *moles* of a gas take part. You can also work out its *volume*, since a mole of any gas occupies the same volume, at a given temperature and pressure. In these examples, all volumes are measured at rtp.

Example 1   What volume of hydrogen will react with 24 dm³ of oxygen to form water?
1   The equation for the reaction is: $2H_2(g) + O_2(g) \longrightarrow 2H_2O(l)$
2   So 2 volumes of hydrogen react with 1 of oxygen, or
  2×24 dm³ of hydrogen react with 24 dm³ of oxygen.
  48 dm³ of hydrogen will react.

Example 2   Coal contains the impurity sulphur. When sulphur burns in air it forms sulphur dioxide. What volume of this polluting gas is produced when 1 g of sulphur burns? (RAM: S = 32)

1   The equation for the reaction is: $S(s) + O_2(g) \longrightarrow SO_2(g)$
2   32 g of sulphur atoms = 1 mole so $1\,g = \dfrac{1}{32}$ mole or 0.03125 moles.

3   1 mole of sulphur atoms gives 1 mole of sulphur dioxide molecules so 0.03125 moles of sulphur atoms gives 0.03125 moles of sulphur dioxide molecules.
4   1 mole of sulphur dioxide molecules has a volume of 24 dm³ at rtp so 0.03125 moles has a volume of 0.03125×24 dm³ at rtp.
  0.75 dm³ or 750 cm³ of sulphur dioxide are produced.

Example 3   What volume of air will provide enough oxygen for the complete combustion of 3.1 g of phosphorus? (RAM: P = 31)

1   The equation for the reaction is: $4P(s) + 5O_2(g) \longrightarrow P_4O_{10}(s)$
2   31 g of phosphorus is 1 mole so 3.1 g is 0.1 moles.
3   4 moles of phosphorus atoms needs 5 moles of oxygen molecules so
  1 mole needs $\dfrac{5}{4}$ mole or 1.25 moles of oxygen molecules and
  0.1 moles needs 0.1×1.25 moles or 0.125 moles of oxygen molecules.
  So the *volume* of oxygen needed is 0.125×24 dm³ or **3 dm³.**
4   Air is about 20% oxygen. This means there is 20 dm³ of oxygen in every 100 dm³ of air or 1 dm³ in every 5 dm³ of air.
  So to obtain 3 dm³ of oxygen, 3×5 dm³ of air is needed.
  The combustion of 3.1 g of phosphorus requires **15 dm³ of air.**

(RAMs: O = 16, N = 14, H = 1, C = 12)
1   What does *rtp* mean? What values does it have?
2   What does *molar gas volume* mean?
3   What is the molar gas volume of neon gas at rtp?
4   For any gas, calculate the volume at rtp of:
  a 7 moles   **b** 0.5 moles   **c** 0.001 moles.
5   Calculate the volume at rtp of:
  a 16 g of oxygen (O_2)   **b** 1.7 g of ammonia (NH_3)

6   You burn 6 g of carbon in plenty of air:
  $C(s) + O_2(g) \longrightarrow CO_2(g)$
  a What volume of gas will form (at rtp)?
  b What volume of oxygen will be used up?
7   If you burn the carbon in limited air, the reaction is different: $2C(s) + O_2(g) \longrightarrow 2CO(g)$
  a What volume of gas will form this time?
  b What volume of oxygen will be used up?

Questions on Section 5

1 Decide whether each change below is a physical change or a chemical change. Give reasons for your answers.
a ice melting
b iron rusting
c petrol burning
d candle wax melting
e a candle burning
f wet hair drying
g milk souring
h perfume evaporating
i a lump of roll sulphur being crushed
j copper being obtained from copper(II) oxide
k clothes being ironed
l custard being made
m a cigarette being smoked
n copper(II) sulphate crystallizing from solution

2 Write a chemical equation for each of the following. You do not need to add state symbols. Example:

$$2H_2 + O_2 \rightarrow 2H_2O$$

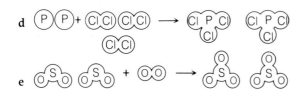

3 Write equations for the following reactions:
a 1 mole of copper atoms combines with 1 mole of sulphur atoms to form 1 mole of copper(II) sulphide, CuS.
b 3 moles of lead atoms combine with 2 moles of oxygen molecules to form 1 mole of lead oxide, Pb_3O_4.
c 1 mole of ethanol molecules, C_2H_5OH, burns in 3 moles of oxygen molecules to form 2 moles of carbon dioxide molecules and 3 moles of water molecules.
d 1 mole of iron(III) oxide, Fe_2O_3, is reduced by 3 moles of hydrogen molecules to form 2 moles of iron atoms and 3 moles of water molecules.

4 Balance these equations:
a $H_2(g) + Br_2(g) \longrightarrow HBr(g)$
b $Cl_2(g) + KBr(aq) \longrightarrow KCl(aq) + Br_2(aq)$
c $C_2H_4(g) + O_2(g) \longrightarrow CO_2(g) + H_2O(l)$
d $Zn(l) + Fe_2O_3(l) \longrightarrow Fe(l) + ZnO(s)$
e $TiCl_4(l) + Mg(l) \longrightarrow Ti(l) + MgCl_2(l)$
f $NH_3(g) + O_2(g) + H_2O(l) \longrightarrow HNO_3(l)$
g $Pb(NO_3)_2(s) \longrightarrow PbO_2(s) + NO_2(g) + O_2(g)$
h $Al(s) + HCl(aq) \longrightarrow AlCl_3(aq) + H_2(g)$
i $C_2H_5OH(l) + O_2(g) \longrightarrow CO_2(g) + H_2O(l)$
j $Na_2CO_3(s) + HCl(aq) \longrightarrow$
$$NaCl(aq) + H_2O(l) + CO_2(g)$$

5 Mercury(II) oxide breaks down into mercury and oxygen when heated. The equation for the reaction is:
$2HgO(s) \longrightarrow 2Hg(l) + O_2(g)$
a Calculate the mass of 1 mole of mercury(II) oxide. Calculate the mass of 1 mole of mercury(II) oxide. (O = 16, Hg = 201)
b Find how much mercury and oxygen are produced when 21.7 g of mercury(II) oxide is heated.

6 Iron(II) sulphide is formed when iron and sulphur react together:
$Fe(s) + S(s) \longrightarrow FeS(s)$
a How many grams of sulphur will react with 56 g of iron? (The RAMs on page 63 will help you.)
b If 7 g of iron and 10 g of sulphur are used, which substance is in excess?
c If 7 g of iron and 10 g of sulphur are used, name the substances present when the reaction is complete, and find the mass of each.
d What mass of iron would react completely with 10 g of sulphur?

7 The following equation represents a reaction in which iron is obtained from iron(III) oxide:
$Fe_2O_3(s) + 3CO(g) \longrightarrow 2Fe(s) + 3CO_2(g)$
a Write a word equation for the reaction.
b What is the formula mass of iron(III) oxide? (The RAMs are: Fe = 56, O = 16.)
c How many moles of Fe_2O_3 are present in 320 kg of iron(III) oxide? (1 kg = 1000 g)
d How many moles of Fe are obtained from 1 mole of Fe_2O_3?
e From **c** and **d**, find the number of moles of iron atoms obtained from 320 kg of iron(III) oxide.
f Find the mass of iron obtained from 320 kg of iron(III) oxide.

8 What is the volume at rtp in dm^3 and cm^3, of:
a 2 moles of hydrogen, H_2?
b 0.5 moles of carbon?
c 0.01 moles of nitrogen, N_2?
d 0.3 moles of oxygen, O_2?
e 0.1 moles of neon, Ne?
f 0.003 moles of ammonia, NH_3?

9 The volumes of some different gases were measured at rtp. The readings were:
  i hydrogen, $6\,dm^3$
  ii oxygen, $3\,dm^3$
  iii neon, $2400\,cm^3$
  iv carbon monoxide, $600\,dm^3$
  v carbon dioxide, $1.2\,dm^3$
  vi sulphur dioxide, $480\,cm^3$

  a What does rtp mean?
  b Given the volumes, calculate the number of moles of each gas present.
  c Now calculate the number of grams of each gas present.

10 The compound sodium hydrogen carbonate, $NaHCO_3$, decomposes as follows when heated:
  $2NaHCO_3(s) \longrightarrow Na_2CO_3(s) + H_2O(l) + CO_2(g)$
  a Write a word equation for the reaction.
  b i How many moles of sodium hydrogen carbonate are there on the left side of the equation?
    ii What is the mass of this amount?
    $(Na = 23, H = 1, C = 12, O = 16)$
  c i How many moles of carbon dioxide are there on the right side of the equation?
    ii What is the volume of this amount at rtp?
  d What volume at rtp of carbon dioxide would be obtained if:
    i 84 g of sodium hydrogen carbonate were completely decomposed?
    ii 8.4 g of sodium hydrogen carbonate were completely decomposed?

11 When calcium carbonate is heated strongly, the following chemical change occurs:
  $CaCO_3(s) \longrightarrow CaO(s) + CO_2(g)$
  $(Ca = 40, C = 12, O = 16)$
  a Write a word equation for the chemical change.
  b How many moles of $CaCO_3$ are there in 50 g of calcium carbonate?
  c i What mass of calcium oxide is obtained from the thermal decomposition of 50 g of calcium carbonate?
    ii What mass of carbon dioxide would be given off at the same time?
    iii What volume would the gas occupy at rtp?

12 Nitrogen monoxide reacts with oxygen to form nitrogen dioxide. The equation is:
  $2NO(g) + O_2(g) \longrightarrow 2NO_2(g)$
  a How many moles of oxygen molecules react with 1 mole of nitrogen monoxide molecules?
  b What volume of oxygen will react with $50\,cm^3$ of nitrogen monoxide?
  c What is the total volume of the two reactants?
  d What volume of nitrogen dioxide will be obtained?
  e Explain why there is a reduction in volume in the reaction. How could the volume be further reduced?

13 Nitroglycerine, $C_3H_5(NO_3)_3$, is used as an explosive. The equation for the explosion reaction is:
  $4C_3H_5(NO_3)_3(l) \longrightarrow$
  $\qquad 12CO_2(g) + 10H_2O(l) + 6N_2(g) + O_2(g)$
  a How many moles of nitroglycerine are represented in the equation?
  b How many moles of *gas molecules* does this produce?
  c How many moles of gas molecules are produced from 1 mole of nitroglycerine?
  d What is the total volume of gas (at rtp) produced from 1 mole of nitroglycerine?
  e What is the mass of 1 mole of nitroglycerine? $(H = 1, C = 12, N = 14, O = 16)$
  f What would be the total volume of gas (at rtp) produced from 1 kg of nitroglycerine explosive?
  g Why is the volume of gas actually produced likely to be much larger than this?

14 Hydrogen peroxide is a colourless liquid which is used as a bleach. It decomposes like this:
  $2H_2O_2(aq) \longrightarrow 2H_2O(l) + O_2(g)$
  It is sold as a 3% solution, which means that $1\,dm^3$ of solution contains 30 g of hydrogen peroxide.
  a Why do bubbles appear in the solution as the hydrogen peroxide decomposes?
  b i What is the mass of 1 mole of hydrogen peroxide? $(H = 1, O = 16)$
    ii What is the concentration in $moles/dm^3$ of a 3% solution?
    iii How many moles of oxygen molecules are formed when $1\,dm^3$ of the solution decomposes?
    iv What volume of oxygen is this at rtp?

15 Magnesium carbonate reacts with hydrochloric acid as follows:
  $MgCO_3(s) + 2HCl(aq) \longrightarrow$
  $\qquad MgCl_2(aq) + H_2O(l) + CO_2(g)$
  a How many moles of magnesium carbonate will react completely with $100\,cm^3$ of 2 M acid?
  b What mass of magnesium carbonate is this? $(Mg = 24, C = 12, O = 16, Cl = 35.5)$
  c What volume of carbon dioxide will be released at rtp?
  d What mass of magnesium chloride will be obtained?

16 2 g (an excess) of iron is added to $50\,cm^3$ of 0.5 M sulphuric acid. After the reaction, when all the hydrogen has bubbled off, the mixture of iron sulphate solution and unreacted iron is filtered. The unreacted iron is dried and weighed. Its mass is 0.6 g.
  a What mass of iron took part in the reaction?
  b How many moles of iron atoms is this? $(Fe = 56)$
  c How many moles of sulphuric acid reacted?
  d Write the equation for the reaction, and deduce the charge on the iron ion that was formed.
  e What volume of hydrogen (calculated at rtp) bubbled off during the reaction?

6.1 Breaking down compounds

Thousands of different reactions are going on as you read this: in laboratories, factories, kitchens, car engines, the atmosphere, the earth beneath you, and within your body. We can divide them into different types of reaction, as you'll see in this and the next three Units.

Decomposition

All the reactions in this Unit have one thing in common: *there is only one reactant*, and it breaks down into two or more simpler products. This is called **decomposition**. It can be brought about by heat, light, electricity, and even enzymes.

Decomposition by heat

Decomposition by heat is called **thermal decomposition**.
An example is the decomposition of calcium carbonate (limestone). This reaction is carried out on a large scale in lime kilns. The quicklime that is produced in the reaction is used on soil to make it less acidic, and to make mortar:

calcium carbonate $\xrightarrow{\text{heat}}$ calcium oxide $+$ carbon dioxide
limestone (quicklime)

$$CaCO_3\,(s) \xrightarrow{\text{heat}} CaO\,(s) \quad + \quad CO_2\,(g)$$

To make cement, limestone is mixed with clay and heated in a giant kiln. The first reaction to take place is the thermal decomposition of limestone to calcium oxide. This then reacts with silica in the clay to form calcium silicate. You can find out more about making cement on page 165.

Decomposition by light

Silver chloride is a white solid. If you expose it to daylight, it quickly breaks down to give tiny black crystals of silver:

silver chloride $\xrightarrow{\text{light}}$ silver $+$ chlorine

$$2AgCl\,(s) \xrightarrow{\text{light}} 2Ag\,(s) \quad + \quad Cl_2\,(g)$$

Silver bromide and silver iodide decompose in the same way. These reactions are important in photography, as you'll see on page 94.

Electrolysis

A powerful way to decompose a substance is to pass electricity through it. For example, if you connect two graphite rods to a battery and stand them in molten lithium chloride, the lithium chloride decomposes. You get beads of silvery lithium at one rod and bubbles of chlorine gas at the other:

lithium chloride $\xrightarrow{\text{electricity}}$ lithium $+$ chlorine

$$2LiCl\,(l) \xrightarrow{\text{electricity}} 2Li\,(l) \quad + \quad Cl_2\,(g)$$

The process is called **electrolysis** and it is very important in industry. For example, it is used to extract aluminium from its ore, and to get chlorine from sodium chloride. You can find out more about it in Section 7.

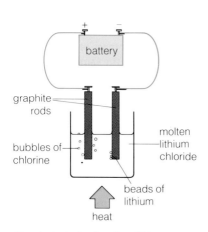

The electrolysis of molten lithium chloride

Fermentation

Like all other living organisms, yeast and bacteria need to feed. They do so by producing enzymes that cause decomposition. For example, yeast produces an enzyme that breaks down the sugar glucose from fruit and grains into ethanol (alcohol) and carbon dioxide. This process is called **fermentation**. It releases energy which the yeast cells use to multiply:

$$\text{glucose} \xrightarrow{\text{yeast}} \underset{\text{(alcohol)}}{\text{ethanol}} + \text{carbon dioxide}$$

$$C_6H_{12}O_6\,(aq) \xrightarrow{\text{yeast}} 2C_2H_5OH\,(aq) + 2CO_2\,(g)$$

Without fermentation, the beer and wine industries wouldn't exist.

The home wine-making kit. Note the water trap to keep air out. This is necessary because fermentation is **anaerobic** – it takes place in the absence of oxygen.

Cracking

Cracking is another important decomposition reaction in industry. A big molecule is cracked into smaller ones using heat and a catalyst (so it is thermal decomposition). Cracking is used for compounds obtained from crude oil. For example, big molecules that aren't so commercially useful are cracked into smaller ones that can be used in petrol.

Crude oil is a mixture of hundreds of different compounds called **hydrocarbons**, which contain only carbon and hydrogen. They consist of chains or rings of carbon atoms with hydrogen bonded on. The diagram below shows how a hydrocarbon could be cracked:

decane

pentane + propene + ethene

Note that two of the products have double bonds between carbon atoms. This makes them reactive. Cracking always gives products with double bonds. These can then be turned into plastics and other useful products, as you'll see in the next Unit.

Cracking ethane

Ethane is a hydrocarbon with *small* molecules. But even it can be cracked:

ethane $\xrightarrow[\text{>800°C}]{\text{steam}}$ ethene + hydrogen

There is more about this on page 246.

---

1 What does *decomposition* mean in chemistry?
2 Give one example of a reaction where decomposition is brought about by:
   a heat **b** light **c** electricity
3 What name is given to decomposition brought about by: **a** heat? **b** electricity?

4 Some decomposition reactions depend on living organisms. Give an example of one which:
   a is commercially useful
   b takes place in your body.
5 What is *cracking*? Give an example.
6 Why is cracking important in industry?

6.2 Building up compounds

In the last Unit, all the reactants were broken down into simpler substances. In this unit they are built into more complex ones.

Combination or synthesis

Often two or more substances react together to form *just one product*. The reaction is called a **combination** or **synthesis**.

iron + sulphur $\longrightarrow$ iron sulphide
$Fe\ (s) + S\ (s) \longrightarrow FeS\ (s)$

Polymerisation

In **polymerization,** small molecules join up to form a long chain. That is how all plastics are made! There are two kinds of polymerization reaction: **addition** and **condensation**.

1 Addition polymerization Look what happens when the gas ethene is heated at very high pressure:

ethene molecules (monomers)

polymerization

part of a polythene molecule (a polymer)

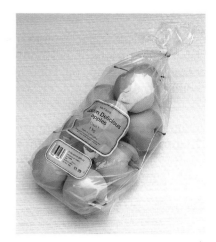

One of the many uses for polythene

The double bonds break, and the ethene molecules join up to make very long molecules with thousands of carbon atoms. The result is **polythene**, the plastic used for plastic bags.

Polythene is a **polymer**. The small starting molecules are called **monomers**. The reaction is called an **addition polymerization** because the monomers just add on to each other.

Below are two more examples. Note the shorthand equations!

The monomer	Part of the polymer molecule	The equation for the reaction
chloroethene (vinyl chloride)	polyvinyl chloride (PVC)	n stands for a large number!
propene	polypropene	

2 Condensation polymerization Nylon and terylene are also polymers. To make them, two *different* monomers join. This table shows the monomers and their reactions. (You don't have to remember their names!) The parts in colour do not change during the reaction so you can draw them as simple blocks.

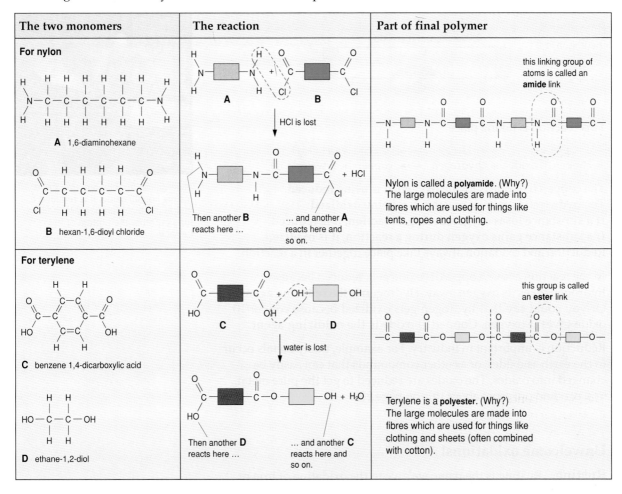

The two monomers	The reaction	Part of final polymer
For nylon **A** 1,6-diaminohexane **B** hexan-1,6-dioyl chloride	HCl is lost Then another **B** reacts here … … and another **A** reacts here and so on.	this linking group of atoms is called an **amide** link Nylon is called a **polyamide**. (Why?) The large molecules are made into fibres which are used for things like tents, ropes and clothing.
For terylene **C** benzene 1,4-dicarboxylic acid **D** ethane-1,2-diol	water is lost Then another **D** reacts here … … and another **C** reacts here and so on.	this group is called an **ester** link Terylene is a **polyester**. (Why?) The large molecules are made into fibres which are used for things like clothing and sheets (often combined with cotton).

Look at the reactions. In each, a small molecule is released during the polymerization: hydrogen chloride in the first, water in the second.
If a small molecule is released during polymerization, the reaction is a condensation polymerization.

Condensation polymerizations are common in nature. For example in plants, when glucose is changed into starch. (See page 302 for more.)

1 What happens during *synthesis*? Give an example.
2 What is: **a** a monomer? **b** a polymer?
3 Draw a diagram to show what happens when vinyl chloride molecules polymerize to form PVC.
4 Write an equation for the polymerization of ethene.

5 **a** Draw a diagram to show how nylon is formed.
  b It is called a *condensation* polymerization. Why?
  c Nylon is called a *polyamide*. Why?
6 **a** Draw a diagram to show the structure of terylene.
  b Explain why terylene is called a polyester.

6.3 Gaining and losing oxygen

Oxidation and reduction

In this Unit, we look at reactions where oxygen is added or removed. Let's start with the reaction between hydrogen and heated black copper(II) oxide, shown below. The black powder turns pink:

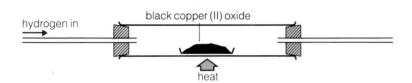

This reaction is taking place:

$$CuO\,(s) \quad + \quad H_2\,(g) \quad \longrightarrow \quad Cu\,(s) \ + \ H_2O\,(g)$$
copper(II) oxide + hydrogen ⟶ copper + water

The copper(II) oxide is losing oxygen. It is being **reduced**.
The hydrogen is gaining oxygen. It is being **oxidized**.
If a substance loses oxygen during a reaction, it is reduced.
If a substance gains oxygen during a reaction, it is oxidized.
Reduction and oxidation always take place together in a reaction.

In the reaction above, copper(II) oxide gets reduced because hydrogen takes its oxygen away. Hydrogen is the **reducing agent**. Or you could say that hydrogen gets oxidized because copper(II) oxide gives it oxygen. Copper(II) oxide is the **oxidizing agent**.

Reduction is important in industry. For example, many metals occur in the earth as oxides, or as other compounds that can easily be changed into oxides. The oxides are reduced to get the pure metal. You can find out more about this on page 203.

Making iron from its ore. Iron ore is mainly iron(III) oxide. It is reduced to iron in the blast furnace. Molten iron runs out from the bottom of the furnace.

Unwelcome oxidations!

Rusting Rusting is the name we give to the oxidation of iron or steel in damp air. It is also called **corrosion**:

$$4Fe\,(s) \ + \ 3O_2\,(g) \ + \quad 2H_2O(g) \quad \longrightarrow \qquad 2Fe_2O_3.H_2O$$
iron + oxygen + water vapour ⟶ hydrated iron(III) oxide (rust)

Rust weakens structures such as car bodies, iron railings, and ships' hulls, and shortens their useful life. Preventing it can cost a lot of money. You can find out more about it on page 210.

Rancidity Oxidation can also cause things to 'go off' in the kitchen. For example, butter, margarine, and cooking oils all contain **unsaturated fats**, which are fats containing carbon–carbon double bonds. The double bonds are unstable so oxygen reacts with them giving substances with unpleasant flavours and smells. The oil, or butter, or margarine, turns **rancid**.

Manufacturers sometimes add **anti-oxidants** to fats, oils, and fatty foods to prevent oxidation taking place. You can also slow down oxidation by storing foods in a cool dark place.

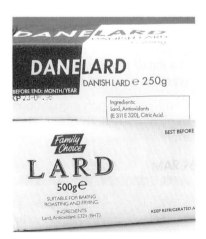

Lard contains unsaturated fats, so should be stored in the fridge. (Why?)

Combustion

Combustion is often just called **burning**. It usually means the reaction of a substance with oxygen from the air. During it:
- the substance that burns is oxidized to produce oxides;
- energy is given out in the form of heat, light, and sound.

Here are some examples of combustion.

1 The combustion of magnesium in the lab:

$$2Mg\,(s) \quad + \quad O_2\,(g) \quad \longrightarrow \quad 2MgO\,(s)$$
magnesium + oxygen $\longrightarrow$ magnesium oxide

The reaction gives out heat, brilliant white light, and a fizzing noise.

2 The combustion of North Sea gas in a central heating system. North Sea gas is mainly **methane**. It burns like this:

$$CH_4\,(g) \quad + \quad 2O_2\,(g) \quad \longrightarrow \quad CO_2\,(g) \quad + \quad 2H_2O\,(g)$$
methane + oxygen $\longrightarrow$ carbon dioxide + water vapour

The reaction gives out a little light and sound in the boiler, and a great deal of heat which is carried around the house. Note that this time *two* different oxides are produced! They are both waste gases and are carried out into the air through the boiler flue.

Controlled combustion in safe conditions

Help, fire!

Sometimes combustion is unwelcome – like when your house goes on fire!

A fire needs three things to keep it going: oxygen, fuel, and heat. This is shown in the **fire triangle**.

Remove any one of them, and the fire will go out.

So to put out a fire you can:
1 **Cut off the fuel.** Turn off the gas or electricity. Cover puddles of oil or petrol with sand or soil.
2 **Get rid of the heat.** Cool things down with water. But don't use water on burning petrol or oil because they will just float and spread the fire further. And don't use it on electrical appliances because it can conduct electricity and give you a shock.
3 **Cut off the air supply.** Cover burning things with foam, carbon dioxide, or a fire blanket. But don't use foam on electrical appliances because it too can conduct and give shocks.

With the air supply cut off, the chip-pan fire will go out. No panic!

---

1 $2Mg\,(s) + SO_2\,(g) \longrightarrow 2MgO\,(s) + S\,(s)$
In this reaction, which substance is:
a oxidized? b the oxidizing agent?
c reduced? d the reducing agent?

2 Copy and complete: Combustion usually means burning in .... The burning substance is ...... and one or more ...... are produced. ...... is also given out. Combustion in body cells is called ......

6.4 Redox reactions

Look again at the reaction between copper(II) oxide and hydrogen from the last Unit:

$$CuO\,(s) + H_2\,(g) \longrightarrow Cu\,(s) + H_2O\,(g)$$

The copper(II) oxide is reduced: oxygen is removed from it.
The hydrogen is oxidized: oxygen is added to it.
Oxidation and reduction are partners in the reaction. They *always* take place together.
When one substance in a reaction is oxidized, another is reduced. The reaction is called a redox reaction.
So combustion and rusting are redox reactions.

Another definition for oxidation and reduction

When magnesium burns in oxygen, magnesium oxide is formed:

$$2Mg\,(s) + O_2(g) \longrightarrow 2MgO\,(s)$$

It's obvious that the magnesium has been oxidized. Oxygen is the only other reactant, so it must have been reduced. But how? We need to look at what's happening to the electrons:

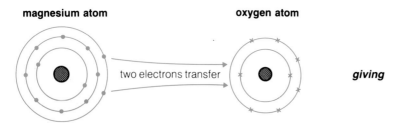

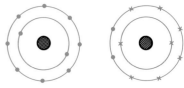

magnesium atom   **oxygen atom**   two electrons transfer   *giving*   **magnesium ion, Mg^{2+}**   **oxide ion, O^{2-}**

During the reaction, a magnesium atom loses two electrons and an oxygen atom gains two. This leads us to a new definition:
If a substance loses electrons during a reaction it has been oxidized. If it gains electrons it has been reduced.

> **Remember OIL RIG!**
>
> **O**xidation **I**s **L**oss of electrons.
> **R**eduction **I**s **G**ain of electrons.

Reactions that don't involve oxygen

The above definition means that a reaction can be a redox reaction *even if it doesn't involve oxygen*. All that's needed is a transfer of electrons. Let's look at some examples.

1 Sodium and chlorine. As you saw on page 36, sodium burns in chlorine to form sodium chloride:

$$2Na + Cl_2 \longrightarrow 2NaCl$$

You can also write this as an **ionic equation**:

$$2Na + Cl_2 \longrightarrow 2Na^+ + 2Cl^-$$

Sodium has lost electrons. It has been oxidized. Chlorine has gained electrons. It has been reduced. The reaction is a redox reaction. (Note that it is also a combination reaction.)

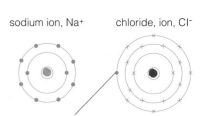

sodium ion, Na$^+$ chloride, ion, Cl$^-$

the sodium atom has lost an electron to the chlorine atom

2 Iron(II) chloride and chlorine. The (II) in iron(II) chloride means it contains Fe^{2+} ions. Let's see how it reacts with chlorine:

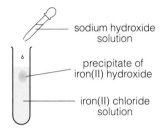

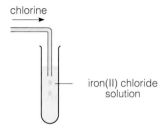

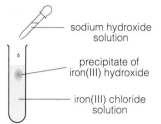

A solution of iron(II) chloride. Add some sodium hydroxide solution and a green precipitate of iron(II) hydroxide forms, proving Fe^{2+} ions are present.

Chlorine is bubbled through the iron(II) chloride solution. This is done in a fume cupboard. (Why?) The solution doesn't look much different afterwards.

But adding sodium hydroxide solution now gives a red-brown precipitate of iron(III) hydroxide. This proves Fe^{3+} ions are present.

The reaction with chlorine is:

$$\underset{\text{iron(II) chloride}}{2FeCl_2(aq)} + Cl_2(g) \longrightarrow \underset{\text{iron(III) chloride}}{2FeCl_3(aq)}$$

The ionic equation for this reaction is:

$$2Fe^{2+} + Cl_2 \longrightarrow 2Fe^{3+} + 2Cl^-$$

The iron(II) ions have given up electrons to chlorine. They have been oxidized. Chlorine has been reduced.

3 Potassium bromide and chlorine. When chlorine is bubbled through potassium bromide solution, the liquid turns red-brown because bromine forms. The reaction is:

$$Cl_2(g) + 2KBr(aq) \longrightarrow 2KCl(aq) + Br_2(aq)$$

The ionic equation is:

$$Cl_2(g) + 2Br^-(aq) \longrightarrow 2Cl^-(aq) + Br_2(aq)$$

Bromine has given up electrons to chlorine. The bromine has been oxidized and the chlorine reduced.

In effect chlorine has pushed bromine out of the potassium compound and taken its place. It has **displaced** bromine. Redox reactions like these, where one element displaces another from a dissolved compound, are called **displacement reactions**. You'll find other examples on pages 195 and 233.

A reactive element 'likes' to exist as ions. So which is more reactive, bromine or chlorine? Check on page 233 to see if you're right.

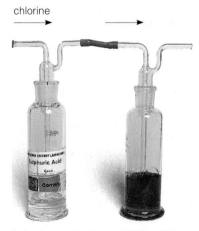

Colourless potassium bromide solution turns orange as chlorine is bubbled in. What is the purpose of the sulphuric acid? (Hint: look on page 231.)

---

1 Where does the word *redox* come from?
2 Is it possible to have oxidation without reduction?
3 Write a definition of oxidation and reduction in terms of electron transfer.
4 What is a *displacement* reaction? Give an example.

5 When molten lithium chloride is electrolysed (page 86) it decomposes:
$$2LiCl(l) \longrightarrow 2Li(l) + Cl_2(g)$$
a Write the ionic equation for this reaction.
b Is the decomposition a redox reaction? Explain.

6.5 Precipitation and neutralization

Water is the best solvent we've got, and it's still quite cheap. For this reason, large numbers of reactions in industry and in the lab are carried out in aqueous solution. The reactions within your body take place in aqueous solution too.

In the Units so far in this Section, some reactions have been in solution, and some have not. In this Unit we look at two further types of reaction. Both of them usually take place in solution.

Precipitation

When two aqueous solutions are mixed, they may react to give a product that is *not* soluble in water.

For example, look what happens when solutions of potassium bromide and silver nitrate are mixed:

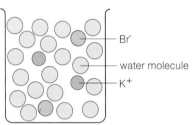

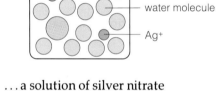

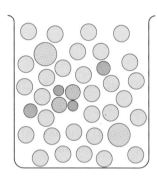

A solution of potassium bromide contains potassium ions and bromide ions, and . . .

. . . a solution of silver nitrate contains silver ions and nitrate ions. When the two are mixed . . .

. . . the silver and bromide ions attract each other strongly. An off-white solid forms.

The off-white solid is silver bromide. It forms as a **precipitate**. The reaction is a **precipitation reaction**.

The equation for the reaction is:

$$AgNO_3\,(aq) \;+\; KBr\,(aq) \;\longrightarrow\; AgBr\,(s) \;+\; KNO_3\,(aq)$$
silver nitrate   potassium bromide   silver bromide   potassium nitrate

How would you obtain dry silver bromide from the final mixture? You would need to do this in the dark. Why?

Some uses of precipitation

1 In photography   The reaction above is used in making film for your camera. Solutions of silver nitrate and potassium bromide are mixed with gelatine, and a precipitate of tiny crystals of silver bromide forms. The mixture is coated on to clear film.

When light strikes the film, the silver bromide breaks down to silver, as you saw on page 86. The silver ions are reduced to silver:

$$2AgBr \longrightarrow 2Ag + Br_2 \quad \text{or} \quad 2Ag^+ + 2e^- \longrightarrow 2Ag$$

The amount that breaks down depends on how much light gets through the camera lens. When the film is developed, the remaining silver bromide is converted to a soluble substance and washed away. This leaves a 'negative' image on the film in silver. It is converted to a 'positive' image by shining a light through it on to photographic paper.

Silver iodide is also used in film, and silver chloride in photographic printing paper. Both break down in light in the same way as silver bromide.

2 To remove ions you don't want   For example, when tap water is pumped from a bore hole it may contain too many iron compounds and not enough dissolved oxygen. It will taste stale. So it is **aerated** by bubbling oxygen through it. At the same time this precipitates Fe^{3+} ions as brown iron(III) oxide, which is removed by filtering the water through sand.

3 To test for the presence of ions   Precipitation reactions help you to play detective in the lab. For example, suppose you suspect that a solution contains lead ions. Just add a few drops of sodium chloride solution. If lead ions are present, lead chloride will precipitate. Unlike silver chloride, it will not turn black in light. (See page 294 for this and other ion tests.)

4 To make insoluble salts in the lab   You can find out more about this in Unit 9.6 on page 148.

In the dark room. Every year, around 1500 tonnes of silver are used for photography, in the form of silver chloride, bromide, and iodide.

Neutralization

When acids react with substances called **bases**, their acidity is cancelled out. The reaction is called a **neutralization**. It always produces a salt and water. The base may be soluble, like sodium hydroxide (when it is called an **alkali**), or insoluble, like copper(II) oxide:

$$\text{copper(II) oxide} + \text{sulphuric acid} \longrightarrow \text{copper(II) sulphate} + \text{water}$$
$$CuO\,(s) \quad + \quad H_2SO_4\,(aq) \quad \longrightarrow \quad CuSO_4\,(aq) \quad + H_2O\,(l)$$

Some uses of neutralization

Neutralization is useful in all kinds of ways. For example, it is used for:
- making fertilizers and other salts in industry
- making salts in the lab
- reducing the acidity of soil
- treating insect stings and upset stomachs.

You can find out more about neutralization and its uses in Section 9.

Spreading a nitrate fertilizer – made by neutralization

1 Describe two uses of precipitation reactions.
2 Silver chloride is used to coat photographic paper. What solutions could you mix to precipitate it?

3 What does *neutralization* mean?
4 How would you obtain crystals of blue copper(II) sulphate using the neutralization reaction above?

6.6 Energy changes in reactions

You have met many different reactions so far in this section. But they all have one thing in common: they involve an energy change.

When iron and sulphur react together, it's easy to see that energy is given out. The reacting mixture glows.

Mixing silver nitrate and sodium chloride solutions gives a white precipitate – and a temperature rise!

But adding a mixture of citric acid and sodium hydrogen carbonate to water produces bubbles and a fall in temperature.

The first two reactions give out heat energy. They are **exothermic**. But in the third reaction, the temperature of the water falls because the reaction takes heat energy from it. It is **endothermic**.

Other examples

Exothermic reactions   These are all exothermic:
- the neutralization of an acid by an alkali.
- the combustion of fuels. If they didn't give out heat energy they'd be no good as fuels!
- respiration in your body cells. It provides the energy to keep your heart and lungs working, and for warmth and movement.

Endothermic reactions   These are all endothermic:
- the reactions that take place within food during cooking.
- the polymerization of ethene to poythene.
- the reduction of silver ions to silver in photography (page 94).
- electrolysis. Energy is provided in the form of electricity.

Just as in physics, the unit for energy in chemistry is the **joule** (J). 1 kilojoule (kJ) = 1000 joules.

Measuring the heat change in a reaction

As you saw above, the neutralization of an acid by an alkali is always exothermic. Suppose you add hydrochloric acid to sodium hydroxide solution. The neutralization reaction is:

$$HCl\,(aq) + NaOH\,(aq) \longrightarrow NaCl\,(aq) + H_2O\,(l)$$
$$\text{acid} + \text{alkali} \longrightarrow \text{salt} + \text{water}$$

How could you measure the energy change? Read on ...

Reactions in aqueous solution

① In an exothermic reaction, heat energy is transferred from the reactants to the water. The temperature of the solution rises.

② In an endothermic reaction, heat energy is transferred from the water to the reactants. The temperature of the solution falls.

Sherbert is a mixture of citric acid and sodium hydrogen carbonate. So what happens on your tongue?

The experiment   This is what you could do:
- Measure out 50 cm³ of a 2M solution of sodium hydroxide, and 50 cm³ of a 2M solution of hydrochloric acid.
- Put the sodium hydroxide solution in a polystyrene cup. (Why polystyrene?) Measure its temperature.
- Add the acid. Stir the solution, note how the temperature changes, and record the highest temperature reached.

The results   Your results could be something like this:

Initial temperature:	18 °C
Highest temperature reached:	31 °C
Temperature rise:	13 °C

The calculation   First, you need this information:

> - **1 cm³ of water has a mass of 1 g.**
> - **It takes 4.2 joules of energy to raise the temperature of 1 g of water by 1 °C.**

The solution is not pure water, but we can assume it is. So:
a rise of 1 °C in 1 g of water takes 4.2 joules of energy, and
a rise of 13 °C in 100 g takes 100 × 13 × 4.2 joules or 5460 J.

The neutralization has given out **5460 J** of energy, or **5.46 kJ**.

Calculating the heat of neutralization

The **heat of neutralization** for this reaction is the heat change when 1 mole of hydrochloric acid is neutralized. It is shortened to $\Delta H_{neutralization}$. ($\Delta$ or *delta* means *change in*.)

1 To find the number of moles of hydrochloric acid used:
number of moles = volume (dm³) × concentration (mol/dm³)
$$= 0.05 \times 2$$
$$= \mathbf{0.1 \ mole}$$

2 5460 joules is given out in the neutralization of 0.1 mole of acid. 54 600 joules, or 54.6 kilojoules, is given out for 1 mole of acid. $\Delta H_{neutralization}$ is **−54.6 kJ/mol**.
The minus sign shows that the reaction is exothermic.
(The next Unit explains why.)

This is a lower value than you'd find if you looked up data tables. That's because the solution loses some heat to the polystyrene cup. Although polystyrene is a good insulator, it in turn loses a little heat to the air.

> $\Delta H_{neutralization}$
>
> - $\Delta H_{neutralization}$ is in fact the heat change when *1 mole of H⁺ ions* is neutralized.
>
> - But for this reaction, 1 mole of hydrochloric acid is the same as a mole of H⁺ ions:
>
> $HCl \, (aq) \longrightarrow H^+ \, (aq) + Cl^- \, (aq)$
> 1 mole   1 mole

1 What is: **a** an endothermic reaction?
b an exothermic reaction?
Give three examples of each.
2 What unit is used to measure energy?

3 During a reaction, the temperature of 50 cm³ of solution *fell* by 10 °C. How much energy did the reaction take in?
4 What is *heat of neutralization*? What is the symbol for it?

6.7 Explaining energy changes

A mixture of hydrogen and chlorine will explode in sunshine. The equation for the reaction is:

hydrogen + chlorine $\longrightarrow$ hydrogen chloride
$H_2(g)$ + $Cl_2(g)$ $\longrightarrow$ $2HCl(g)$

The explosion is a sign that a lot of energy is given out. But where does it come from? Let's look more closely at what happens:

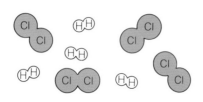 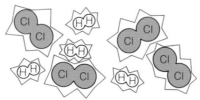 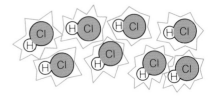

These are hydrogen and chlorine molecules. The atoms are held together by single bonds.

The bonds must be broken before a reaction can take place. This step needs energy. It is **endothermic**.

Now the hydrogen and chlorine atoms react to form new bonds. This step gives out energy. It is **exothermic**.

You can look up **bond energy tables** to find the energy needed to break bonds, and given out when bonds form. For this reaction:

Energy in to break bonds:
for a mole of hydrogen molecules	436 kJ
for a mole of chlorine molecules	242 kJ
Total energy in	678 kJ

Energy out from forming bonds:
for 2 moles of hydrogen chloride	862 kJ

Energy in − energy out = 678 kJ − 862 kJ = −184 kJ

Overall the reaction takes in 184 kJ less energy than it gives out. This can be written as:
$\Delta H_{reaction} = -184$ kJ. The reaction is exothermic.
If the energy taken in to break bonds is *less* than the energy given out when new bonds form, the reaction is exothermic.

> **Heat of reaction, $\Delta H_{reaction}$**
>
> - The heat of reaction is the heat change when the number of moles of reactants shown in the equation react together.
> - For the reaction between hydrogen and chlorine, it is the heat change when one mole of hydrogen molecules reacts with one mole of chlorine molecules to form hydrogen chloride.
> - It is shortened to $\Delta H_{reaction}$.

An endothermic reaction

If you blow steam through white-hot coke, the carbon is oxidized:

carbon + water vapour $\longrightarrow$ carbon monoxide + hydrogen
$C(s)$ + $H_2O(g)$ $\longrightarrow$ $CO(g)$ + $H_2(g)$

As before, energy is *taken in* to break the bonds in the reactants. Energy is *given out* when bonds form in the products. The energy calculation is shown on the right.

The reaction takes in 132 kJ more energy than it gives out.
$\Delta H_{reaction} = +132$ kJ. **The + sign shows energy is taken in overall.** The reaction is endothermic.
If the energy taken in to break bonds is *greater* than the energy given out when new bonds form, the reaction is endothermic.

> **Energy in to break bonds:**
>
> | | |
> |---|---|
> | 1 mole of carbon | 717 kJ |
> | 1 mole of water | 928 kJ |
> | **Total energy in** | **1645 kJ** |
>
> **Energy out from forming bonds:**
>
> | | |
> |---|---|
> | 1 mole of carbon monoxide | 1077 kJ |
> | 1 mole of hydrogen | 436 kJ |
> | **Total energy out** | **1513 kJ** |
>
> **Energy in − energy out =** **132 kJ**

Bond energies

As you saw opposite, it takes 242 kJ to break the bonds in a mole of chlorine molecules. This is called the **bond energy** of chlorine. If the chlorine atoms were to bond to each other again, exactly the same amount of energy would be given out.

Bond energy is the energy needed to break a mole of bonds. The same amount of energy is given out when a mole of these bonds forms.
Bond energy is measured in kilojoules per mole or kJ/mol.
A *high* bond energy value means a bond is *strong*. More energy is needed to break it and more is given out when it forms.

Bond energies in kJ/mole	
C—C	346
C=C	612
C—O	358
O=O	498
O—H	464
H—H	436
N≡N	946
N—H	391

Drawing an energy diagram

You can show the energy change in a reaction on an **energy diagram**.
This compares the energy levels of the reactants and products.

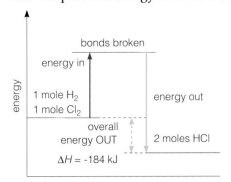

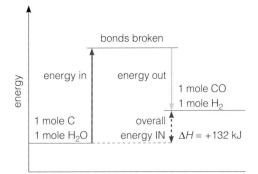

Overall, the reaction between hydrogen and chlorine gives out energy. So hydrogen chloride must contain less energy than the reactants did. This is shown by the negative sign in front of ΔH.

Overall, the reaction between carbon and steam takes in energy. So between them, the products must contain more energy than the reactants did. This is shown by the positive sign in front of ΔH.

Activation energy

All reactions need energy to start them off, even the exothermic ones. This is called the **activation energy**. It is the minimum energy needed to break enough bonds to get a reaction started.

Activation energy is different for different reactions. You need heat from a bunsen flame to start iron filings reacting with sulphur. You need heat from a match to get paper to burn. But the heat in cold water is more than enough to start sodium and water reacting.

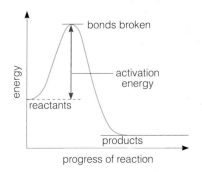

1 When chlorine and hydrogen react:
   a which part of the reaction is endothermic?
   b which is exothermic?
   c why does the reaction *give out* energy overall?
2 Write the equation for the reaction between hydrogen and oxygen. Now sketch the molecules taking part, showing any double bonds.

3 a Using the bond energies given above, calculate the energy change for the reaction in question 2.
   b Is the reaction endothermic or exothermic?
   c What is $\Delta H_{\text{reaction}}$ for it?
4 Draw an energy diagram for the combustion of North Sea gas in a gas cooker. How would you provide the activation energy for this reaction?

6.8 Energy from burning fuels

Measuring the heat given out by fuels

We burn **fuels** to provide us with heat energy. Two examples are **ethanol** and **butane**. Ethanol is widely used as motor fuel in Brazil. Butane is used for camping gas and lighter fuel.

The more heat a fuel gives out the better. The amount of heat given out when a mole of fuel burns is called the **heat of combustion**. This is often written as $\Delta H_{combustion}$.

You measure this in the lab *indirectly* by burning the fuel to heat water. Simple apparatus is shown below right. The basic idea is:

heat gained by the water = heat given out by the fuel.

Can you see any problems with this idea?

> **What happens when they burn?**
>
> Compare these reactions:
>
> ethanol
> $$C_2H_5OH(l) + 3O_2(g) \longrightarrow 2CO_2(g) + 3H_2O(g)$$
>
> butane
> $$2C_4H_{10}(l) + 13O_2(g) \longrightarrow 8CO_2(g) + 10H_2O(g)$$
>
> - What do you notice about the products?
> - Which fuel needs more oxygen per mole of fuel?

The method   These are the steps:

- Pour a measured volume of water into the tin. Since you know its volume you also know its mass. (1 cm^3 of water has a mass of 1 g.)
- Weigh the fuel and its container.
- Measure the temperature of the water.
- Light the fuel and and let it burn for a few minutes.
- Measure the water temperature again, to find the increase.
- Reweigh the fuel and container to find how much fuel was burned.

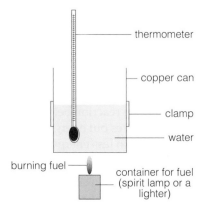

Apparatus for measuring the heat given out by a burning fuel

The calculations   It takes 4.2 J of energy to raise the temperature of 1 g of water by 1 °C. So you can calculate the energy given out when the fuel burns by using this equation:

energy given out (J) =
   4.2 × mass of water used (g) × its rise in temperature (°C)

Then, since you know what mass of fuel you burned, you can work out the energy that would be given out by burning one mole of it.

Example   The experiment gave these results for ethanol and butane. Check that you understand the calculations:

Ethanol (burned in a spirit lamp)	Butane (burned in a butane cigarette lighter)
Results	*Results*
mass of ethanol used: 0.9 g	mass of butane used: 0.32 g
mass of water used: 200 g	mass of water heated: 200 g
temperature rise: 20 °C	temperature rise: 12 °C
Calculations	*Calculations*
Heat given out $= 4.2 \times 200 \times 20\,J = 16800\,J$ or **16.8 kJ**	Heat given out $= 4.2 \times 200 \times 12\,J = 10080\,J$ or **10.08 kJ**
The formula mass of ethanol (C_2H_5OH) is 46.	The formula mass of butane (C_4H_{10}) is 58.
0.9 g gives out 16.8 kJ of energy so	0.32 g gives out 10.08 kJ of energy so
46 g gives out $\dfrac{16.8}{0.9} \times 46 = -859$ kJ of energy.	58 g gives out $\dfrac{10.08}{0.32} \times 58 = -1827$ kJ of energy.
So $\Delta H_{combustion}$ for ethanol is $-$**859 kJ/mol**.	So $\Delta H_{combustion}$ for butane is $-$**1827 kJ/mol**.

How reliable is the experiment?

This table compares the experimental results with values from a data book. Note the big difference! The experimental results are almost 40% lower for both fuels. There are two reasons for this.

1 **Heat loss** Not all the heat from the burning fuel is transferred to the water. Some is lost to the air, and some to the copper can.
2 **Incomplete combustion** In *complete* combustion *all* the carbon in a fuel is converted to carbon dioxide. But here combustion is *incomplete*. Some carbon is deposited as soot on the bottom of the can, and some converted to carbon monoxide. For example, when butane burns, a mixture of all these reactions may take place:

$$2C_4H_{10}(g) \; + \; 13O_2(g) \longrightarrow \; 8CO_2(g) \; + \; 10H_2O(g)$$
$$\text{carbon dioxide}$$

$$2C_4H_{10}(g) \; + \; 9O_2(g) \longrightarrow \; 8CO(g) \; + \; 10H_2O(g)$$
$$\text{carbon monoxide}$$

$$2C_4H_{10}(g) \; + \; 5O_2(g) \longrightarrow \; 8C(s) \; + \; 10H_2O(g)$$
$$\text{carbon}$$

The less oxygen there is, the more carbon monoxide and carbon will form. As you'll see in Unit 11.3, incomplete combustion of petrol (a mixture of octane and other hydrocarbons) in car engines leads to pollution of the atmosphere by carbon monoxide and carbon particles.

Fuel	Heat of combustion (kJ)	
	from the experiment	from a data book
ethanol	−859	−1367
butane	−1827	−2877

heat output decreases

pollution increases

Is it a good fuel?

These are three key factors in deciding what makes a good fuel:
- **how much heat it gives out**. To make comparisons easier you can express this as heat given out per gram of fuel.
- **how much it costs**.
- **how cleanly it burns**. The more cleanly it burns, the more efficient it is as a fuel and the less pollution it causes.

This table compares ethanol and butane for those three factors. Which fuel would *you* choose?

What makes a good fuel?

These are questions to consider:
- How much heat does it give out?
- How much does it cost?
- How cleanly does it burn?
- Does it cause pollution?
- Is it easily available? (If it's cheap it probably is!)
- Does it light easily?
- Does it leave much ash behind?
- Is it easy and safe to store and transport?
- Would it be better used for something else?

Fuel	Heat given out (kJ/g) (by experiment)	Cost per g (pence)	Heat gained for 1p spent (kJ)	Cleanliness
Ethanol	18.7	0.3	62.3	The oxygen in the compound helps combustion. So *less* carbon monoxide and soot form than with butane.
Butane	31.5	5	6.3	Carbon monoxide and soot may form.

1 a What is *heat of combustion*?
  b Will it be positive or negative for a fuel? Explain.
2 Methane (CH_4) is a fuel. When 1 g burns in air the heat given out raises the temperature of 1000 g of water by 13 °C. What is its heat of combustion?

3 Explain why your answer for 2 is likely to be lower than the value you'd find in data tables.
4 Compare ethanol and butane: a Which gives more heat per gram? b Which is better value for money? c Which is better for the environment? Why?

Questions on Section 6

1 Write balanced equations, using the correct symbols, for these chemical reactions:
a the combustion of carbon monoxide to form carbon dioxide
b the decomposition of solid mercury(II) oxide into its elements
c the synthesis of ammonia gas, NH_3, from nitrogen and hydrogen
d the precipitation of barium sulphate when barium chloride and sodium sulphate solutions are mixed
e the reduction of copper(II) oxide by hydrogen
f the polymerization of chloroethene, $CH_2 = CHCl$
g the decomposition of silver chloride by light
h the decomposition of molten aluminium oxide by electricity
i the cracking of ethane to ethene and hydrogen
j the fermentation of glucose to make ethanol.

2 When 6.5 g of zinc (Zn) were added to a solution of copper(II) sulphate ($CuSO_4$), 6.4 g of copper were obtained. (The RAM's are: Zn = 65, Cu = 64.)
a What type of chemical reaction is this?
b How many moles of zinc atoms were used?
c How many moles of copper atoms were obtained?
d Write a word equation for the reaction.
e Use the information from b and c to write a balanced equation for the reaction.

3 When solutions of potassium sulphate and barium chloride are mixed, a white *precipitate* forms. The equation for the reaction is:
$K_2SO_4 (aq) + BaCl_2 (aq) \longrightarrow BaSO_4 (s) + 2KCl (aq)$
a What is a precipitate?
b Which compound above is the precipitate?
c How would you separate the precipitate from the solution?
d i What would remain if the precipitate was removed?
   ii How would you obtain this substance as a dry solid?
e A precipitate also forms if sodium sulphate is used instead of potassium sulphate, above. Write an equation for the reaction.

4 23.3 g of barium sulphate was obtained from the precipitation reaction in question 3.
a What is the formula mass of barium sulphate? (Ba = 137, S = 32, O = 16)
b How many moles of $BaSO_4$ is 23.3 g of barium sulphate?
c How many moles of K_2SO_4 and $BaCl_2$ must have reacted, to form 23.3 g of barium sulphate?
d The concentrations of the two reacting solutions were 0.1 M (that is, 0.1 mole of solute in 1 litre of solution). What volume of each solution was needed to form 23.3 g of barium sulphate?

5 Crude oil is an important raw material. Fractional distillation of crude oil produces a number of useful products. The distillation process, however, produces too much of the 'heavy' fractions. There is greater demand for 'lighter' fractions. The heavy fractions can be converted to lighter fractions by cracking. The products obtained are used as fuels and monomers. The fuels are burned to provide energy. The monomers are used for the production of plastics.
a Why is there more demand for the lighter fractions?
b Which process mentioned above involves a physical rather than chemical change?
c What type of chemical reaction is cracking?
d What type of chemical reaction is involved in the production of plastics?
e i Which of the chemical changes mentioned in the passage is an oxidation?
   ii What other name is used for this type of reaction?

6 Water at 25 °C was used to dissolve two compounds. The temperature of each solution was measured immediately after the compound had dissolved.

Compound	Temperature of solution/°C
NH_4NO_3	21
$CaCl_2$	45

a Name the two compounds.
b Calculate the temperature change for each.
c Which compound dissolved exothermically?
d Which dissolved endothermically?
e For each solution, estimate the temperature of the the solution if:
   i the amount of water was halved but the same mass of compound was used
   ii the mass of the compound was halved but the volume of water was unchanged
   iii both the mass of the compound and the volume of water were halved.

7 Sodium hydroxide and hydrochloric acid react as shown in this equation:
$NaOH (aq) + HCl (aq) \longrightarrow NaCl (aq) + H_2O (l)$
The reaction is exothermic.
a What type of chemical reaction is it?
b Is heat given out or taken in during the reaction, or neither?
c In that case, what happens to the temperature of the solutions as they react?
d What volume of 0.5 M hydrochloric acid would react completely with 50 cm^3 of 1.0 M sodium hydroxide solution?
e If *twice* the volume of acid was added to the sodium hydroxide solution, what would you expect to happen to the temperature? Explain.

8 Calor gas is a hydrocarbon called propane, C_3H_8. In an experiment, 1 gram of propane was used to heat $1000\,cm^3$ of water. The temperature of the water rose from 20 °C to 32 °C.

a By how much did the water temperature rise?

b Is the burning of Calor gas exothermic or endothermic?

c i Calculate the heat energy given out when 1 gram of propane burns. (4.2 kJ raises the temperature of $1000\,cm^3$ of water by 1 °C.)

   ii What is the heat energy given out when 1 mole of propane burns? (C = 12, H = 1)

Lighter fuel is another hydrocarbon, called butane (C_4H_{10}). When 1 mole of butane burns, 2900 kJ of heat energy is given out.

d Use the idea of making and breaking bonds to explain why burning 1 mole of butane produces more energy than burning 1 mole of propane.

9 When ammonium chloride (NH_4Cl) dissolves in water, the temperature drops by a few degrees.

a i Is the dissolving exothermic or endothermic?

   ii What can you say about the bonds that are formed with water in the solution?

When 1 mole of ammonium chloride dissolves in water, 15 kJ of energy is taken in.

b How much energy is taken in when: i 107 g

   ii 1 kg of ammonium chloride dissolves in water?

10 Methane is the main component of natural gas. The reaction between methane and oxygen is exothermic:
$$CH_4(s) + O_2(g) \longrightarrow CO_2(g) + 2H_2O(g)$$

a Explain in terms of bond breaking and bond making why this reaction is exothermic.

b i Copy the diagram below and complete it to show the energy diagram for this reaction.

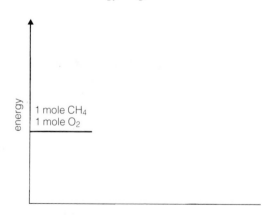

   ii Methane will not burn in air until a spark or flame is applied. Why not?

When 1 mole of methane burns in oxygen, 890 kJ of energy is given out.

c How much energy is given out when 1 g of methane burns? (C = 12, H = 1)

11 This energy diagram represents the reaction between citric acid solution and sodium hydrogen carbonate:

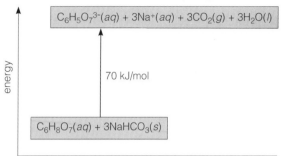

a Write the chemical equation for the reaction.

b What is the energy change, ΔH, for the reaction?

c Is the reaction endothermic or exothermic?

d Which energy change in the reaction is greater, the breaking or the forming of bonds?

e Which have stronger bonds, the reactants or the products? Explain your answer.

12 Hydrazine, N_2H_4, burns in oxygen as follows:

$$\begin{array}{c} H \quad\quad H \\ \backslash\quad / \\ N-N \quad (g) + O=O(g) \longrightarrow N\equiv N(g) + 2\ \overset{H}{\underset{H}{\diagup}}O(g) \\ / \quad\quad \backslash \\ H \quad\quad H \end{array}$$

a Count and list the bonds broken in this reaction.

b Count and list the new bonds formed.

c Calculate the total energy
   i required to break the bonds
   ii released when the new bonds are made.

(The bond energies in kJ/mol are: N–H 391; N–N 158; N≡N 945; O–H 464; O=O 498)

d Calculate the value of $\Delta H_{reaction}$.

e Is energy released or absorbed overall?

f Draw an energy diagram for the reaction.

g How much energy is transferred if 100 g of hydrazine is burnt in oxygen? (RAMs: N = 14, H = 1.)

h Where is the energy transferred from and to?

i Would hydrazine be suitable as a fuel? Why?

13 The energy change for the following reaction involving methane is +1664 kJ/mol.

$$\begin{array}{c} H \\ | \\ H-C-H(g) \longrightarrow C(g) + 4H(g) \\ | \\ H \end{array}$$

a How many bonds are broken in this reaction?

b How many bonds are formed?

c From the information on the energy change, calculate the bond energy of one C–H bond.

d For a similar reaction involving ethane (C_2H_6), the energy change is +2826 kJ/mol. Write an equation like the one above, to show this reaction.

e List the bonds broken during the reaction.

f Use the data you have available to calculate the bond energy of the C–C bond in ethane.

7.1 Conductors and insulators

Batteries and electric current

The photograph below shows a battery, a bulb and a rod of graphite (carbon) joined or **connected** to each other by copper wires. The arrangement is called an **electric circuit**. The bulb is lit: this shows that electricity must be flowing in the circuit.
Electricity is a stream of moving electrons.
Do you remember what an electron is? It is a tiny particle with a negative charge and almost no mass.

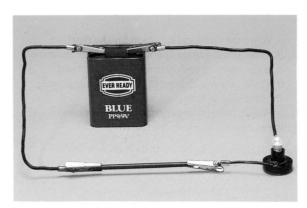

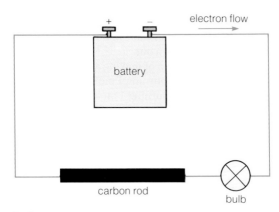

The diagram above right shows how the electrons move through the circuit. The battery acts like an electron pump. Electrons leave it through one terminal, called the **negative terminal**. They are pumped through the wires, the bulb and the rod, and enter the battery again through the **positive terminal**. When the electrons stream through the fine wire in the bulb, they cause the wire to heat up so much that it gets white-hot and gives out light.

Conductors

In the circuit above, the light will go out if:
- you disconnect a wire, so that everything no longer joins up. *Or*
- you connect something into the circuit that prevents electricity from flowing through.

The copper wires and graphite rod obviously do allow electricity to flow through them – they **conduct** electricity. Copper and graphite are therefore called **conductors**.
A conductor of electricity is a substance that allows electricity to flow through it.
A substance that does not conduct electricity is called a **non-conductor** or **insulator**.

Testing substances to see if they conduct

The circuit above can be used to test any substance to see if it conducts electricity. The substance is simply connected into the circuit, like the graphite rod above. Some examples are given on the next page.

Copper is the conductor inside this electric drill. But plastic (an insulator) is used to connect the bit to the motor, and for the outer case. Why?

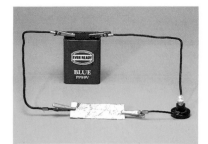

Testing tin to see if it conducts electricity. A strip of tin is connected into the circuit. The bulb lights, so tin must be a conductor.

Testing ethanol. The liquid is connected into the circuit by dipping graphite rods into it. The bulb does not light, so ethanol is a non-conductor.

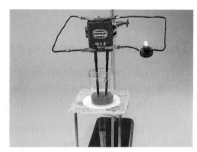

Testing molten lead bromide. A Bunsen is used to melt it. The molten compound conducts, and at the same time gives off a choking brown vapour.

The results   These are the results from a range of tests:

1 **The only solids that conduct are the metals and graphite.**
These conduct because of their free electrons (pages 52 and 56). The free electrons get pumped out of one end of the solid by the battery. Electrons then flow in the other end, and through the spaces left behind.
For the same reason, *molten* metals conduct. (It is not possible to test molten graphite, because graphite sublimes.)

2 **Molecular or covalent substances are non-conductors.**
This is because they contain no free electrons, or other charged particles, that can flow through them.
The ethanol above is one example of a molecular substance. Others are petrol, paraffin, sulphur, sugar and plastic. These never conduct, whether solid or molten.

3 **Ionic substances do not conduct when solid. However, they conduct when melted or dissolved in water, and they decompose at the same time.**
An ionic substance contains no free electrons. However, it does contain ions, which are also charged particles. The ions become free to move when the substance is melted or dissolved, and it is they that conduct the electricity.
The lead bromide above is an example. It is a non-conductor when solid. But it begins to conduct the moment it is melted, and a brown vapour bubbles off at the same time. The vapour is bromine, and it forms because electricity causes the lead bromide to **decompose**.
Decomposition caused by electricity is called electrolysis, and the liquid that decomposes is called an electrolyte.
Molten lead bromide is therefore an electrolyte. Ethanol is a **non-electrolyte** because it does not conduct at all.

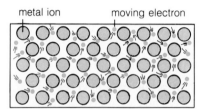

metal ion   moving electron

Metals conduct thanks to their free electrons.

Like other metals, aluminium conducts electricity. It is used for electricity cables because it is so light.

1 What is a **conductor** of electricity?
2 Draw a circuit to show how you would test whether mercury conducts.
3 Explain why metals are able to conduct electricity.
4 Naphthalene is a molecular substance. Do you think it conducts when molten? Explain why.
5 What is: **a** an electrolyte? **b** a non-electrolyte? Give *three* examples of each.

7.2 A closer look at electrolysis

The electrolysis of lead bromide

On the last page, you saw that molten lead bromide decomposes when it conducts electricity. Decomposition caused by electricity is called **electrolysis**, and molten lead bromide is an **electrolyte**.

The apparatus This is shown on the right. The graphite rods carry the current into and out of the molten lead bromide. Conducting rods like these are called **electrodes**.

The electrode joined to the negative terminal of the battery is called the **cathode**. It is also negative, because the electrons from the battery flow to it. The other electrode is positive, and is called the **anode**.

Notice the switch. When it is open, no electricity can flow.

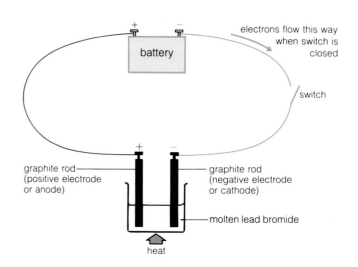

The electrolysis Once the switch is closed, **bromine vapour** starts to bubble out of the molten lead bromide, around the anode. After some time a bead of **molten lead** forms below the cathode. The electrical energy from the battery has caused a **chemical change**:

$$\text{lead bromide} \longrightarrow \text{lead} + \text{bromine}$$
$$\text{PbBr}_2\,(l) \longrightarrow \text{Pb}\,(l) + \text{Br}_2\,(g)$$

Why the molten lead bromide decomposes When lead bromide melts, its lead ions (Pb^{2+}) and bromide ions (Br^-) become free to move about. When the switch is closed, the electrodes become charged, and the ions are immediately attracted to them:

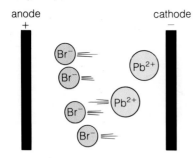

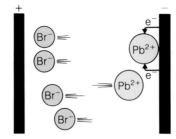

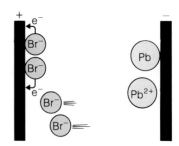

Opposite charges attract, so the lead ions are attracted to the cathode, and the bromide ions to the anode.

At the cathode, the lead ions each receive 2 electrons and become lead atoms:
$$Pb^{2+} + 2e^- \longrightarrow Pb$$
The lead atoms collect together on the cathode, and in time fall to the bottom of the beaker.

At the anode, the bromide ions each give up 1 electron to become bromine atoms. These pair together as molecules:
$$2\,Br^- \longrightarrow Br_2 + 2e^-$$
The bromine bubbles off as a gas.

Why the molten lead bromide conducts During the electrolysis, each lead ion takes two electrons from the cathode, as shown on the right. At the same time two bromide ions each give an electron to the anode. The effect is the same as if two electrons *flowed through the liquid* from the cathode to the anode. In other words, the lead bromide is acting as a conductor of electricity.

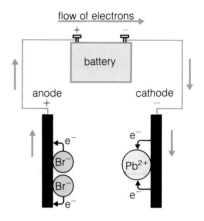

The electrolysis of other compounds

All ionic compounds can be electrolysed, when they are molten. (Another word for molten is **fused**.) These are some points to remember about the process:

1 The electrolyte always **decomposes**. So electrical energy is causing a **chemical change**.
2 The electrodes are usually made of graphite or platinum. These substances are unreactive or **inert**. That means they will not react with the electrolyte or the products of the electrolysis.
3 Metals always form positive ions. Positive ions always go to the cathode. So they are also called **cations**.
4 Nonmetals (except hydrogen) always form negative ions. They always go to the anode so are called **anions**.
5 **So when a molten ionic compound is electrolysed, a metal is always formed at the cathode and a nonmetal at the anode.**
   You can see more examples in this table:

> **Remember**
>
> • Cations (+) go to the cathode (−).
>
> • Anions (−) go to the anode (+).
>
> • Ca^{2+} is a *ca*tion.

Electrolyte	The decomposition	At the cathode	At the anode
Sodium chloride NaCl	sodium chloride $\longrightarrow$ sodium + chlorine $2NaCl(l) \longrightarrow 2Na(l) + Cl_2(g)$	$2Na^+ + 2e^- \longrightarrow 2Na$	$2Cl^- \longrightarrow Cl_2 + 2e^-$
Potassium iodide KI	potassium iodide $\longrightarrow$ potassium + iodine $2KI(l) \longrightarrow 2K(l) + I_2(g)$	$2K^+ + 2e^- \longrightarrow 2K$	$2I^- \longrightarrow I_2 + 2e^-$
Copper(II) bromide $CuBr_2$	copper(II) bromide $\longrightarrow$ copper + bromine $CuBr_2(l) \longrightarrow Cu(l) + Br_2(g)$	$Cu^{2+} + 2e^- \longrightarrow Cu$	$2Br^- \longrightarrow Br_2 + 2e^-$

6 Electrolysis is the most powerful way to decompose an ionic compound. So it is used in industry to extract metals such as sodium and aluminium from their ores. (Sodium is extracted from molten rock salt or sodium chloride. Aluminium is extracted from molten aluminium oxide.) But electrolysis guzzles up electricity, which makes it very expensive. So it is used only for very stable compounds which are difficult to decompose in any other way. You can find out more on page 202.

> **Oxidation and reduction in electrolysis**
>
> • At the cathode, ions gain electrons. They are *reduced*.
>
> • At the anode, ions lose electrons. They are *oxidized*.

---

1 Explain what each of these words means:
  electrolysis   anode   cathode
2 For the electrolysis of molten lead bromide, draw diagrams to show:
  a how the ions move when the switch is closed;
  b what happens at the anode;
  c what happens at the cathode.
3 What is: a cation?   an anion?
4 Molten copper and molten copper(II) bromide both conduct. What changes would you expect to see when they conduct?
5 Write equations for the overall reaction, and the reaction at each electrode, when fused magnesium chloride ($MgCl_2$) is electrolysed.

7.3 The electrolysis of solutions

When a salt such as sodium chloride is dissolved in water, its ions become free to move. So the solution can be electrolysed. But the products may be different from when you electrolyse the *molten* salt, *because water itself also produces ions.* Although water is molecular, a tiny fraction of its molecules is split into ions:

some water molecules $\longrightarrow$ hydrogen ions + hydroxide ions
$$H_2O\,(l) \longrightarrow H^+\,(aq) + OH^-\,(aq)$$

During electrolysis, these H^+ and OH^- ions compete with the metal and nonmetal ions from the dissolved salt, to receive or give up electrons. So which ions win? These are the rules.

At the cathode:
1 The more reactive a metal, the more it 'likes' to exist as ions. So if a metal is very reactive, its ions remain in solution. The H^+ ions accept electrons, and hydrogen molecules are formed.
2 The ions of less reactive metals will accept electrons and form metal atoms, leaving the H^+ ions in solution.

At the anode:
3 If ions of a halogen are present (Cl^-, Br^- or I^-), they will give up electrons more readily than the OH^- ions do. Molecules of chlorine, bromine or iodine are formed.
4 If no halogen ions are present, OH^- ions will give up electrons more readily than other nonmetal ions do, and oxygen is formed.

Now let's look at some examples.

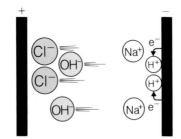

Use apparatus like this when you want to collect the gases from electrolysis.

Sodium chloride solution

A concentrated solution of sodium chloride is electrolysed in the lab using the equipment shown above right. This is what happens:

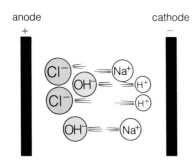

The solution contains Na^+ ions and Cl^- ions from the salt, and H^+ and OH^- ions from water. The positive ions go to the cathode and the negative ions to the anode.

At the cathode, it is the H^+ ions which accept electrons since sodium is more reactive than hydrogen:
$$2H^+ + 2e^- \longrightarrow H_2$$
Hydrogen gas bubbles off while Na^+ ions remain in solution.

At the anode, the Cl^- ions give up electrons more readily than the OH^- ions do. Chlorine gas bubbles off:
$$2Cl^- \longrightarrow Cl_2 + 2e^-$$
The OH^- ions remain in solution.

When the hydrogen and chlorine bubble off, Na^+ and OH^- ions are left behind: a solution of sodium hydroxide is formed.

Other salt solutions

Here are the results of electrolysing some other salt solutions.
Check them out. Do they obey the rules given opposite?

When the electrolyte is a solution of ...	At the cathode you get ...	At the anode you get ...
potassium bromide KBr (aq)	hydrogen $2H^+ + 2e^- \longrightarrow H_2$	bromine $2Br^- \longrightarrow Br_2 + 2e^-$
sodium iodide NaI (aq)	hydrogen $2H^+ + 2e^- \longrightarrow H_2$	iodine $2I^- \longrightarrow I_2 + 2e^-$
magnesium sulphate $MgSO_4$ (aq)	hydrogen $4H^+ + 4e^- \longrightarrow 2H_2$	oxygen $4OH^- \longrightarrow 2H_2O + O_2 + 4e^-$
lead(II) nitrate $Pb(NO_3)_2$ (aq)	lead $2Pb^{2+} + 4e^- \longrightarrow 2Pb$	oxygen $4OH^- \longrightarrow 2H_2O + O_2 + 4e^-$
copper(II) chloride $CuCl_2$ (aq)	copper $Cu^{2+} + 2e^- \longrightarrow 2Cu$	chlorine $2Cl^- \longrightarrow Cl_2 + 2e^-$
silver nitrate $AgNO_3$ (aq)	silver $4Ag^+ + 4e^- \longrightarrow 4Ag$	oxygen $4OH^- \longrightarrow 2H_2O + O_2 + 4e^-$

metals increasingly reactive

Dilute sulphuric acid

Sulphuric acid has the formula H_2SO_4. In water it forms ions:
$$H_2SO_4 (aq) \longrightarrow 2H^+(aq) + SO_4^{2-}(aq)$$
As you've seen, water also produces ions. So a dilute solution of the
acid contains H^+ ions from both water and acid, OH^- ions from the
water and SO_4^{2-} ions from the acid. It can be electrolysed using
apparatus like that shown on the right or on the opposite page.

At the cathode This time there are no metal ions to compete with.
So hydrogen gas is formed and bubbles off:
$$4H^+ + 4e^- \longrightarrow 2H_2$$

At the anode The OH^- and SO_4^{2-} ions compete to give up
electrons. As you'd expect from rule 4, the OH^- ions win:
$$4OH^- \longrightarrow 2H_2O + O_2 + 4e^-$$
The oxygen bubbles off. The SO_4^{2-} ions are left behind in solution.

The overall result is that the water decomposes rather than the acid:
$$\text{water} \longrightarrow \text{hydrogen} + \text{oxygen}$$
$$2H_2O (l) \longrightarrow 2H_2 (g) + O_2 (g)$$
So this electrolysis is often called **the electrolysis of acidified water**.

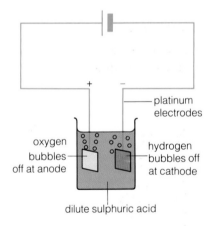

Use apparatus like this when you don't
want to collect the gases from
electrolysis.

1 Explain why water conducts electricity very slightly.
2 Describe what happens during the electrolysis of a
concentrated solution of sodium chloride.
3 Which ions are there in a solution of sodium iodide?

4 Write an equation for the reaction you expect at each
electrode, when you electrolyse a solution of:
a sodium nitrate, $NaNO_3$ b hydrochloric acid, HCl
c potassium iodide, KI d copper sulphate, $CuSO_4$

7.4 Using electrolysis (I)

Extracting metals

Reactive metals 'like' to exist as compounds, so it is difficult to extract them from their compounds. Electrolysis delivers the energy that's needed. For example, it is used to extract sodium from molten sodium chloride:

$$2NaCl\,(l) \longrightarrow 2\,Na\,(l) + Cl_2\,(g)$$

Since electricity is expensive, electrolysis is not used for less reactive metals. These can be extracted by cheaper methods. See page 202.

Purifying metals

When a solution of copper(II) sulphate is electrolysed using carbon or platinum electrodes, copper is obtained at the cathode and oxygen at the anode, as you'd expect from the rules on page 108. But when *copper* electrodes are used, there's a different result.

At the cathode Copper ions become atoms, as you would expect:
$$Cu^{2+} + 2e^- \longrightarrow Cu$$
The copper atoms cling to the cathode.

At the anode The copper anode *dissolves*, forming copper ions:
$$Cu \longrightarrow Cu^{2+} + 2e^-$$
So the anode wears away while the cathode grows thicker.

This process is used in industry to purify copper, which must be very pure for electrical wiring. This is how it's carried out:

> **Magnesium from sea water**
>
> We aren't likely to run out of magnesium in a hurry.
>
> Every 10000 grams of sea water contains about 13 grams of it (or 0.13%).
>
> In the latest process it is obtained from sea water as magnesium chloride by evaporation. Electrolysis is then used to extract magnesium metal from the molten salt.

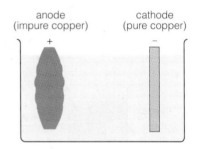

anode (impure copper)   cathode (pure copper)

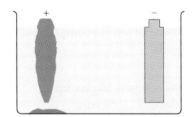

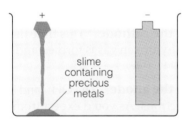

slime containing precious metals

The anode is made of impure copper. The cathode is pure copper. The electrolyte is copper(II) sulphate solution.

The anode dissolves and the impurities drop to the floor of the cell. A layer of pure copper builds up on the cathode.

Eventually the cathode is removed. The slime on the cell floor is checked for silver and other precious metals.

The cathodes are sold off to copper manufacturers. They are over 99.9 % pure. The precious metals are recovered from the slime and sold. They help to make the business profitable.

A similar process is used for stripping the tin off 'tinned' steel. 'Tins' are steel cans with a fine coating of tin. This must be removed before the steel is recycled. So crushed used cans are made into an anode. The cathode is 'clean' steel. Sodium hydroxide solution is the electrolyte. The tin leaves the anode and plates the cathode, from which it can be scraped off. Both the tin and steel are recycled.

Electroplating

Electrolysis is also used to coat one metal with another, to make it look better or to prevent corrosion. For example, it is used to coat steel car bumpers with chromium, or steel cans with tin, or cheap metal jewellery with silver. This process is called **electroplating**.

The diagram on the right shows how a steel jug is electroplated with silver. The jug becomes the cathode of an electrolytic cell. The anode is made of silver. The electrolyte is a solution of a silver compound, for example silver nitrate.

At the anode   The silver dissolves, forming ions in solution:
$$Ag \longrightarrow Ag^+ + e^-$$

At the cathode   Silver ions receive electrons, forming a coat of silver on the jug:
$$Ag^+ + e^- \longrightarrow Ag$$

When the layer of silver is thick enough, the jug is removed.

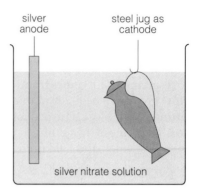
silver anode   steel jug as cathode

silver nitrate solution

Silverplating: electroplating with silver

In general, to electroplate any object with metal X, the set-up is:
cathode   – object to be electroplated
anode   – metal X
electrolyte – solution of a soluble compound of X

Chromium-plated car bumpers

Chromium does not stick well to steel. So the steel is first electroplated with copper or nickel, and then chromium.

The chromium compound used in the electrolyte is chromium(VI) oxide.

Anodizing aluminium

Although aluminium is quite reactive, corrosion in air isn't a problem. This is because a thin layer of aluminium oxide quickly forms and acts as a barrier to oxygen (page 211). The layer can be made thicker, to give even more protection, by a process called **anodizing**.

The aluminium is used as the anode of a cell. Dilute sulphuric acid is the electrolyte. When this is electrolysed, as you saw on page 109, oxygen forms at the anode. It reacts with the aluminium:

$$4Al\,(s) + 3O_2\,(g) \longrightarrow 2Al_2O_3\,(s)$$

The anodized aluminium absorbs dyes and pigments very readily. It is used for a whole range of things including masts for yachts and windsurfing boards, facings for buildings, and window frames.

Hanging on to his anodized aluminium. Think of two reasons why it's used for masts.

---

1 Name *two* metals extracted by electrolysis.
2 Why is electrolysis *not* used in extracting iron?
3 If you want to purify a metal by electrolysis, will you make it the anode or the cathode? Why?
4 What does *electroplating* mean?
5 Why might you want to electroplate steel cutlery with nickel?

6 You decide to electroplate steel cutlery with nickel.
  a What will the anode be?
  b What will the cathode be?
  c Suggest a suitable electrolyte. (See page 148!)
7 What does *anodizing* mean?
8 Give two reasons why anodized aluminium is used for the masts for windsurfing boards.

7.5 Using electrolysis (II)

Making important chemicals from brine

Brine is a concentrated solution of sodium chloride or common salt. It is the starting point for a whole range of chemicals.

Most salt in Britain comes from Cheshire, where it lies 200 metres below ground in beds 200 m thick. Some is mined as rock salt for gritting roads. Most is turned into brine by pumping water into the beds. The brine is then pumped out of the ground and into reservoirs, where it is held until it is needed for electrolysis:

$$\text{brine} \xrightarrow{\text{electrolysis}} \text{sodium hydroxide} + \text{chlorine} + \text{hydrogen}$$
$$2NaCl\,(aq) + 2H_2O\,(l) \longrightarrow 2NaOH\,(aq) + Cl_2\,(g) + H_2\,(g)$$

The electrolysis   It is a bit like the one you can do in the lab (page 108), but on a much larger scale.

The diagram shows a **membrane cell**, one of several types of cell that are used.

The anode is made of titanium and the cathode of nickel. An **ion-exchange membrane** up the middle lets sodium ions through but keeps the gases apart. The sodium ions move freely to the cathode.

At the cathode   Hydrogen bubbles off:
$$2H^+ + 2e^- \longrightarrow H_2$$

At the anode   Chlorine bubbles off:
$$2Cl^- \longrightarrow Cl_2 + 2e^-$$

Na^+ and OH^- ions are left behind, which means a solution of sodium hydroxide forms. Some is evaporated to a more concentrated solution, and some evaporated completely to give solid sodium hydroxide.

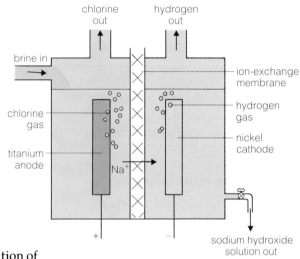

The membrane cell for the electrolysis of brine

Tanks of chlorine at a waterworks. What's it doing here?

All three products from the electrolysis of brine must be transported with care. Why? What precautions should be taken?

Chlorine, a poisonous yellow-green gas

Used for making...
- PVC (nearly $\frac{1}{3}$ of it used for this)
- solvents such as trichloroethane for degreasing and dry-cleaning
- paints and dyestuffs
- bleaches, weedkillers, pesticides
- pharmaceuticals
- titanium dioxide, a white pigment used in paints, ceramics, cosmetics, and paper
- hydrogen chloride and hydrochloric acid

It is also used for...
- killing bacteria in tap water (at waterworks)
- killing bacteria in swimming pools

Sodium hydroxide solution, alkaline and corrosive

Used for making ...

- soaps
- detergents
- viscose (rayon) and other textiles
- paper (such as used in this book)
- ceramics
- dyes
- medical drugs

Hydrogen, a colourless flammable gas

Used for...
- making nylon
- making hydrogen peroxide
- 'hardening' vegetable oils for margarine
- burning to make steam for other processes

A whole section of the chemical industry is based on the electrolysis of brine. It's called the **chlor-alkali** industry. As you can see above, the products from the electrolysis have a wide range of uses.

Producing hydrogen from water

As you will see on page 216, hydrogen may become a much more important fuel in the future. It can be produced by the electrolysis of acidified water (page 109). The problem with doing this on a large scale is that electricity is expensive, because it's made by burning expensive fossil fuels. But that is changing:

- Solar cells made of silicon are being used increasingly to generate electricity from sunlight. When sunlight strikes the silicon an electric current is produced. It can be used for lighting and heating homes *and* to split water.
- The use of windpower to generate electricity is becoming increasingly important. Windpower is a renewable resource.

In the future, hydrogen from the electrolysis of water may be piped into homes and used as a fuel for cooking and heating. It will also be used in batteries called **fuel cells** to power homes and electric cars.

1 How is brine made in Cheshire?
2 Write a word equation for the electrolysis of brine.
3 Draw a rough sketch of the membrane cell.
4 Give three uses for each of the products formed.
5 Where does the term 'chlor-alkali' come from?

6 What are the advantages of hydrogen as a fuel?
7 Explain how hydrogen is obtained by the electrolysis of water. Write equations for the reactions that take place at the electrodes.
8 Explain why electricity may get cheaper.

7.6 Calculations on electrolysis

Depositing copper by electrolysis

Suppose you set up an experiment to electrolyse copper(II) sulphate solution, using copper electrodes. As you saw on page 110, a layer of copper will build up on the cathode while the anode dissolves.

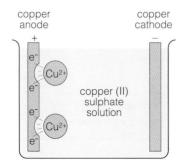

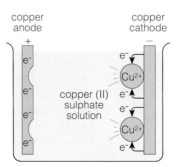

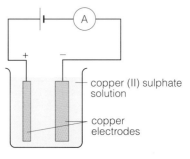

The anode dissolves because its atoms give up electrons to form ions: $Cu \longrightarrow Cu^{2+} + 2e^-$

At the cathode the copper ions grab electrons and deposit as atoms: $Cu^{2+} + 2e^- \longrightarrow Cu$

So the mass of copper gained by the cathode exactly equals the mass lost from the anode.

Calculations from the experiment

In the experiment above, you will find that:
- the bigger the current, the more copper is deposited
- the longer the current flows, the more copper is deposited
  This is shown by the graphs on the right.

In fact, if you select a current, and a time, you can calculate the amount of copper that will be deposited. You can do this for *any* substance formed during electrolysis, because:
The amount of a substance formed during electrolysis is directly proportional to the size of the current and how long it flows.

But first, you need to know a bit more about current.

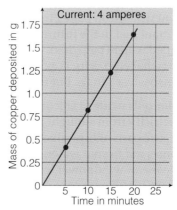

A closer look at current

A current is a flow of electrons, each with a tiny electric charge. Current is measured in units called **amperes** (A). The total charge carried by the current is measured in **coulombs** (C).
A current of 1 ampere flowing for 1 second carries a charge of 1 coulomb.

To find the total charge carried by a current, use this formula:
current (A) × time (s) = charge in coulombs (C)

How is the charge on an electron related to the coulomb? Like this:
1 mole of electrons has a charge of 96 500 coulombs.

The charge on a mole of electrons is called a **Faraday**.

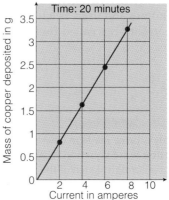

Doing the calculations

Example 1 A solution of copper(II) sulphate is electrolysed using copper electrodes. How much copper will be deposited by a current of 2 amps flowing for 20 minutes?

Step 1 Calculate the charge of the current.
Charge (coulombs) = time (seconds) × current (amps)
= 20 × 60 × 2 = 2400 coulombs

Step 2 Convert the charge to moles of electrons.
96 500 coulombs = 1 mole of electrons

so 1 coulomb $= \dfrac{1}{96\,500}$ moles of electrons

2400 coulombs $= \dfrac{1}{96\,500} \times 2400$ moles of electrons

= 0.025 moles of electrons

Step 3 Now use the equation for the reaction at the cathode to find how much copper is deposited.
$Cu^{2+} + 2e^- \longrightarrow Cu$
So 2 moles of electrons give 1 mole of copper atoms.
Changing moles of copper atoms to grams:
2 moles of electrons give 64 g of copper so
1 mole of electrons gives 32 g and
0.025 moles of electrons give (0.025 × 32) g or 0.8 g.
0.8 grams of copper will be deposited.

Example 2 1 kg of sodium is produced by electrolysing sodium chloride. What volume of chlorine is produced at the same time?

Step 1 Calculate the number of moles of sodium produced.
23 g of sodium = 1 mole

1 kg of sodium $= 1000\,g = \dfrac{1000}{23}$ moles = 43.48 moles.

Step 2 Calculate the number of moles of chlorine obtained.
At the cathode: $Na^+ (l) + e^- \longrightarrow Na\,(l)$
At the anode: $2Cl^- (l) \longrightarrow Cl_2(g) + 2e^-$
1 mole of chlorine molecules gives up 2 moles of electrons, which is enough to produce 2 moles of sodium atoms.
But 43.48 moles of sodium atoms are produced. So
$\dfrac{43.48}{2}$ moles or 21.74 moles of chlorine molecules form.

Step 3 Convert moles to volume of gas.
1 mole of gas molecules has a volume of 24 dm³ at rtp.
So the volume of chlorine obtained is 24 × 21.74 dm³ or 521.76 dm³.

The Faraday

As you saw on the opposite page, the Faraday is the charge on 1 mole of electrons: 96 500 coulombs.

It is named after the famous British scientist Michael Faraday (1791–1867).

Remember Avogadro's Law?

1 mole of molecules of any gas occupies a volume of 24 dm³ at 20 °C and 1 atmosphere pressure.

20 °C and 1 atmosphere pressure is also called **room temperature and pressure** or **rtp**.

See page 82 for more.

---

1 What two things affect the amount of a substance produced during electrolysis?
2 In an electrolysis, 200 g of metal is obtained per hour. How would you double this to 400 g per hour?
3 In the electrolysis of molten sodium chloride, what current gives 23 g (1 mole) of sodium in an hour?

4 Copper(II) sulphate solution is electrolysed for 15 minutes using copper electrodes and a 4 A current.
   a What charge is carried by the current?
   b How many moles of electrons does this equal?
   c How many moles of copper are deposited?
   d What mass of copper is this?

Questions on Section 7

1 a What does the term *electrolysis* mean?
b Copy the diagram below, and label it using the words in this list, which are all connected with electrolysis:
anode, cathode, electrolyte, anion, cation.

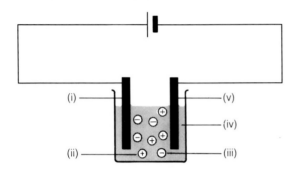

2 In which of these would the bulb light?

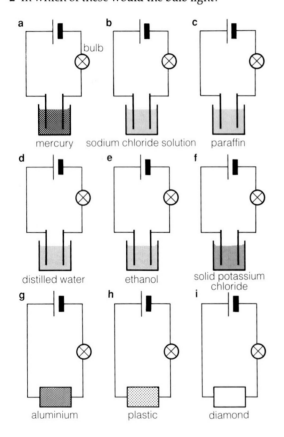

3 a Which of the substances in question 2 are:
   i conductors?   ii non-conductors?
   iii electrolytes?   iv non-electrolytes?
b What is the difference between a conductor and an electrolyte?
c For which substances above would you expect to see changes taking place at the electrodes?

4 The electrolysis of lead bromide can be investigated using the following apparatus.

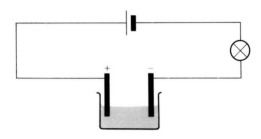

a What must be done to the lead bromide before the bulb will light?
b What would be *seen* at the positive electrode during the experiment?
c Name the substance in b.
d What is formed at the negative electrode?
e Write an equation for the reaction at each electrode.

5 This question is about the electrolysis of *molten* lithium chloride. Lithium chloride is ionic, and contains lithium ions (Li^+) and chloride ions (Cl^-).
a Which ion is the anion?
b Which ion is the cation?
c Copy the following diagram and use arrows to show which way:
   i the ions flow when the switch is closed;
   ii the electrons flow in the wires.

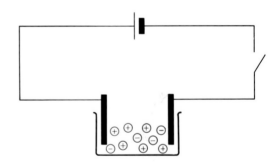

d Write equations for the reaction at each electrode, and the overall reaction.

6 This question is about the electrolysis of an aqueous solution of lithium chloride.
a Write down the names and symbols of all the ions present in the solution.
b Lithium is a reactive metal, like sodium. What will be formed at the cathode?
c What will be formed at the anode?
d Write an equation for the reaction at each electrode.
e Name two other electrolytes that will give the same electrolysis products as this one.

7 Write an equation for:
  a the overall decomposition
  b the reaction at each electrode
  when molten sodium chloride is electrolysed.

8 **a** List the anions and cations present in:
    i sodium chloride solution
    ii copper(II) chloride solution
  b Write down the reaction you would expect at:
    i the anode **ii** the cathode
  when each solution in **a** is electrolysed, using platinum electrodes.
  c Explain why the anode reactions in **b** are both the same.
  d Explain why copper is obtained at the cathode, but not sodium.

9 Six substances A to F were dissolved in water, and connected in turn into the circuit below. The symbol ⒜ represents an ammeter, which is an instrument for measuring the current.

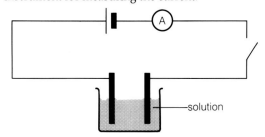

The results are shown in this table:

Substance	Current (amperes)	Cathode (−)	Anode (+)
A	0.8	copper	chlorine
B	1.0	hydrogen	chlorine
C	0.0	——	——
D	0.8	copper	oxygen
E	1.2	hydrogen	oxygen
F	0.7	silver	oxygen

  a Which solution conducts best?
  b Which solution is a non-electrolyte?
  c Which solution could be:
    i silver nitrate?
    ii copper(II) sulphate?
    iii copper(II) chloride?
    iv sodium hydroxide?
    v sugar?
    vi potassium chloride?

10 Hydrogen chloride is a molecular substance. However, it dissolves in water to form *hydrochloric acid*, which exists as ions:
  $$HCl(g) \longrightarrow H^+(aq) + Cl^-(aq)$$
  List the ions present in a solution of hydrochloric acid. What result would you expect, when the solution is electrolysed with platinum electrodes?

11 Aluminium is extracted by electrolysis of molten aluminium oxide. The aluminium ion is Al^{3+}.
  a Write an equation for the reaction at the cathode.
  b How many moles of electrons are required to obtain 1 mole of aluminium atoms?
  c What mass of aluminium will be obtained if a current of 25 000 A flows for 24 hours?
  (The RAM of $Al = 27$; the charge due to 1 mole of electrons is 96 500 coulombs.)

12 Using platinum electrodes the apparatus below was set up to electrolyse three different solutions.

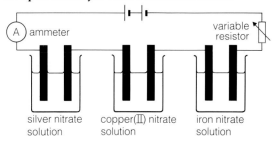

  silver nitrate solution   copper(II) nitrate solution   iron nitrate solution

  a Why is a variable resistor needed in the circuit?
  b Write an electron transfer equation to show how silver is formed at the negative electrode.
  c In the experiment, 0.403 g of silver was formed when a current was passed for 30 minutes.
  The RAM of $Ag = 108$; the charge due to 1 mole of electrons is 96 500 coulombs.
    i Calculate the current used.
    ii Calculate the mass of copper formed in the middle cell. (The RAM of $Cu = 64$.)
  d In the third cell, 0.070 g of iron was obtained. The RAM of $Fe = 56$. Calculate whether the solution used was iron(II) nitrate (containing Fe^{2+} ions), or iron(III) nitrate (containing Fe^{3+} ions).
  e What would be obtained at the positive electrode in each beaker?

13 When sodium chloride solution is electrolysed, the gases hydrogen and chlorine are obtained.
  a i Write down the formulae for the hydrogen and chloride ions, then say which gas is obtained at each electrode.
    ii Explain why hydrogen is released instead of sodium metal.
  b i Write the equation for the formation of chlorine gas at the electrode.
    ii How many moles of electrons are required to release 1 mole of chlorine gas (Cl_2)?
    iii How many coulombs is this? (The charge due to 1 mole of electrons is 96 500 coulombs.)
  c A current of 2 A was passed for 20 minutes.
    i Calculate the volume of chlorine released at room temperature and pressure (rtp). (The volume of 1 mole of any gas is 24 dm^3 or 24 000 cm^3 at rtp.)
    ii Explain why the volume of hydrogen released is the same as the volume of chlorine.

8.1 Rates of reaction

Fast and slow

Some reactions are **fast** and some are **slow**. Look at these examples:

Silver chloride precipitating, when solutions of silver nitrate and sodium chloride are mixed. This is a very fast reaction.

Concrete setting. This reaction is quite slow. It will take a couple of days for the concrete to harden.

Rust forming on a heap of scrap iron in a scrapyard. This is usually a very slow reaction.

It is not always enough to know just that a reaction is fast or slow. For example, in a factory that makes products from chemicals, the chemical engineers need to know *exactly* how fast each reaction is going, and how long it takes to complete. In other words, they need to know the **rate** of each reaction.

What is rate?

Rate is a measure of how fast or slow something is. Here are some everyday examples:

This plane has just flown 2000 kilometres in 1 hour. It flew at a **rate** of 2000 kilometres per hour.

This petrol pump can pump petrol at a **rate** of 50 litres per minute.

This machine can print newspapers at a **rate** of 10 copies per second.

From these examples you can see that:

Rate is a measure of the change that happens in a single unit of time.

Any suitable unit of time can be used – a second, a minute, an hour, even a day.

Rate of a chemical reaction

When zinc is added to dilute sulphuric acid, they react together. The zinc disappears slowly, and a gas bubbles off.

After a time, the bubbles of gas form less quickly. The reaction is slowing down.

Finally, no more bubbles appear. The reaction is over, because all the acid has been used up. Some zinc remains behind.

In this example, the gas that forms is hydrogen. The equation for the reaction is:

$$zinc \ + \ sulphuric\ acid \longrightarrow zinc\ sulphate \ + \ hydrogen$$
$$Zn\,(s) \ + \ H_2SO_4\,(aq) \longrightarrow ZnSO_4\,(aq) \ + \ H_2\,(g)$$

Both zinc and sulphuric acid get used up in the reaction. At the same time, zinc sulphate and hydrogen form.

You could measure the rate of the reaction, by measuring either:

- the amount of zinc used up per minute *or*
- the amount of sulphuric used up per minute *or*
- the amount of zinc sulphate produced per minute *or*
- the amount of hydrogen produced per minute.

For this reaction, it is easiest to measure the amount of hydrogen produced per minute – the hydrogen can be collected as it bubbles off, and its volume can then be measured. On the next page you will find out how this is done.

In general, to find the rate of a reaction, you should measure:
the amount of a reactant used up per unit of time *or*
the amount of a product produced per unit of time.

---

1 Here are some reactions that take place in the home. Put them in order of decreasing rate (the fastest one first).
  a Gloss paint drying.
  b Fruit going rotten.
  c Cooking gas burning.
  d A cake baking.
  e A metal bath rusting.

2 Which of these rates of travel is slowest?
  5 kilometres per second.
  20 kilometres per minute.
  60 kilometres per hour.

3 Suppose you had to measure the rate at which zinc is used up in the reaction above. Which of these units would be suitable?
  a litres per minute;
  b grams per minute;
  c centimetres per minute.
  Explain your choice.

4 Iron reacts with sulphuric acid like this:
  $$Fe\,(s) + H_2SO_4\,(aq) \longrightarrow FeSO_4\,(aq) + H_2\,(g)$$
  a Write a word equation for this reaction.
  b Write down four different ways in which the rate of the reaction could be measured.

8.2 Measuring the rate of a reaction

On the last page you saw that the rate of a reaction is found by measuring the amount of a **reactant** used up per unit of time or the amount of a **product** produced per unit of time.
Take, for example, the reaction between magnesium and excess dilute hydrochloric acid. Its equation is:

magnesium + hydrochloric acid $\longrightarrow$ magnesium chloride + hydrogen
$$Mg\,(s) \quad + \quad 2HCl\,(aq) \quad \longrightarrow \quad MgCl_2\,(aq) \quad + \quad H_2\,(g)$$

In this reaction, hydrogen is the easiest substance to measure. This is because it is the only gas in the reaction. It bubbles off and can be collected in a **gas syringe**, where its volume is measured.

Some reactions are so fast that their rates would be very difficult to measure, like this detonation of an old mine.

The method   This apparatus is suitable:

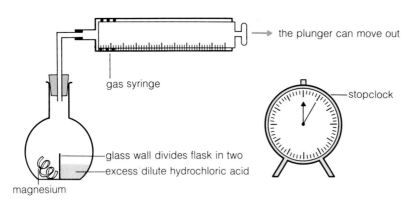

The magnesium is cleaned with sandpaper and put into one part of the flask. Dilute hydrochloric acid is put into the other part. The flask is tipped up to let the two reactants mix, and the clock is started at the same time. Hydrogen begins to bubble off. It rises up the flask, and pushes its way into the gas syringe. The plunger is forced to move out:

At the start the plunger is fully in. No gas has yet been collected.

Now the plunger has moved out to the 20 cm³ mark. 20 cm³ of gas been been collected.

The volume of gas in the syringe is noted at intervals, for example at the end of each half-minute. How will you know when the reaction is complete?

The results   Here are some typical results:

Time/minutes	0	$\frac{1}{2}$	1	$1\frac{1}{2}$	2	$2\frac{1}{2}$	3	$3\frac{1}{2}$	4	$4\frac{1}{2}$	5	$5\frac{1}{2}$	6	$6\frac{1}{2}$
Volume of hydrogen/cm³	0	8	14	20	25	29	33	36	38	39	40	40	40	40

These results can be plotted on a graph, as shown on the next page.

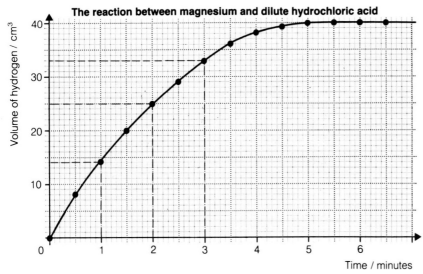

The reaction between magnesium and dilute hydrochloric acid

(Graph: Volume of hydrogen / cm³ on the y-axis against Time / minutes on the x-axis)

Notice these things about the results:

1 In the first minute, 14 cm³ of hydrogen is produced.
  So the rate for the first minute is 14 cm³ of hydrogen per minute.
  In the second minute, only 11 cm³ is produced. (25 – 14 = 11)
  So the rate for the second minute is 11 cm³ of hydrogen per minute.
  The rate for the third minute is 8 cm³ of hydrogen per minute.
  So you can see that the rate decreases as time goes on.
  The rate changes all through the reaction. It is greatest at the start, but gets less as the reaction proceeds.

2 The reaction is fastest in the first minute, and the curve is steepest then. It gets less steep as the reaction gets slower.
  The faster the reaction, the steeper the curve.

3 After 5 minutes, no more hydrogen is produced, so the volume no longer changes. The reaction is over, and the curve goes flat.
  When the reaction is over, the curve goes flat.

4 Altogether, 40 cm³ of hydrogen is produced in 5 minutes.

  $$\text{The } average \text{ rate for the reaction} = \frac{\text{total volume of hydrogen}}{\text{total time for the reaction}}$$

  $$= \frac{40 \text{ cm}^3}{5 \text{ minutes}}$$

  $$= \textbf{8 cm}^3 \textbf{ of hydrogen per minute.}$$

Note that this method can be used to measure the rate of *any* reaction in which one product is a gas – like the reaction shown on page 119.

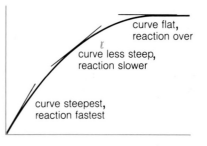

1 For this experiment, can you explain why:
  a a divided flask is used?
  b the magnesium ribbon is first cleaned?
  c the clock is started the moment the reactants are mixed?

2 From the graph above, how can you tell when the reaction is over?

3 This question is about the graph above.
  a How much hydrogen is produced in:
    i 2.5 minutes?   **ii** 4.5 minutes?
  b How many minutes does it take to produce:
    i 10 cm³ **ii** 20 cm³   of hydrogen?
  c What is the rate of the reaction during:
    i the fourth minute?   **ii** the fifth minute?

8.3 Changing the rate of a reaction (I)

The effect of concentration

A reaction can be made to go faster or slower by changing the **concentration** of a reactant.

Suppose the experiment with magnesium and excess hydrochloric acid is repeated twice (A and B below). Everything is kept the same each time, *except* the concentration of the acid:

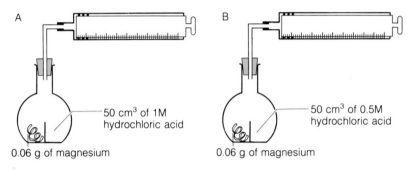

A 50 cm³ of 1M hydrochloric acid
0.06 g of magnesium

B 50 cm³ of 0.5M hydrochloric acid
0.06 g of magnesium

The acid in A is *twice as concentrated* as the acid in B.
Here are both sets of results shown on the same graph.

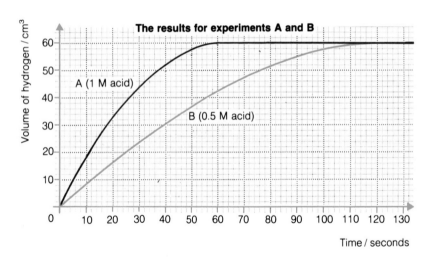

The results for experiments A and B

A (1 M acid)

B (0.5 M acid)

Volume of hydrogen / cm³

Time / seconds

Notice these things about the results:
1 Curve A is steeper than curve B. From this you can tell straight away that the reaction was faster in A than in B.
2 In A, the reaction lasts 60 seconds. In B it lasts for 120 seconds.
3 Both reactions produced 60 cm³ of hydrogen. Do you agree?
   In A it was produced in 60 seconds, so the average rate was 1 cm³ of hydrogen per second. In B it was produced in 120 seconds, so the average rate was 0.5 cm³ of hydrogen per second.
   The average rate in A was twice the average rate in B.

These results show that:
A reaction goes faster when the concentration of a reactant is increased.
For the reaction above, the rate doubles when the concentration of acid is doubled.

This stain could be removed with a solution of bleach. The more concentrated the solution, the faster the stain will disappear.

The effect of temperature

A reaction can also be made to go faster or slower by changing the **temperature** of the reactants.

This time, a different reaction is used: when dilute hydrochloric acid is mixed with sodium thiosulphate solution, a fine yellow precipitate of sulphur forms. The rate can be followed like this:

1 A cross is marked on a piece of paper.
2 A beaker containing some sodium thiosulphate solution is put on top of the paper. The cross should be easy to see through the solution, from above.
3 Hydrochloric acid is added quickly, and a clock started at the same time. The cross grows fainter as the precipitate forms.
4 The clock is stopped the moment the cross can no longer be seen from above.

View from above the beaker:

The cross grows fainter with time

The experiment is repeated several times. The quantity of each reactant is kept exactly the same each time. Only the temperature of the reactants is changed. This table shows the results:

Temperature/°C	20	30	40	50	60
Time for cross to disappear/seconds	200	125	50	33	24

The higher the temperature, the faster the cross disappears

The cross disappears when enough sulphur forms to blot it out. Notice that this takes 200 seconds at 20 °C, but only 50 seconds at 40 °C. So the reaction is *four times faster* at 40 °C than at 20 °C.
A reaction goes faster when the temperature is raised. When the temperature increases by 10 °C, the rate approximately doubles.
This fact is used a great deal in everyday life. For example, food is kept in the fridge to slow down decomposition reactions and keep it fresh for longer. Can you think of any other examples?

The low temperature in the fridge slows down decomposition reactions.

---

1 Look at the graph on the opposite page.
  a After two minutes, how much hydrogen was produced in:
    i experiment A?   ii experiment B?
  b From the shape of the curves, how can you tell which reaction was faster?
2 Explain why experiments A and B both produce the same amount of hydrogen.
3 Copy and complete: A reaction goes . . . . when the concentration of a . . . . is increased. It also goes . . . . when the . . . . is raised.
4 Why does the cross disappear, in the experiment with sodium thiosulphate and hydrochloric acid?
5 What will happen to the rate of a reaction when the temperature is *lowered*? Use this to explain why milk is stored in a fridge.

8.4 Changing the rate of a reaction (II)

The effect of surface area

In many reactions, one of the reactants is a solid. The reaction between hydrochloric acid and calcium carbonate (marble chips) is one example. Carbon dioxide gas is produced:

$$CaCO_3 + 2HCl\,(aq) \longrightarrow CaCl_2\,(aq) \longrightarrow + H_2O\,(l) + CO_2\,(g)$$

The rate can be measured using the apparatus on the right.

The method   Marble chips and acid are placed in the flask, which is then plugged with cotton wool. This prevents any liquid from splashing out during the reaction. Next the flask is weighed. Then it is tipped up, to let the reactants mix, and a clock is started at the same time. The mass is noted at regular intervals, until the reaction is complete.

Since carbon dioxide can escape through the cotton wool, the flask gets lighter as the reaction proceeds. So by weighing the flask, you can follow the rate of the reaction.

The experiment is repeated twice. Everything is kept exactly the same each time, except the **surface area** of the marble chips:

For experiment 1, large chips of marble are used. The surface area is the total area of the surface of these chips.

For experiment 2, the same *mass* of marble is used. But this time it is in small chips, so its surface area is greater.

The results   The results of the two experiments are plotted below:

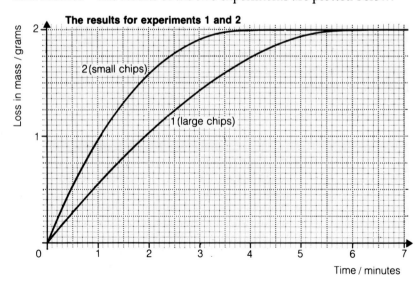

> **How to draw the graph:**
>
> First you have to find the *loss in mass* at different times:
>   loss in mass at a given time =
>   mass at start − mass at that time
> Then you plot the values for loss in mass against the times.

You should notice these things about the results:

1 Curve 2 is steeper than curve 1. This shows immediately that the reaction is faster for the small chips.
2 In both experiments, the final loss in mass is 2.0 grams. In other words, 2.0 grams of carbon dioxide is produced each time.
3 For the small chips, the reaction is complete in 4 minutes. For the large chips, it lasts for 6 minutes.

These results show that:

The rate of a reaction increases when the surface area of a solid reactant is increased.

The effect of a catalyst

Hydrogen peroxide is a clear, colourless liquid with the formula H_2O_2. It can decompose to water and oxygen:

hydrogen peroxide $\longrightarrow$ water + oxygen
$$2H_2O_2\,(aq) \longrightarrow 2H_2O\,(l) + O_2\,(g)$$

The rate of the reaction can be followed by collecting the oxygen:

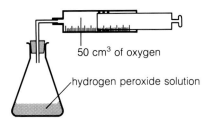

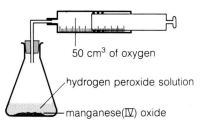

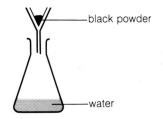

The reaction is in fact very slow. It could take 500 days to collect 50 cm³ of oxygen.

However, if 1 gram of black manganese(IV) oxide is added, the reaction goes much faster. 50 cm³ of oxygen is produced in a few minutes.

After the reaction, the black powder is removed by filtering. It is then dried, weighed and tested. It is still manganese(IV) oxide, and still weighs 1 gram.

The manganese(IV) oxide speeds up the reaction without being used up itself. It is called a **catalyst** for the reaction.
A catalyst is a substance that changes the rate of a chemical reaction but remains chemically unchanged itself.

Catalysts have been discovered for many reactions. They are usually **transition metals** or **compounds of transition metals.** For example, iron speeds up the reaction between nitrogen and hydrogen, to make ammonia (page 218). There are also many biological catalysts, called **enzymes.** For example, the pancreatic juice made by your pancreas contains enzymes that speed up digestion.

---

1 This question is about the graph on the opposite page. For each experiment find:
  a the loss in mass in the first minute;
  b the mass of carbon dioxide produced during the first minute;
  c the average rate of production of the gas.

2 What is a catalyst? Give two examples of reactions and their catalysts.
3 Look again at the decomposition of hydrogen peroxide, above. How would you show that:
  a the reaction goes even faster *if more than* 1 gram of catalyst is used?   b the catalyst is not used up?

8.5 Explaining rates

A closer look at a reaction

On page 120, you saw that magnesium and dilute hydrochloric acid react together:

magnesium + hydrochloric acid ⟶ magnesium chloride + hydrogen
Mg (s)   +   2HCl (aq)   ⟶   MgCl₂ (aq)   +   H₂ (g)

In order for the magnesium and acid particles to react together:
- **they must collide with each other.**
- **the collision must have enough energy.**

This is shown by the drawings below.

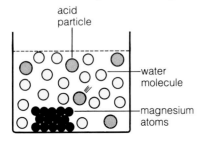

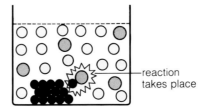

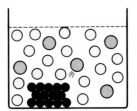

The particles in the liquid move around continuously. Here an acid particle is about to collide with a magnesium atom.

If the collision has enough energy, reaction takes place. Magnesium chloride and hydrogen are formed.

If the collision does not have enough energy, no reaction occurs. The acid particle bounces away again.

If there are lots of successful collisions in a given minute, then a lot of hydrogen is produced in that minute. In other words, the reaction goes quickly – its rate is high. If there are not many, its rate is low. **The rate of a reaction depends on how many successful collisions there are in a given unit of time.**

> **In a successful collision**
> - bonds are broken (this needs energy)
> - new bonds are formed (this releases energy).

Changing the rate of a reaction

Why rate increases with concentration   If the concentration of the acid is increased, the reaction goes faster. It is easy to see why:

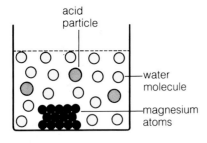

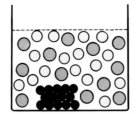

In dilute acid, there are not so many acid particles. This means there is not much chance of an acid particle hitting a magnesium atom.

Here the acid is more concentrated – there are more acid particles in it. There is now more chance of a successful collision occurring.

The more successful collisions there are, the faster the reaction.

> **What about reactions between gases?**
> - Increasing the pressure on two reacting gases has the same effect as increasing the concentration.
> - It means you squeeze more gas molecules into a given space.
> - The more molecules there are in a given space, the greater the chance of successful collisions.
> - So if pressure ↑ then rate ↑ for a gaseous reaction.

This idea also explains why the rate decreases with time:

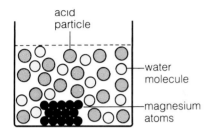

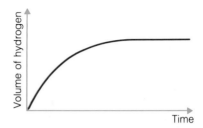

At the start, there are plenty of magnesium atoms and acid particles. But they get used up during successful collisions.

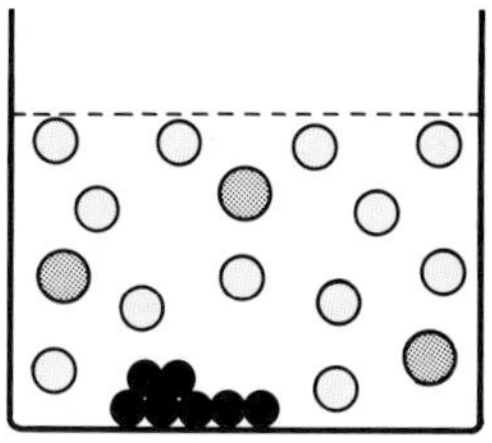

After a time, there are fewer magnesium atoms, and the acid is less concentrated. So the reaction slows down.

This means that the slope of the reaction curve decreases with time, as shown above.

Why rate increases with temperature   When reacting substances are heated, the particles take in energy. They move faster, which means they collide more often and with more energy. So there are more successful collisions. The rate goes up.

For the same reason reactions that depend on *light energy* speed up in stronger light. Photosynthesis is a good example.

Why rate increases with surface area   The reaction between the magnesium and acid is much faster when the metal is powdered:

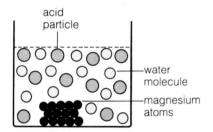

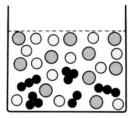

The acid particles can collide only with those magnesium atoms in the outer layer of the metal ribbon.

When the metal is powdered, many more atoms are exposed. So there is a greater chance of successful collisions.

Why a catalyst increases the rate   Some reactions can be speeded up by adding a catalyst. *In the presence of a catalyst, a collision needs less energy in order to be successful.* The result is that more collisions become successful, so the reaction goes faster. Catalysts are very important in industry, because they speed up reactions even at low temperatures. This means that less fuel is needed, so money is saved.

This airborne detonation test of an industrial explosive shows clearly the rapid rate of reaction.

---

1 Copy and complete: Two particles can only react together if they . . . . and if the . . . . has enough . . . .
2 What is:
  a a successful collision?
  b an unsuccessful collision?

3 In your own words, explain why the reaction between magnesium and acid goes faster when:
  a the temperature is raised;
  b the magnesium is powdered.
4 Explain why a catalyst can speed up a reaction, even at low temperatures.

8.6 More about catalysts

How catalysts work

This gas jar contains a mixture of hydrogen and oxygen. Even if you leave it for hours, the two gases won't react together.

But hold a platinum wire in the mouth of the jar, and the gases explode immediately with a pop, producing water.

So platinum is a **catalyst** for the reaction between hydrogen and oxygen. It speeds it up, while remaining chemically unchanged itself. The diagram on the right shows how it works.

1   This curve shows the situation in jar ①. The reaction can't start until the reacting molecules collide with enough energy to break bonds. The energy they need is called the **activation energy** (page 99). In jar ① they don't have enough energy – so there's no reaction.

2   When platinum is present, the activation energy is lowered. This means some molecules already have enough energy to react. The reaction is exothermic. Once it starts, the energy given out breaks further bonds. The reaction goes so fast it's explosive.

A catalyst lowers the energy needed for a reaction to take place. In other words, it lowers the activation energy.

A closer look at the platinum catalyst

In the reaction above, platinum acts as a **surface catalyst**:

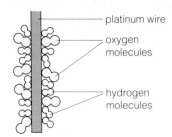

Instead of meeting on collision, gas molecules are adsorbed on to the metal surface. They get very close together, making reaction more likely.

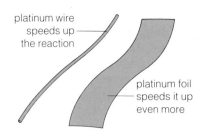

The larger its surface area the better the platinum works, because more gas is adsorbed. Many other catalysts work in the same way. For example . . .

. . . in modern car exhausts, harmful gases are adsorbed on to a rhodium catalyst, where they react with oxygen to make less harmful products.

Catalysts in industry

In industry, many reactions are endothermic. Energy must be put in to make them happen. The energy may come from oil, gas, coal, or electricity. But in any case it's expensive. It's one of the biggest costs industry faces.

With a catalyst, a reaction goes faster *at a given temperature*. That means you get your product faster, saving time and therefore money. Even better, it may go fast enough at a lower temperature. And that means a lower fuel bill and higher profits.

So catalysts can turn an uneconomic process into a profitable one. This makes them very important in industry – so important that over 10 million tonnes of catalysts a year are produced worldwide.

Different reactions need different catalysts. For example:
- **Iron** acts as a catalyst for the reaction between nitrogen and hydrogen to make ammonia.
- A mixture of **platinum** and **rhodium** acts as a catalyst for the oxidation of ammonia to nitric acid.

Although a catalyst is not itself chemically changed in a reaction, it does adsorb impurities. This eventually stops it being active. So the beds of catalyst in a chemical plant are replaced from time to time.

Catalysts are made in a variety of shapes and sizes.

Fitting a platinum–rhodium catalyst gauze to a reactor where ammonia will be oxidized to make nitric acid.

A close-up of the platinum–rhodium gauze. Can you think of three advantages of using the catalyst in this form?

1 Which does a catalyst *not* change?
   a the speed of a reaction   b the products formed
   c the total amount of each product formed.
2 Explain how a surface catalyst works.

3 Give two reasons why a catalyst can turn an uneconomic process into a profitable one.
4 Try to think of a reason why the same catalyst will not work for all reactions.

8.7 Enzymes

Enzymes are proteins produced by living organisms. They are catalysts for biological reactions.

You have thousands of enzymes in your body. Each catalyses a particular reaction. For example, the enzyme **amylase** in your saliva catalyses the breakdown of starch (from bread and other foods) into sugar. Without enzymes, digestion of food and other body reactions would be too slow at body temperature to keep you alive.

How enzymes work

This is how an enzyme catalyses the breakdown of a molecule:

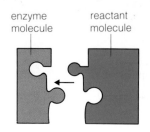

enzyme molecule   reactant molecule

First, the molecules fit together like jigsaw pieces. For this, the reactant molecule has to be the right shape.

reactant molecule breaking down

The 'complex' that forms makes it easier for the reactant molecule to break down. When decomposition is complete . . .

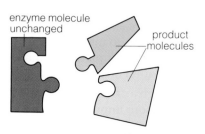

enzyme molecule unchanged   product molecules

. . . the product molecules break away. Another molecule of reactant will take their place . . . as long as its shape is right!

Note these important points:
- Since shape is so important, an enzyme will catalyse only one specific reaction where the 'fit' is right.
- An enzyme works best between about 25°C and 45°C. At higher temperatures it loses its shape and stops working. It becomes **denatured**. At low temperatures it becomes **inactive**.
- The graph on the right shows how the rate of an enzyme-catalysed reaction changes with temperature. This enzyme is at its most active at 40°C. That's its **optimum** temperature.
- An enzyme also works best in a particular pH range, depending on the enzyme. You can denature it by adding acid or alkali.

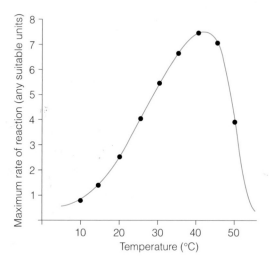

Enzymes in the kitchen

- Enzymes in foods bring about discoloration and decay. Keeping foods in the fridge slows the process. Why?
- Enzymes are completely inactive (but not destroyed) around −18°C. At 0°C they are very slightly active. So raw vegetables are often blanched (immersed quickly in boiling water) before freezing. Why?
- Foods can be preserved by pickling them in vinegar (a solution of ethanoic acid). Why?

Enzymes in industry

Enzymes are widely used in industry. For example:
- in making bread, yogurt, cheese, and chocolate
- in making beer and wine
- to tenderize meat
- to remove hair and bristle from hides, and to tan leather
- in biological detergents

Making bread   Yeast is a fungus consisting of millions of tiny living cells. These feed on sugar, using an enzyme to break it down into ethanol and carbon dioxide. The process is called **fermentation**.

$$C_6H_{12}O_6\,(aq) \longrightarrow 2C_2H_5OH\,(l) + 2CO_2\,(g) + energy$$
$$\text{sugar} \qquad\qquad \text{ethanol} \qquad \text{carbon dioxide}$$

Bread dough contains yeast and sugar. When the dough is left to stand in a warm place for an hour or so, the yeast feeds on the sugar and carbon dioxide is produced. This makes the dough rise. Then it is baked in a hot oven. The heat makes the carbon dioxide expand, so the dough rises still further. The heat also kills off the yeast.

Making wine and beer   Yeast is also used in making wine and beer. The fermentation process is the same as for bread. The sugar comes from fruit or barley. The ethanol makes the wine or beer alcoholic, and carbon dioxide bubbles off. Yeast stops working when the alcohol content reaches 14%, or it may be removed before then.

Making yogurt   Yogurt is made by adding bacteria to milk. First the milk is sterilized or **pasteurized** by heating, to kill harmful bacteria. Then it is inoculated with a **starter culture** of special bacteria. These produce enzymes which convert the sugar (lactose) in the milk into lactic acid and other substances, and make its proteins thicken.

Making cheese   Traditionally, a liquid called **rennet** from calves' stomachs was used in making cheese. Rennet contains two enzymes, chymosin and pepsin. Today some calves' rennet is still used, but rennet substitutes have been developed that give the same results.

First, a starter culture is added to pasteurized milk to sour it. Then rennet is added. It causes the protein **casein** in the milk to form a solid gel with fat and calcium trapped inside. When the gel is cut a liquid called **whey** runs out, leaving behind a rubbery solid called **curd**. This is compressed and stored at a controlled temperature. The enzymes act on the protein and fat in it, and turn it into cheese.

Yogurt with **bio-** in the name contains the bacterium *lactobacillus acidophilus*, which aids digestion. Other yogurt is heat-treated to kill off all bacteria.

1 What is an *enzyme*?
2 The process that makes bread rise is the same as the one that makes wine alcoholic. Explain.
3 How is yogurt made?
4 How is cheese made?
5 Explain how an enzyme breaks down a molecule of fat.
6 Why wouldn't you use a biological detergent:
  a in very hot water?   b to wash a woollen sweater?

8.8 Reversible reactions

Water of crystallization

- The water in blue copper(II) sulphate crystals is called **water of crystallization**.
- The blue compound is **hydrated**.
- The white compound is **anhydrous**.

When you heat blue crystals of copper(II) sulphate, they break down into *anhydrous* copper(II) sulphate, a white powder:

$$CuSO_4.5H_2O\,(s) \longrightarrow$$
$$CuSO_4\,(s) + 5H_2O\,(l)$$

The reaction is easy to reverse: just add water! The white powder turns blue again. In fact this is used as a test for water:

$$CuSO_4\,(s) + 5H_2O\,(l) \longrightarrow$$
$$CuSO_4.5H_2O\,(s)$$

The above reaction is **reversible**: it can go in either direction. Reaction 1 is the **forward** reaction. Reaction 2 is the **back** reaction. We use the symbol $\rightleftharpoons$ to show that the reaction is reversible:

$$CuSO_4.5H_2O\,(s) \rightleftharpoons CuSO_4\,(s) + 5H_2O\,(g)$$

Many chemical reactions are reversible. Here are two more examples.

The reaction between chlorine and iodine   It goes like this:

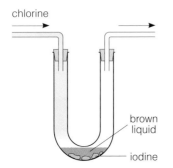

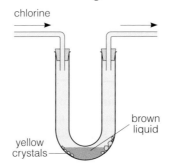

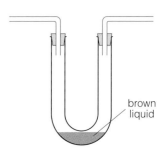

When chlorine is passed over iodine in a U-tube, a brown liquid forms. It is **iodine monochloride**, ICl:

$$Cl_2\,(g) + I_2\,(s) \longrightarrow 2ICl\,(l)$$
$$\text{brown}$$

If chlorine keeps flowing, yellow crystals will form on the sides of the tube. These are **iodine trichloride**, ICl_3:

$$ICl\,(l) + Cl_2\,(g) \longrightarrow ICl_3\,(s)$$
$$\text{brown} \qquad\qquad \text{yellow}$$

When the flow of chlorine stops, the yellow crystals will turn back into a brown liquid:

$$ICl_3\,(s) \longrightarrow ICl\,(l) + Cl_2\,(g)$$
$$\text{yellow} \qquad \text{brown}$$

So the reaction between iodine monochloride and iodine trichloride is reversible. To get more yellow crystals just add more chlorine:

$$ICl\,(l) + Cl_2\,(g) \rightleftharpoons ICl_3\,(s)$$

The thermal decomposition of ammonium choride   When you heat ammonium chloride (a solid) in the bottom of a test tube, it breaks down into ammonia and hydrogen chloride (gases). The gases recombine at the top of the tube, where it's cool:

$$NH_4Cl\,(s) \underset{\text{cool}}{\overset{\text{heat}}{\rightleftharpoons}} NH_3\,(g) + HCl\,(g)$$

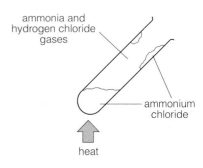

ammonia and hydrogen chloride gases

ammonium chloride

heat

Reversible reactions and dynamic equilibrium

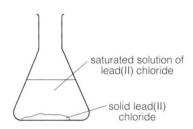

saturated solution of lead(II) chloride

solid lead(II) chloride

closed system

ions move continually between solid and solution

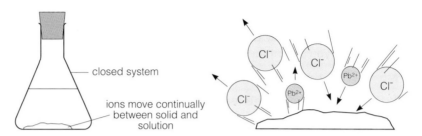

Lead(II) chloride is not very soluble. 1.1 g forms a saturated solution in 100 g of water:

$$PbCl_2\ (s) \longrightarrow Pb^{2+}\ (aq) + 2Cl^-\ (aq)$$

If you add more lead chloride it will sink to the bottom.

With a stopper on, the flask becomes a closed sytem. Inside, it's all action. Ions move continually from the solid to the solution and from the solution to the solid.

This is happening *at the same rate in both directions*, so no extra lead(II) chloride is dissolving. We can describe the situation like this:

$$PbCl_2\ (s) \rightleftharpoons Pb^{2+}\ (aq) + 2Cl^-\ (aq)$$

Inside the flask, the solid and its saturated solution have reached a state of **dynamic equilibrium** (often just called **equilibrium**).

Equilibrium means there is no overall change. But *dynamic* means that change is in fact taking place continuously. You can prove this by adding lead(II) chloride containing radioactive lead ions to the saturated solution in the closed flask. The solid falls to the bottom as you'd expect. After 20 minutes its mass is unchanged, but there are radioactive lead ions in the solution.

Exactly the same thing happens with a reversible chemical reaction.

In a closed system, a reversible reaction eventually reaches a state of dynamic equilibrium. The forward and back reactions take place at the same rate, so no overall change occurs.

Dynamic equilibrium and industry

Many very important reactions in industry are reversible. For example, the reaction between nitrogen and hydrogen to make ammonia:

$$N_2\ (g)\ +\ 3H_2\ (g)\ \rightleftharpoons\ 2NH_3\ (g)$$

This reaction eventually reaches a state of dynamic equilibrium: while some molecules of ammonia are forming, others are breaking down. So the reaction *never goes to completion*.

This is a problem for companies who make ammonia. They want the yield to be as large as possible. So what steps can they take to make more ammonia form? You will find out in the next Unit.

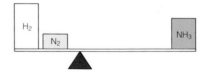

Nitrogen, hydrogen, and ammonia in equilibrium. The mixture is in balance.

1 What is a *reversible* reaction?

2 Write an equation for the reversible reaction between copper sulphate and water.

3 What does *dynamic equilibrium* mean?

4 Explain how you could establish a state of dynamic equilibrium between table salt and water.

8.9 Shifting the equilibrium

The manufacture of ammonia

Imagine you run a factory that makes ammonia from nitrogen and hydrogen. As you saw in the last Unit, the reaction is **reversible**:

$$N_2(g) + 3H_2(g) \rightleftharpoons 2NH_3(g)$$

Let's look more closely at this reaction:

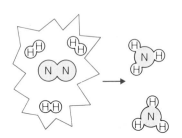

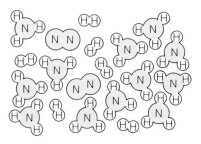

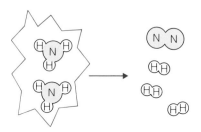

Three molecules of hydrogen react with one of nitrogen to form two of ammonia. So if you mix the right amounts of nitrogen and hydrogen...

...will it all turn into ammonia? No! Once a certain amount of ammonia is formed the system reaches a state of dynamic equilibrium. From then on...

...every time two ammonia molecules form, another two break down into nitrogen and hydrogen. So the level of ammonia remains unchanged.

But the more ammonia that forms, the better for your profits. So how can you increase the yield?

This idea, called Le Chatelier's principle, will help you:
When a reversible reaction is in equilibrium and you make a change, it will do what it can to oppose that change.

You can't make a reversible reaction go to completion. It *always* ends up in a state of equilibrium. But by changing the conditions you can *shift equilibrium to the right* and obtain more product. So let's see what changes you can make to obtain more ammonia.

Shifting the equilibrium to the right will give you more ammonia.

1 Increasing the temperature

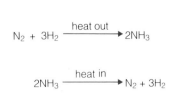

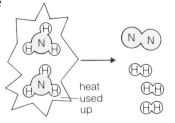

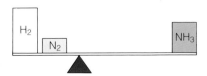

As you saw, heat speeds up *any* reaction. But here the forward reaction is exothermic – it gives out heat. The back reaction is endothermic – it takes it in.

If you heat the equilibrium mixture it will act to oppose the change. More ammonia will break down in order to use up the heat you've added.

The result is that the reaction reaches equilibrium faster, which is good, but the level of ammonia *decreases*. So you're worse off than before.

But if you decide to improve the yield by running at a very *low* temperature, the reaction will take too long to reach equilibrium!

2 Increasing the pressure

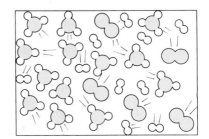

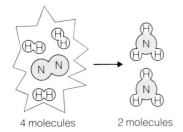

4 molecules   2 molecules

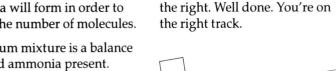

Pressure is caused by collisions between gas molecules and the walls of the container. So the fewer molecules present, the lower the pressure.

If you apply more pressure, the equilibrium mixture will act to oppose the change. More ammonia will form in order to reduce the number of molecules.

The result is that the level of ammonia in the mixture *increases*. Equilibrium shifts to the right. Well done. You're on the right track.

When you remove ammonia, nitrogen and hydrogen react to restore equilibrium.

3 Removing the ammonia The equilibrium mixture is a balance between the levels of nitrogen, hydrogen, and ammonia present. Suppose you cool the mixture. Ammonia condenses first so you can run it off as a liquid. Then warm the remaining nitrogen and hydrogen again. *Et voilà*, more ammonia!

4 Adding a catalyst Iron acts as a catalyst for this reaction. But note that it speeds up the forward and back reactions *equally*. So although the reaction reaches equilibrium faster, the equilibrium position doesn't change. Still, the catalyst is worth using because it saves time and therefore money. Eventually it will get poisoned and must be replaced.

Choosing the optimum conditions To make production economical, you have to balance all the above factors. You have to:
- use high pressure, and remove ammonia, to shift equilibrium to the right
- use a catalyst to reach equilibrium quickly
- use moderate heat

On page 218 you can see how these conditions are applied.

A summary

Remember, for reversible reactions between gases . . .
If the forward reaction is:
- exothermic, temperature ↑ means yield ↓ and vice versa
- endothermic, temperature ↑ means yield ↑ and vice versa

If there are *fewer* moles of product than reactant, in the equation:
pressure ↑ means yield ↑ and vice versa

---

1 a Explain what goes on in an equilibrium mixture of nitrogen, hydrogen, and ammonia.
   b Why does this cause a problem in industry?
2 What is Le Chatelier's principle? Write it down.
3 In manufacturing ammonia, explain why:
   a high pressure is used   b ammonia is removed
   c only moderate heat is used.

4 Sulphur dioxide (SO_2) and oxygen react to form sulphur trioxide (SO_3). The reaction is exothermic and reversible.
   a Write a balanced equation for it.
   b What happens to the yield of sulphur trioxide:
      i if you increase the pressure?
      ii if you increase the temperature?

Questions on Section 8

1 The rate of the reaction between magnesium and dilute hydrochloric acid could be measured using this apparatus:

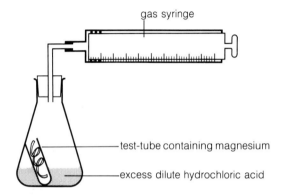

gas syringe

test-tube containing magnesium

excess dilute hydrochloric acid

a What is the purpose of:
  i the test-tube?
  ii the gas syringe?
b How would you get the reaction to start?

2 Some magnesium and an *excess* of dilute hydrochloric acid were reacted together. The volume of hydrogen produced was recorded every minute, as shown in the table:

Time/min	0	1	2	3	4	5	6	7
Volume of hydrogen/cm³	0	14	23	31	38	40	40	40

a What does an *excess* of acid mean?
b Plot a graph of the results, labelling the axes as on page 121.
c how much hydrogen was produced in:
  i the first minute?
  ii the second minute?
  iii the third minute?
  iv the fourth minute?
  v the fifth minute?
d What is the *rate of the reaction* (cm³ of hydrogen per minute) during each minute?
e What is the total volume of hydrogen produced in the reaction?
f How many minutes pass before the reaction finishes?
g What is the *average rate* of the reaction?
h A similar reaction had a rate of 15 cm³ of hydrogen in the first minute. Is this a slower or faster reaction than the one above?
i How could you make the above reaction go slower, while still using the same quantities of metal and acid?
j How could you make it go faster?

3 For this question you will need the graph you drew for question 2.
The experiment with magnesium and an excess of dilute hydrochloric acid was repeated. This time a different concentration of hydrochloric acid was used. The results were:

Time/min	0	1	2	3	4	5	6
Volume of hydrogen/cm³	0	22	34	39	40	40	40

a Plot these results on the graph you drew for question 2.
b Which reaction was faster? How can you tell?
c In which experiment was the acid more concentrated? Give a reason for your answer.
d The same volume of hydrogen was produced in each experiment. What does that tell you about the mass of magnesium used?

4 Name three factors that affect the rate of a reaction, and describe the effect of changing each factor.

5 Suggest a reason for each of the following observations:
a Magnesium powder reacts faster than magnesium ribbon, with dilute sulphuric acid.
b Hydrogen peroxide decomposes much faster if you add a piece of raw liver. But boiled liver has no effect on it.
c The reaction between manganese carbonate and dilute hydrochloric acid speeds up when some concentrated hydrochloric acid is added.
d Zinc powder burns much more vigorously in oxygen than zinc foil does.
e The reaction between sodium thiosulphate and hydrochloric acid takes a very long time if carried out in an ice bath.
f Zinc and dilute sulphuric acid react much more quickly when a few drops of copper(II) sulphate solution are added.
g Drenching with water prevents too much damage from spilt acid.
h A car's exhaust pipe will rust faster if the car is used a lot.
i In fireworks, powdered magnesium is used rather than magnesium ribbon.
j In this country, dead animals decay quite quickly. But in Siberia, bodies of mammoths that died 30 000 years ago have been found fully preserved in ice.
k The more sweet things you eat, the faster your teeth decay.
l Food cooks much faster in a pressure cooker than in an ordinary saucepan.
m A concentrated solution of bleach is used for removing stains.

6 When sodium thiosulphate reacts with hydrochloric acid, a precipitate forms. In an investigation, the time taken for the solution to become opaque was recorded. (*Opaque* means that you cannot see through it.) Four experiments (A to D) were carried out. Only the concentration of the sodium thiosulphate solution was changed each time. The results were:

Experiment	A	B	C	D
Time taken/seconds	42	71	124	63

a Draw a diagram of suitable apparatus for this experiment.
b Name the precipitate that forms.
c What would be *observed* during the experiment?
d In which experiment was the reaction:
  i fastest? ii slowest?
e In which experiment was the sodium thiosulphate solution most concentrated? How can you tell?
f Suggest two other ways of speeding up this reaction.

7 Copper(II) oxide catalyses the decomposition of hydrogen peroxide. 0.5 g of the oxide was added to a flask containing 100 cm³ of hydrogen peroxide solution. A gas was released. It was collected and its volume noted every 10 seconds. This table shows the results:

Time/sec	0	10	20	30	40	50	60	70	80	90
Volume/cm³	0	18	30	40	48	53	57	58	58	58

a What is a catalyst?
b Draw a diagram of suitable apparatus for this experiment.
c Name the gas that is formed.
d Write a balanced equation for the decomposition of hydrogen peroxide.
e Plot a graph of the volume of gas (vertical axis) against time (horizontal axis).
f Describe how the rate changes during the reaction.
g What happens to the concentration of hydrogen peroxide as the reaction proceeds?
h What chemicals are present in the flask after 90 seconds?
i What mass of copper(II) oxide would be left in the flask at the end of the reaction?
j Sketch on your graph the curve that might be obtained for 1.0 g of copper(II) oxide.
k Name one other chemical that catalyses this decomposition.
l A catalyst works by lowering the *activation energy* for a reaction. Explain what that means.

8 When chlorine gas is passed over iodine, the following reactions take place.
  Reaction 1: $I_2(s) + Cl_2(g) \longrightarrow 2\,ICl(l)$
  brown

  Reaction 2: $ICl(l) + Cl_2(g) \rightleftharpoons ICl_3(s)$
  brown   yellow

a What would you see when:
  i chlorine is first passed over the iodine?
  ii more chlorine is passed over it?
b Which of the reactions can be *reversed*?
c What change would you see as the chlorine gas supply is turned off? Explain your answer.

9 Hydrogen and bromine react together like this:
  $H_2(g) + Br_2(g) \rightleftharpoons 2HBr(g)$
a Which of the following will favour the formation of more hydrogen bromide?
  i Adding more hydrogen.
  ii Removing bromine.
  iii Removing the product as it is formed.
b Explain why increasing the pressure has no effect on the amount of HBr formed.

10 Ammonia is manufactured from nitrogen and hydrogen. $\Delta H_{reaction} = -92\,kJ/mol$.
a Write the equation for the reaction.
b What does $\Delta H_{reaction}$ mean?
c Is the forward reaction endothermic or exothermic? How can you tell?
d Explain why the yield of ammonia:
  i rises if you increase the pressure
  ii falls if you increase the temperature
e Why is the reaction carried out at 450 °C rather than a lower temperature?

11 The dichromate and chromate ions, $Cr_2O_7^{2-}$ and CrO_4^{2-}, exist in equilibrium as follows:
  $Cr_2O_7^{2-}(aq) + H_2O(l) \rightleftharpoons 2CrO_4^{2-}(aq) + 2H^+(aq)$
  orange   yellow

a What would you see if you added dilute acid to a solution containing chromate ions?
b How would you reverse the change?
c Use Le Chatelier's principle to explain why adding hydroxide ions shifts the equilibrium.

12 The enzyme *polyphenoxidase* is involved in the oxidation reaction that causes sliced apple to turn brown in air. Explain the following observations.
a When the apple is first cut open, the flesh is not brown.
b Browning is much slower when the sliced apple is placed in the fridge.
c No browning takes place if the apple is placed in boiling water for 30 seconds straight after slicing.
d Apple that has been pulped in a food mixer turns brown much faster than sliced apple does.
e The browning reaction does not take place if the sliced apple is dipped in lemon juice straight away.

9.1 Acids and alkalis

Acids

One important group of chemicals is called **acids**:

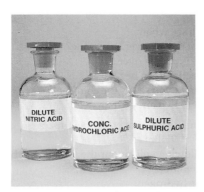

Here are some acids. You have probably seen them in the lab. They are all liquids. In fact they are **solutions** of pure compounds in water.

They must be handled carefully, especially the concentrated acids, for they are **corrosive**. They can eat away metals, skin, and cloth.

But some acids are not so corrosive, even when they are concentrated. They are called **weak** acids. Ethanoic acid is one example. It's found in vinegar.

You can tell if something is an acid, by its effect on **litmus**. Litmus is a purple dye. It can be used as a solution, or on paper:

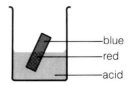

Litmus solution is purple. Litmus paper for testing acids is blue.

Acids will turn litmus solution red.

They will also turn blue litmus paper red.

Some common acids The main ones are:

hydrochloric acid	HCl (aq)
sulphuric acid	H_2SO_4 (aq)
nitric acid	HNO_3 (aq)
ethanoic acid	CH_3COOH (aq)

But there are plenty of others. For example, lemon juice contains **citric acid**, ant and nettle stings contain **methanoic acid** and tea contains **tannic acid**.

Alkalis

There is another group of chemicals that also affect litmus, but in a different way to acids. They are the **alkalis**.
Alkalis turn litmus solution blue, and red litmus paper blue.
Like acids, they must be handled carefully, because they can burn skin.

Some common alkalis Most pure alkalis are solids. But they are usually used in laboratory as aqueous solutions. The main ones are:

sodium hydroxide   NaOH (*aq*)
potassium hydroxide   KOH (*aq*)
calcium hydroxide   Ca(OH)$_2$ (*aq*)
ammonia   NH$_3$ (*aq*)

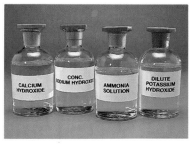
Common laboratory alkalis

Neutral substances

Many substances do not affect the colour of litmus solution, so they are not acids or alkalis. They are **neutral**. Examples are pure water and aqueous solutions of sodium chloride and sugar.

The pH scale

You saw on page 138 that some acids are weaker than others. It is the same with alkalis.
The strength of an acid or an alkali is shown using a scale of numbers called the **pH scale**. The numbers go from 0 to 14:

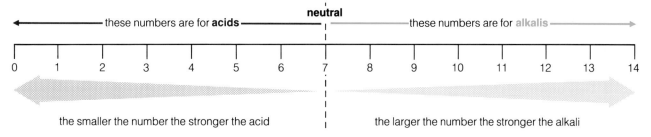

On this scale:
An acidic solution has a pH number less than 7.
An alkaline solution has a pH number greater than 7.
A neutral solution has a pH number of exactly 7.

You can find the pH of any solution by using **universal indicator**. Universal indicator is a mixture of dyes. Like litmus, it can be used as a solution, or as universal indicator paper. It goes a different colour at different pH values, as shown in this diagram:

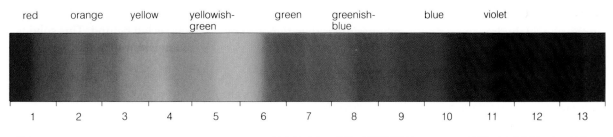

1 What does *corrosive* mean?
2 How would you test a substance, to see if it is an acid?
3 Write down the formula for:
   sulphuric acid   nitric acid
   calcium hydroxide   ammonia solution

4 What effect do alkalis have on litmus solution?
5 Say whether a solution is acidic, alkaline or neutral, if its pH number is:
   9   4   7   1   10   3
6 What colour would universal indicator show, in an aqueous solution of sugar? Why?

9.2 A closer look at acids

increasing acidity

| 4.0 | 5.5 | 6.8 |

| pH 1.0 | 2.0 | 2.5 | 2.9 | 3.2 | 3.8 | 5.2 | 5.9 | 6.8 |

The liquids above are all **acidic**, as their pH numbers show. Which is the most acidic? How can you tell?

The properties of acids

1 Acids have a sour taste. Think of the taste of vinegar.
   But *never* taste the laboratory acids, because they could burn you.
2 They turn litmus red.
3 They have pH numbers less than 7.
4 They usually react with **metals**, forming hydrogen and a **salt**:

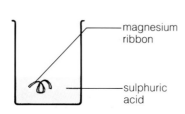

magnesium ribbon

sulphuric acid

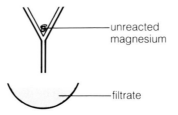

unreacted magnesium

filtrate

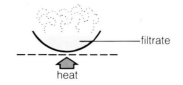

filtrate

heat

When magnesium is dropped into dilute sulphuric acid, hydrogen quickly bubbles off.

The bubbles stop when the reaction is over. The unreacted magnesium is then removed by filtering.

The filtrate is heated, to evaporate the water. A white solid is left behind. This solid is **magnesium sulphate**.

The equation for the reaction is:

$$\text{magnesium} + \text{sulphuric acid} \longrightarrow \text{magnesium sulphate} + \text{hydrogen}$$
$$Mg\,(s) \quad + \quad H_2SO_4\,(aq) \quad \longrightarrow \quad MgSO_4\,(aq) \quad + \quad H_2\,(g)$$

The magnesium has driven the hydrogen out of the acid, and taken its place. Magnesium sulphate is called a **salt**.
When a metal takes the place of hydrogen in an acid, the compound that forms is called a salt.

The salts of sulphuric acid are always called **sulphates**.
The salts of hydrochloric acid are called **chlorides**.
The salts of nitric acid are called **nitrates**.

5 Acids react with **carbonates**, forming a salt, water and carbon dioxide. Hydrochloric acid reacts with calcium carbonate like this:

calcium carbonate + hydrochloric acid $\longrightarrow$ calcium chloride + water + carbon dioxide

$$CaCO_3\,(s) + 2HCl\,(aq) \longrightarrow CaCl_2\,(aq) + H_2O\,(l) + CO_2\,(g)$$

6 They react with **alkalis**, forming a salt and water. For example:

sodium hydroxide + nitric acid $\longrightarrow$ sodium nitrate + water

$$NaOH\,(aq) + HNO_3\,(aq) \longrightarrow NaNO_3\,(aq) + H_2O(l)$$

7 They also react with **metal oxides**, forming a salt and water:

zinc oxide + hydrochloric acid $\longrightarrow$ zinc chloride + water

$$ZnO\,(s) + 2HCl\,(aq) \longrightarrow ZnCl_2\,(aq) + H_2O\,(l)$$

When an acid reacts with a carbonate, carbon dioxide fizzes off.

What causes acidity?

Acids have a lot in common, as you have just seen.
There must be something *in* them all, that makes them act alike. That 'something' is **hydrogen ions**.
Acids contain hydrogen ions.
Acids are solutions of pure compounds in water. The pure compounds are molecular. But in water, the molecules break up to form ions. They *always* give hydrogen ions. For example, in hydrochloric acid:

$$HCl\,(aq) \longrightarrow H^+\,(aq) + Cl^-\,(aq)$$

The more H^+ ions there are in a solution, the more acidic it is. In other words the more H^+ ions there are, the lower the pH number.

H^+ – just a proton!

- A hydrogen atom has just 1 proton and 1 electron.
- This means the H^+ ion is just a proton!
- In fact, although we refer to it as the H^+ ion, it attaches itself to a water molecule, forming the hydroxonium ion H_3O^+.

Strong and weak acids

Dilute hydrochloric acid reacts quickly with magnesium ribbon. The reaction could be over in minutes. But when ethanoic acid of the same concentration is used instead, the reaction is much slower. It could take all day.

The hydrochloric acid reacts faster *because it contains more hydrogen ions.*
In hydrochloric acid, *nearly all* the acid molecules break up to form ions. It is called a **strong** acid. But in ethanoic acid, only some of the acid molecules form ions, so it is called a **weak** acid.
In a strong acid, nearly all the acid molecules form ions. In a weak acid, only some of the acid molecules form ions.
So a strong acid *always* has a lower pH number than a weak acid of the same concentration.
Some strong and weak acids are shown in the box on the right.

Strong acids
Hydrochloric acid
Sulphuric acid
Nitric acid

Weak acids
Ethanoic acid
Citric acid
Carbonic acid

---

1 Name the acid found in:
lemon juice   wine   tea   vinegar
2 Write a word equation for the reaction of dilute sulphuric acid with: **a** zinc
b magnesium oxide   **c** sodium carbonate

3 Bath salts contain sodium carbonate. Explain why they fizz when you put vinegar on them.
4 What causes the acidity in acids?
5 Why is ethanoic acid called a weak acid?
6 Name two other weak acids, and two strong ones.

9.3 A closer look at alkalis

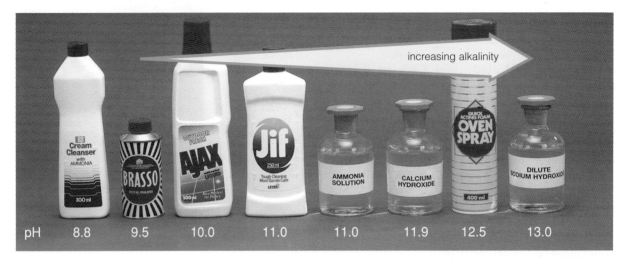

| pH | 8.8 | 9.5 | 10.0 | 11.0 | 11.0 | 11.9 | 12.5 | 13.0 |

increasing alkalinity

All the substances above are alkaline – you can tell by their pH numbers. Note the kitchen cleaners! They are alkaline because they contain ammonia or sodium hydroxide, which attack grease.

What makes things alkaline?

You saw on page 141 that all acid solutions contain hydrogen ions, H^+. The alkalis also have something in common:
All alkaline solutions contain hydroxide ions, OH^-.
In sodium hydroxide solution, the ions are produced like this:

$$NaOH \ (aq) \longrightarrow Na^+ \ (aq) + OH^- \ (aq)$$

In ammonia solution, ammonia molecules react with water molecules to form ions:

$$NH_3 \ (aq) + H_2O \ (l) \longrightarrow NH_4^+ \ (aq) + OH^- \ (aq)$$

The more OH^- ions there are in a solution, the more alkaline it will be, and the higher its pH number.

Strong and weak alkalis

Like acids, alkalis can also be strong or weak.
Sodium hydroxide is a **strong** alkali, because it exists almost completely as ions, in solution. Ammonia solution is a **weak** alkali because only some ammonia molecules form ions in solution.

Properties of alkalis

1 Alkalis feel soapy. But don't try – they can burn flesh.
2 Their solutions turn litmus blue.
3 Their solutions have pH numbers greater than 7.
4 They react with acids to form a salt and water. The reaction is called a **neutralization**. See next page for more.
5 All the alkalis except ammonia will react with ammonium compounds, driving ammonia out. For example:

| calcium hydroxide | + | ammonium chloride | → | calcium chloride | + | steam | + ammonia |
| $Ca(OH)_2 \ (s)$ | + | $2NH_4Cl \ (s)$ | → | $CaCl_2 \ (s)$ | + | $2H_2O \ (g)$ | + $2NH_3 \ (g)$ |

This reaction is used for making ammonia in the laboratory.

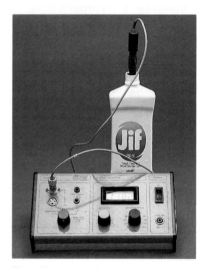

The pH number of a kitchen cleaner can be measured with a pH meter.

Strong alkalis
Sodium hydroxide
Potassium hydroxide
Calcium hydroxide

Weak alkali
Ammonia

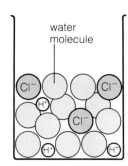

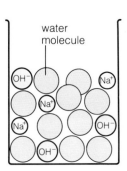

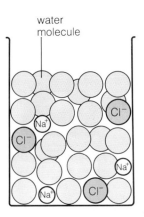

This is a solution of hydrochloric acid. It contains H^+ ions and Cl^- ions. It will turn litmus red.

This is a solution of sodium hydroxide. It contains Na^+ ions and OH^- ions. It will turn litmus blue. When you add it to the acid ...

... the H^+ and OH^- ions join to form **water molecules**. The result is a neutral solution of sodium chloride, with no effect on litmus.

The equation for this reaction is:

$$HCl\ (aq) + NaOH\ (aq) \longrightarrow NaCl\ (aq) + H_2O\ (l)$$

You could also write it showing only the ions that combine:

$$H^+\ (aq) + OH^-\ (aq) \longrightarrow H_2O\ (l)$$

The alkali has **neutralized** the acid by removing its H^+ ions, and turning them into water.
During a neutralization reaction, the H^+ ions of the acid are turned into water.
You could obtain solid sodium chloride by heating the last solution above. The water evaporates leaving the solid behind.

Tracking a neutralization

When you add an alkali to an acid, you can track the neutralization in several ways.
1 **pH.** As H^+ ions are removed the pH of the solution rises. You can follow this using **universal indicator** or a **pH meter**. When the pH rises rapidly you'll know that neutralization is complete.
2 **Conductivity.** As H^+ ions get removed, the solution is less able to conduct electricity. Conductivity reaches its lowest when neutralization is complete. Track it using a **conductivity meter**.
3 **Temperature.** Neutralization is exothermic. So the temperature of the solution rises until the reaction is complete. Track this using a **thermometer**.

---

1 Look at the substances shown at the top of page 142. Which ion do they all have in common?
2 A 0.1 M solution of potassium hydroxide has a higher pH number than a 0.1 M solution of ammonia. Why is this? Which one will be a better conductor of electricity?

3 What happens to: **a** the H^+ ions
b the temperature   **c** the conductivity
of the solution when an acid is neutralized?
4 Write a balanced equation for the reaction between sodium hydroxide and:
a sulphuric acid   **b** nitric acid

9.4 Bases and neutralization

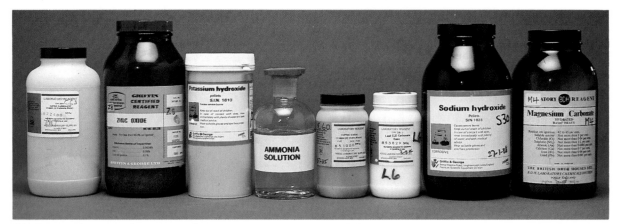

All these substances have something in common, as you will find out below.

Bases

An alkali can **neutralize** an acid, and destroy its acidity. It does this by removing the H^+ ions and converting them to water. But alkalis are not the only compounds that can neutralize acids:

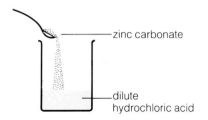

zinc carbonate

dilute
hydrochloric acid

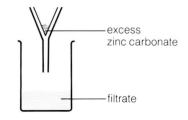

excess
zinc carbonate

filtrate

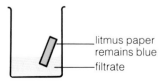

litmus paper
remains blue
filtrate

1 When zinc carbonate is added to dilute hydrochloric acid, it dissolves, and carbon dioxide bubbles off.

2 More is added until no more will dissolve, even with heating. The excess zinc carbonate is removed by filtering.

3 Then the filtrate is tested with blue litmus. There is no colour change. The acidity of the acid has been destroyed.

The equation for the reaction is:
$$ZnCO_3\,(s) + 2HCl\,(aq) \longrightarrow ZnCl_2\,(aq) + H_2O\,(l) + CO_2\,(g)$$
So the H^+ ions of the acid have been turned into water. The acid has been neutralized by the zinc carbonate.
In fact, acid can be neutralized by any of these compounds:

> **metal oxides**
> **metal hydroxides**
> **metal carbonates**
> **metal hydrogen carbonates**
> **ammonia solution**

These compounds are all called **bases**.
Any compound that can neutralize an acid is called a base.

Alkalis are bases

The photograph at the top of the opposite page shows some bases. Notice that it includes alkalis – they can neutralize acids, so they are bases. They are bases that are soluble in water.
Alkalis are soluble bases.

Neutralization always produces a salt

Neutralization always produces a salt, as these general equations show:

acid + metal oxide $\longrightarrow$ metal salt + water

acid + metal hydroxide $\longrightarrow$ metal salt + water

acid + metal carbonate $\longrightarrow$ metal salt + water + carbon dioxide

acid + metal hydrogen carbonate $\longrightarrow$ metal salt + water + carbon dioxide

acid + ammonia solution $\longrightarrow$ ammonium salt + water

Some neutralizations in everyday life

Insect stings   When a bee stings, it injects an acidic liquid into the skin. The sting can be neutralized by rubbing on **calamine lotion**, which contains zinc carbonate, or **baking soda**, which is sodium hydrogen carbonate.
Wasp stings are alkaline, and can be neutralized with vinegar. Why? Ant stings and nettle stings contain methanoic acid. How would you treat them?

Bee stings are acidic.

Indigestion   To digest food properly, the liquid in your stomach must be acidic. But too much acidity leads to **indigestion**, which can be very painful. To cure indigestion, you must take an **antacid** to neutralize the excess acid. For example, a drink of sodium hydrogen carbonate solution (baking soda), or an indigestion tablet.

Soil treatment   Most plants grow best when the pH of the soil is close to 7. If the soil is too acidic, or too alkaline, the plants grow badly or not at all.
Chemicals can be added to soil to adjust its pH. Most often, soil is too acid, so it is treated with **quicklime** (calcium oxide), **slaked lime** (calcium hydroxide), or **chalk** (calcium carbonate). These are all bases, and are quite cheap.

Factory waste   Liquid waste from factories often contains acid. If it reaches a river, the acid will kill fish and other river life. This can be prevented by adding slaked lime to the waste, to neutralize it.

Slaked lime being spread on fields where the soil is too acidic

---

1 Look at the reaction shown on page 144. What salt does the filtrate contain? How would you obtain the dry salt from the filtrate?
2 What is a base? Name six bases.
3 What special property do alkalis have?
4 Which bases react with acids to give carbon dioxide?

5 Write a word equation for the reaction between:
  a dilute hydrochloric acid and copper(II) oxide;
  b dilute sulphuric acid and potassium carbonate.
6 Write a word equation for the reaction that takes place in your stomach, when you take baking soda to cure indigestion.

9.5 Making salts (I)

Making salts from acids

Acid + metal   Zinc sulphate can be made by reacting dilute sulphuric acid with zinc:

$$Zn\,(s) + H_2SO_4\,(aq) \longrightarrow ZnSO_4\,(aq) + H_2\,(g)$$

These are the steps:

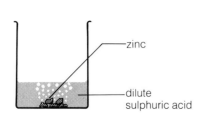

1 Some dilute sulphuric acid is put in a beaker, and zinc is added. The zinc begins to dissolve, and hydrogen bubbles off. The bubbles stop when all the acid has been used up.

2 Some zinc is still left. It is removed by filtering, which leaves an aqueous solution of zinc sulphate.

3 The solution is heated to evaporate some of the water. Then it is left to cool. Crystals of zinc sulphate start to form.

The method above is not suitable for *all* metals, or *all* acids.
It is fine for magnesium, aluminium, zinc and iron.
But the reactions of sodium, potassium and calcium with acid are dangerously violent. The reaction of lead is too slow, and copper, silver and gold do not react at all (page 196).

Acid + insoluble base   Copper(II) oxide is an insoluble base. Although copper will not react with dilute sulphuric acid, copper(II) oxide will. The salt that forms is copper(II) sulphate:

$$CuO\,(s) + H_2SO_4\,(aq) \longrightarrow CuSO_4\,(aq) + H_2O\,(l)$$

The method is quite like the one above.

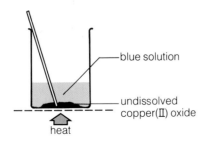
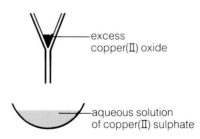
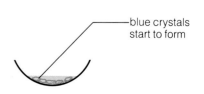

1 Some copper(II) oxide is added to dilute sulphuric acid. On warming, it dissolves and the solution turns blue. More is added until no more dissolves.

2 That means all the acid has been used up. The excess solid is removed by filtering. This leaves a blue solution of copper(II) sulphate in water.

3 The solution is heated to evaporate some of the water. Then it is left to cool. Blue crystals of copper(II) sulphate start to form.

Acid + alkali (soluble base) The reaction of sodium with acids is very dangerous. So sodium salts are usually made by starting with sodium hydroxide. This reaction can be used to make sodium chloride:

$$NaOH\,(aq) + HCl\,(aq) \longrightarrow NaCl\,(aq) + H_2O\,(l)$$

Both reactants are soluble, and no gas is given off during the reaction. So it is difficult to know when the reaction is over. You have to use an **indicator**. Universal indicator or litmus could be used, but even better is **phenolphthalein**. This is pink in alkaline solution, but colourless in neutral and acid solutions:

Titration

- The process shown below is called **titration**.
- Titration is used to find the amount of acid needed to neutralize a known amount of alkali – or vice versa.

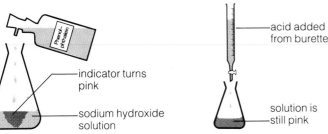

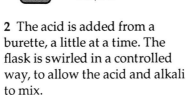

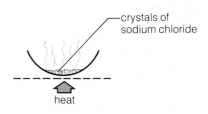

on adding one more drop, pink colour suddenly disappears

1 25 cm³ of sodium hydroxide solution is measured into a flask, using a pipette. Then two drops of phenolphthalein are added. The indicator turns pink.

2 The acid is added from a burette, a little at a time. The flask is swirled in a controlled way, to allow the acid and alkali to mix.

3 When all the alkali has been used up, the indicator suddenly turns colourless, showing that the solution is neutral. There is no need to add more acid.

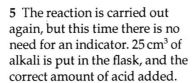

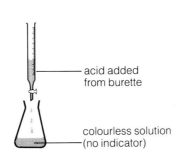

crystals of sodium chloride

heat

4 You can tell how much acid was added, using the scale on the burette. So now you know how much acid is needed to neutralize 25 cm³ of alkali.

5 The reaction is carried out again, but this time there is no need for an indicator. 25 cm³ of alkali is put in the flask, and the correct amount of acid added.

6 The solution from the flask is heated, to let the water evaporate. You will find dry crystals of sodium chloride are left behind.

In step 5 the reaction has to be carried out again, *without* indicator, because indicator would make the salt impure.
A similar method can be used for making potassium salts from potassium hydroxide, and ammonium salts from ammonia solution.

---

1 Name the acid and metal you would use for making:
   a zinc chloride   **b** magnesium sulphate
2 Why would you *not* make potassium chloride from potassium and hydrochloric acid?

3 How would you obtain lead(II) nitrate, starting with the insoluble compound lead(II) carbonate?
4 Write instructions for making potassium chloride, starting with solid potassium hydroxide.

9.6 Making salts (II)

Making salts by precipitation

The salts made so far have all been soluble. They were obtained as crystals by evaporating solutions. But not all salts are soluble:

Soluble	Insoluble
All sodium, potassium, and ammonium salts	
All nitrates	
Chlorides... *except*	silver and lead chloride
Sulphate... *except*	calcium, barium and lead sulphate
Sodium, potassium, and ammonium carbonates...	but all other carbonates are insoluble

Insoluble salts can be made by **precipitation**.
For example, insoluble barium sulphate is precipitated when solutions of barium chloride and magnesium sulphate are mixed:

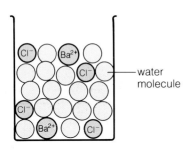

This is a solution of barium chloride, $BaCl_2$. It contains barium ions and chloride ions.

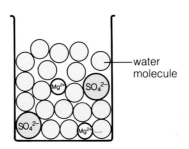

This is a solution of magnesium sulphate, $MgSO_4$. It contains magnesium ions and sulphate ions.

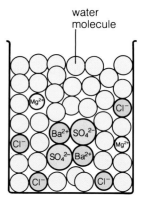

When the two solutions are mixed, the barium ions and sulphate ions bond together, because they are strongly attracted to each other. Solid barium sulphate precipitates.

The equation for the reaction is:

$$BaCl_2\,(aq) + MgSO_4\,(aq) \longrightarrow BaSO_4\,(s) + MgCl_2\,(aq)$$

or you could write it in a shorter way as:

$$Ba^{2+}\,(aq) + SO_4^{2-}\,(aq) \longrightarrow BaSO_4\,(s)$$

These are the steps for obtaining the barium sulphate:

1 Solutions of barium chloride and magnesium sulphate are mixed. A white precipitate of barium sulphate forms at once.
2 The mixture is filtered. The barium sulphate gets trapped in the filter paper.
3 It is rinsed with distilled water.
4 Then it is put in a warm oven to dry.

The precipitation of barium sulphate

Barium sulphate could also be made from barium nitrate and sodium sulphate, for example, since these are both soluble. As long as barium ions and sulphate ions are present, barium sulphate will be precipitated.

To precipitate an insoluble salt, you must mix a solution that contains its positive ions with one that contains its negative ions.

Making salts by combining elements

Some salts contain just two elements. You can make them by **direct combination** of the elements.

For example you can make iron(III) chloride, $FeCl_3$, by heating iron in a stream of chlorine as shown here. The reaction is exothermic. Once it begins you can turn off the bunsen.

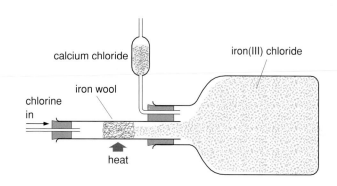

The iron(III) chloride forms as a dark brown vapour. It solidifies on the cold collecting jar. Calcium chloride keeps the apparatus dry. (It absorbs moisture.)

The product is **anhydrous** which means it has no water of crystallization. But in damp air it absorbs moisture and turns yellow. It has become **hydrated** iron(III) chloride, $FeCl_3.6H_2O$.

Making salts in industry

Many useful salts occur naturally and can be dug out of the earth. Sodium chloride is an example. But others must be made in factories, using methods like those you use in the lab.

For example **ammonium nitrate** is an important fertilizer. It is made by neutralizing nitric acid with ammonia solution, in huge tanks:

$$HNO_3\ (aq) + NH_3\ (aq) \longrightarrow NH_4NO_3\ (aq)$$

The water is driven off in an evaporator, leaving molten ammonium nitrate. This is sprayed down a tall tower. By the time it reaches the bottom it has cooled into small pellets that are easy to spray on soil.

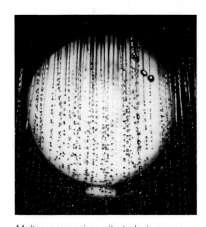

Molten ammonium nitrate being sprayed down a tower. It cools into pellets.

1 Choose two starting compounds you could use to precipitate: **a** calcium sulphate
   b zinc carbonate **c** lead chloride
2 Write a balanced equation for each reaction in question 1.
3 Some salts can be made by *direct combination* of their elements. What does that mean?

4 **a** Suggest how you could make iron(III) bromide by direct combination.
   b The product is *anhydrous*. What does that mean?
   c List the precautions you would take with this experiment, and give reasons.
   d Would you expect the reaction to be faster, or slower, than with chlorine? Explain why.

9.7 Using titration to find concentrations

On page 147, the volume of acid needed to neutralize an alkali was found using indicator, in a process called **titration**. This method can also be used to find the concentration of an acid.

For that, you simply titrate the acid against a solution of alkali of known concentration: a **standard solution**. Then, once you've calculated the concentration of the acid you can use *it* in turn as a standard solution, to find the concentration of a solution of alkali.

Example 1 Standardize a solution of hydrochloric acid

You have a solution of hydrochloric acid with a concentration of around $2 M$, or $2\, mol/dm^3$. You want to find its *exact* concentration: you want to **standardize** it.
How can you do that? By reacting it with a solution of sodium carbonate of known concentration, using methyl orange indicator.

First, make up a standard 1 M solution of sodium carbonate.
- Molar mass of $Na_2CO_3 = (2 \times 23) + 12 + (3 \times 16) = 106$
  A 1 M solution contains 106 grams in $1\, dm^3$ of solution, so
  $250\, cm^3$ of a 1 M solution contains $106 \div 4 = 26.5\, g$.
- Weigh out this amount and place it in a $250\, cm^3$ standard flask.
- Add distilled water and swirl to dissolve the solid. Fill the flask to the $250\, cm^3$ mark. Put the stopper in and shake well.

Then titrate the acid against this standard solution.
- Measure $25\, cm^3$ of the standard solution into a conical flask, using a pipette. Add a few drops of methyl orange indicator.
- Pour the acid into a $50\, cm^3$ burette. Record the level.
- Drip the acid slowly into the conical flask, swirling the flask continuously. Stop adding acid when a single drop finally turns the indicator red. Record the new level of acid in the burette.
- Calculate the volume of acid used, as shown on the right.

How much acid was used?

Final level:	$28.8\, cm^3$
Initial level:	$1.0\, cm^3$
Volume used:	$27.8\, cm^3$

Now calculate the concentration of the acid.
Step 1 Calculate the number of moles of sodium carbonate used.
  $1000\, cm^3$ of 1 M solution contains 1 mole so

  $25\, cm^3$ contains $\dfrac{25}{1000} \times 1$ mole or 0.025 mole.

Step 2 From the equation, find the molar ratio of acid to alkali.
  $2HCl\,(aq) + Na_2CO_3\,(aq) \longrightarrow 2NaCl\,(aq) + H_2O\,(l) + CO_2\,(g)$
  2 mole   1 mole
  The ratio is 2 moles of acid to 1 of alkali.
Step 3 Work out the number of moles of acid neutralized.
  1 mole of alkali neutralizes 2 moles of acid so
  0.025 mole of alkali neutralizes 2×0.025 moles of acid.
  0.05 moles of acid were neutralized.
Step 4 Calculate the concentration of the acid.

  $\text{Concentration} = \dfrac{\text{number of moles}}{\text{volume in } dm^3} = \dfrac{0.05}{0.0278} = 1.8\, mol/dm^3$

  So the concentration of the hydrochloric acid is **1.8 M**.

Use the calculation triangle

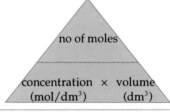

no of moles

concentration × volume
(mol/dm³)   (dm³)

To convert cm^3 to dm^3:
divide by 1000 (or just move the decimal point 3 places to the left). So $27.8\, cm^3 = 0.0278\, dm^3$.

Example 2 Find the concentration of ethanoic acid in vinegar

You can do this by titrating vinegar against a standard solution of sodium hydroxide, using phenolphthalein as indicator. In one experiment, 25 cm^3 of vinegar were neutralized by exactly 20 cm^3 of 1 M sodium hydroxide solution. (25 cm^3 = 0.025 dm^3)

Step 1 Calculate the number of moles of sodium hydroxide used.
1000 cm^3 of 1 M solution contains 1 mole so
20 cm^3 contains $\dfrac{20}{1000} \times 1$ mole or 0.02 mole.

Step 2 From the equation, find the molar ratio of acid to alkali.
$$CH_3COOH\,(aq) + NaOH\,(aq) \longrightarrow CH_3COONa\,(aq) + H_2O\,(l)$$
　　1 mole　　　　　　1 mole
The ratio is 1 mole of acid to 1 mole of alkali.

Step 3 Work out the number of moles of acid neutralized.
1 mole of alkali neutralizes 1 mole of acid so
0.02 mole of alkali neutralizes 0.02 mole of acid.

Step 4 Calculate the concentration of the acid.
$$\text{Concentration} = \frac{\text{number of moles}}{\text{volume in dm}^3} = \frac{0.02}{0.025} = 0.8 \,\text{mol/dm}^3$$
So the concentration of the ethanoic acid is **0.8 M**.

Vinegar is a solution of ethanoic acid in water, which accounts for its tang! Ethanoic acid is made from ethanol, which is the alcohol found in wine and beer.

Example 3 Find the concentration of ammonia in a liquid cleaner

You can titrate an ammonia-based household cleaner against a standard solution of sulphuric acid, using methyl orange as indicator. In one experiment, 10 cm^3 of cleaner were neutralized by exactly 12 cm^3 of 2 M sulphuric acid. (10 cm^3 = 0.01 dm^3)

Step 1 Calculate the number of moles of acid neutralized.
1000 cm^3 of the sulphuric acid contains 2 moles so
12 cm^3 contains $\dfrac{12}{1000} \times 2$ moles or 0.0024 mole.

Step 2 From the equation, find the molar ratio of acid to alkali.
$$H_2SO_4\,(aq) + 2NH_3\,(aq) \longrightarrow (NH_4)_2SO_4\,(aq)$$
　　1 mole　　　2 mole
The ratio is 1 mole of acid to 2 moles of alkali.

Step 3 Work out the number of moles of alkali used.
1 mole of acid is neutralized by 2 moles of alkali so
0.0024 mole of acid is neutralized by 0.0048 mole of alkali.

Step 4 Calculate the concentration of the alkali.
$$\text{Concentration} = \frac{\text{number of moles}}{\text{volume in dm}^3} = \frac{0.048}{0.01} = 4.8 \,\text{mol/dm}^3$$
So the concentration of ammonia in the cleaner is **4.8 M**.

---

1 What volume of 2 M hydrochloric acid will neutralize 25 cm^3 of 2 M sodium carbonate?

2 25 cm^3 of ammonia solution were neutralized by 20 cm^3 of 1 M sulphuric acid. Calculate the concentration of the ammonia solution.

3 3 moles of sodium hydroxide will neutralize 1 mole of citric acid. What volume of 2 M sodium hydroxide solution will neutralize 50 cm^3 of 1 M citric acid solution?

4 What is a *standard* solution?

Questions on Section 9

1 Rewrite the following, choosing the correct word from each pair in brackets.

Acids are compounds which dissolve in water giving (hydrogen/hydroxide) ions. Sulphuric acid is one example. It is a (strong/weak) acid, which can be neutralized by (acids/alkalis) to form salts called (nitrates/sulphates).

Many (metals/nonmetals) react with acids to give (hydrogen/carbon dioxide). Acids react with (chlorides/carbonates) to give (chlorine/carbon dioxide).

Solutions of acids are (good/poor) conductors of electricity. They also affect indicators. For example, phenolphthalein turns (pink/colourless) in acids, while litmus turns (red/blue).

The strength of an acid is shown by its (concentration/pH) number. The (higher/lower) the number, the stronger the acid.

Alkalis are compounds that dissolve in water giving (hydrogen/hydroxide) ions. They have a pH number (greater/less) than 7.

2 This is a brief description of a neutralization reaction.

'25 cm^3 of potassium hydroxide solution was placed in a flask and a few drops of phenolphthalein were added. Dilute hydrochloric acid was added until the indicator changed colour. It was found that 21 cm^3 of acid was used.'

a Draw a labelled diagram of titration apparatus for this neutralization.

b What piece of apparatus should be used to measure 25 cm^3 of sodium hydroxide solution accurately?

c What colour was the solution in the flask at the start of the titration?

d What colour did it turn when the alkali had been neutralized?

e Was the acid more concentrated or less concentrated than the alkali? Explain your answer.

f Name the salt formed in this neutralization.

g Write an equation for the reaction.

h How would you obtain *pure* crystals of the salt?

3 The table below is about the preparation of salts. Copy it and fill in the missing details.

4 Do these properties belong to acids, alkalis, or both?

a Sour taste.

b pH values greater than 7.

c Change the colour of litmus.

d Soapy to touch.

e Soluble in water.

f May be strong or weak.

g Neutralize bases.

h Form ions in water.

i Dangerous to handle.

j Form salts with certain other chemicals.

k Usually react with metals.

5 A and B are white powders. A is insoluble in water but B is soluble and its solution has a pH of 3.

A **mixture** of A and B bubbles or effervesces in water. A gas is given off and a clear solution forms.

a One of the white powders is an acid. Is it A or B?

b The other white powder is a carbonate. What gas is given off in the reaction?

c Although A is insoluble in water, a clear solution forms when the mixture of A and B is added to water. Explain why.

6 The chemical name for aspirin is **2-ethanoyloxybenzoic acid**.

This acid is soluble in hot water.

a How would you expect an aqueous solution of aspirin to affect litmus paper?

b Do you think it is a strong acid or a weak one? Explain why you think so.

c What would you expect to see when baking soda is added to an aqueous solution of aspirin?

7 Insoluble salts can be made by reacting two soluble chemicals together.

a What is this type of reaction called?

b For each pair of solutions, name the salt formed. (You may need to look at the table on page 148.)

   i silver nitrate and sodium chloride

   ii sodium carbonate and lead(II) nitrate

   iii ammonium sulphate and barium chloride

   iv magnesium sulphate and calcium chloride

c Write a balanced equation for each reaction.

Method of preparation	Reactants	Salt formed	Other products
a acid + alkali	calcium hydroxide and nitric acid	calcium nitrate	water
b acid + metal	zinc and hydrochloric acid	....	....
c acid + alkali	.... and potassium hydroxide	potassium sulphate	water only
d acid + carbonate	.... and	sodium chloride	water and ....
e acid + metal	.... and	iron(II) sulphate	....
f acid + ...	nitric acid and sodium hydroxide	....	....
g acid + base	.... and copper(II) oxide	copper(II) sulphate	....
h acid + ...	.... and	copper(II) sulphate	carbon dioxide and
i precipitation	silver nitrate and potassium chloride	....	....
j precipitation	lead nitrate and potassium iodide	....	....

8 Slaked lime, $Ca(OH)_2$, is sparingly soluble in water. It forms an alkaline solution called **lime water**. Its solubility can be found by titrating a saturated solution of lime water with a standard solution of hydrochloric acid, using methyl orange as indicator.
a What is a *saturated* solution?
b What is the chemical name for slaked lime?
c Which ion makes the solution alkaline?
d Which ion from the acid reacts with this ion?
e Write an ionic equation for the neutralization.

9 The neutralization of $10 \, cm^3$ of citric acid ($C_6H_8O_7$) by sodium hydroxide (NaOH) was investigated by titration. The citric acid solution contained $9.6 \, g/dm^3$ of the acid. The sodium hydroxide solution contained $4.0 \, g/dm^3$ of the alkali. The sodium hydroxide solution was placed in the burette. The results of the titration were as follows:

Titration number	1	2	3
2nd burette reading (cm^3)	15.1	30.1	45.0
1st burette reading (cm^3)	0.0	15.1	30.1

a Calculate the concentration of the citric acid solution. (RAMs: C = 12, O = 16, H = 1)
b How many moles of $C_6H_8O_7$ were used?
c Name a suitable indicator for the titration.
d What colour change would indicate the end point of the titration?
e Calculate the concentration of the sodium hydroxide solution. (RAM: Na = 23)
f Work out the average volume of sodium hydroxide solution used in the titration.
g How many moles of NaOH does this represent?
h How many moles of NaOH will neutralize 1 mole of $C_6H_8O_7$?
i Write a balanced equation for the neutralization.

10 a Divide the following salts into *soluble in water* and *insoluble in water*:
sodium chloride
calcium carbonate
potassium chloride
barium sulphate
barium carbonate
silver chloride
sodium citrate
zinc chloride
sodium sulphate
copper(II) sulphate
lead sulphate
lead nitrate
sodium carbonate
ammonium carbonate
b Now write down two starting compounds that could be used to make each *insoluble* salt.

11 Washing soda is crystals of hydrated sodium carbonate, $Na_2CO_3.xH_2O$. The value of x can be found by titration. In the experiment, 2 g of hydrated sodium carbonate neutralized $14 \, cm^3$ of a standard 1 M solution of hydrochloric acid.
a Write a balanced equation for the reaction.
b How many moles of HCl were neutralized?
c How many moles of sodium carbonate, Na_2CO_3, were present in 2 g of the hydrated salt?
d What mass of sodium carbonate, Na_2CO_3, is this? (RAMs: Ca = 40, C = 12, O = 16)
e What mass of the hydrated sodium carbonate was water?
f How many moles of water is this?
g How many moles of water are there in 1 mole of $Na_2CO_3.xH_2O$?
h Write the full formula for washing soda.

12 These diagrams show the stages in the preparation of copper(II) ethanoate, which is a salt of ethanoic acid.

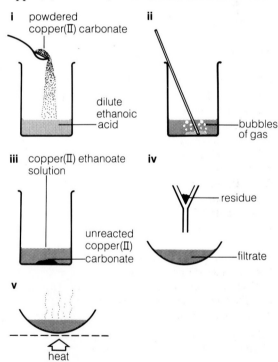

a Which gas is given off in stage **ii**?
b Write a word equation for the reaction in stage **ii**.
c How can you tell when the reaction is complete?
d Which reactant is completely used up in the reaction? Explain your answer.
e Why is copper(II) carbonate powder used, rather than lumps?
f Name the residue in stage **iv**.
g Write a list of instructions for carrying out this preparation in the laboratory.
h Suggest another copper compound that could be used instead of copper(II) carbonate, to make copper(II) ethanoate.

10.1 How Earth began

About 15 billion years ago, our Universe was created in a giant explosion called the **big bang**.

Scientists believe it started as a tiny empty space containing enormous energy. This exploded violently, and within the first second the energy was converted into a cloud of protons, electrons, neutrons, and radiation, at around 10 billion °C. In the first three minutes after the explosion the first atoms were formed.

Hydrogen atoms were first to appear. They are the simplest: just one proton and one electron. Then hydrogen atoms started fusing to form helium atoms, in a process called **nuclear fusion**. But by now the cloud had cooled a lot. Atoms of heavier elements would not be created until much later, in the nuclear furnaces inside stars:

Thanks to gravity, dense clouds of hydrogen and helium contract and swirl into shapes that will become galaxies. Next, part of a cloud separates and starts to . . .

. . . spin and flatten. Its centre gathers into a ball, getting hotter and denser. A star is born! It shines because hydrogen is turning into helium inside it.

It runs out of hydrogen and dies. But in its final intense heat, atoms of new elements such as carbon and oxygen are created by nuclear fusion.

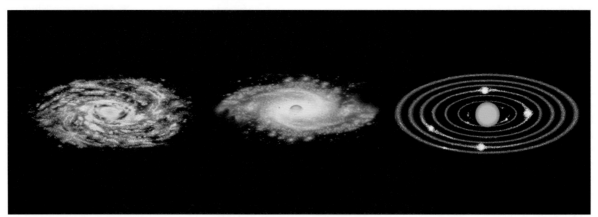

After the birth and death of many stars, the cloud contains not only hydrogen and helium, but some of all the other elements.

Time for the birth of our solar system. At the centre of the spinning disc are hydrogen and some helium, which will form the Sun. Heavier elements . . .

. . . collect around the edge as dust particles. These collide and coalesce to form chunks of material – and the nine planets, including Earth, are born.

Evidence for the big bang

The big bang seems an incredible idea. So what is the evidence for it?
- The universe is still expanding. This was discovered in 1929 by an astronomer called Edwin Hubble.
- Space is still full of cosmic background radiation, the kind a big bang would produce. It was discovered in 1965.
- Hydrogen and helium are by far the two most common elements in the universe. The relative quantities fit the big bang theory.

Today, 15 billion years after the big bang, new stars are still being born, and old stars dying. Our own star, the Sun, is in the prime of life. It consists of 70% hydrogen and 28% helium. It too will die when it runs out of hydrogen. But relax, that won't happen for another 5.5 billion years! Meanwhile the nuclear fusion of hydrogen to helium inside the Sun provides the energy that reaches us as sunlight.

The American astronomer Edwin Hubble (1889–1953). Behind him is the telescope that helped him deduce that our universe is still expanding.

The Earth

For billions of years, Earth was a hot, molten, spinning ball. Gradually a thin outer crust cooled and solidified. Gases bubbled and burst out through volcanoes, giving rise to the atmosphere. But Earth is still not all solid, as this diagram shows.

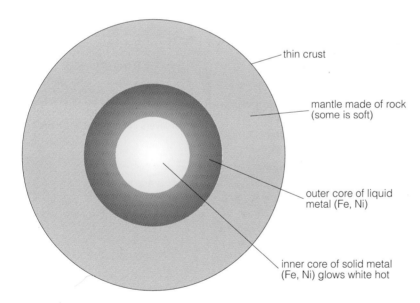

thin crust

mantle made of rock (some is soft)

outer core of liquid metal (Fe, Ni)

inner core of solid metal (Fe, Ni) glows white hot

Note that the two inner layers are made of two metals, iron and nickel. Altogether, Earth contains about 90 different elements. For these we have to thank the big bang, and the births and deaths of millions of stars.

1 What happened in the big bang?
2 The reaction that produces helium from hydrogen is called *nuclear fusion*. Why?
3 How did the other elements come into being?
4 a How is the Sun's energy produced?
  b Why and when will the Sun die?
5 Which part of Earth is:
  a the mantle? b white hot? c still molten?

10.2 The atmosphere past and present

The **atmosphere** is the layer of gas around Earth. It has four parts.
- We live in the **troposphere**, the layer where weather occurs.
- We take flight into the **stratosphere** now and again.
- Only a few of us – the astronauts – have passed through the **mesosphere**.
- The **ionosphere** consists mainly of charged particles.

Because of gravity, the gas is at its most dense at sea level, forming the mixture we call **air**. It thins out rapidly as you rise through the atmosphere. In fact, half of the mass of the atmosphere is in the lowest 6.5 km. Its composition is uniform up to about 80 km. Beyond that you'll find mainly the lighter gas particles.

What's in air?

This pie chart shows the gases that make up air:

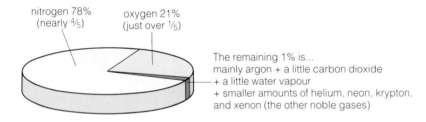

nitrogen 78%
(nearly ⅘)

oxygen 21%
(just over ⅕)

The remaining 1% is...
mainly argon + a little carbon dioxide
+ a little water vapour
+ smaller amounts of helium, neon, krypton,
and xenon (the other noble gases)

The composition of air is not *exactly* the same everywhere. It changes slightly from day to day and place to place. There is more water vapour in the air on a damp day. Over busy cities and industrial areas there is more carbon dioxide, as well as pollutants such as carbon monoxide and sulphur dioxide. But the uneven heating of Earth's surface by the sun causes the air to move continually, resulting in winds. The pollutants get spread around.

The ozone layer

The ozone layer is about 25 kilometres above sea level, in the stratosphere. Ozone has the formula O_3. It is produced when energy from ultraviolet light causes oxygen molecules to break into atoms. These then react with other oxygen molecules:

$$O_2 \xrightarrow[\text{light}]{\text{ultraviolet}} 2O \quad \text{then} \quad 2O_2 + 2O \longrightarrow 2O_3$$

At the same time, some ozone molecules absorb ultraviolet radiation and break down again to oxygen. In this way, the ozone layer protects us. Radiation that would harm us if it reached Earth is absorbed instead by the ozone layer and used to break bonds.

The energy given out when new bonds form is passed on to neighbouring molecules through collisions. So the ozone layer is warmer than the air below it, as you'll see in the diagram above.

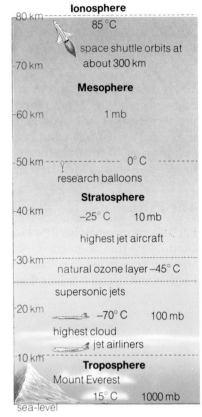

Ionosphere
−80 km
85 °C
space shuttle orbits at about 300 km
−70 km
Mesophere
−60 km
1 mb
−50 km research balloons 0 °C
Stratosphere
−40 km
−25 °C 10 mb
highest jet aircraft
−30 km
natural ozone layer −45 °C
supersonic jets
−20 km −70 °C 100 mb
highest cloud
jet airliners
−10 km
Troposphere
Mount Everest
15 °C 1000 mb
sea-level

The four layers in Earth's atmosphere

Allotropes

As you saw on page 56, some elements can exist in different forms in the same state. These are called **allotropes**.

Oxygen (O_2) and ozone (O_3) are allotropes of oxygen. Diamond and graphite are allotropes of carbon.

Where did the atmosphere come from?

As you saw on page 155, Earth was a molten ball for millions of years. Then a thin, solid crust formed. Inside, there was intense activity. Hot gases seeped out through the crust and burst from volcanoes. Over billions of years, our atmosphere developed from these gases.

4500 million years ago

3500 million years ago: Life starts

2200 million years ago: first green plants

200 million years ago: mammals were established

1–2 million years ago: our ancestors appeared

- We're not certain which gases formed the early atmosphere. Some scientists believe they were mainly carbon dioxide and steam, plus smaller amounts of methane, ammonia, sulphur dioxide, and other gases. Some think there was much more carbon monoxide than carbon dioxide. Some think there was lots of nitrogen. But all agree *there was little or no free oxygen*.
- When the temperature dropped below 100 °C, the steam cooled and condensed. The oceans started to form.
- Carbon dioxide dissolved in the falling rain, forming an acidic solution. This attacked minerals in Earth's crust, forming carbonates that were washed into the ocean. Other gases and minerals dissolved too. The oceans were hot and salty.
- Then, around 3500 million years ago, an amazing event took place: primitive life appeared. It may have started in hot pools, with lightning providing the spark. First, reactions between the dissolved compounds produced *amino acids*. These then polymerized to form *proteins*. And finally, proteins combined to form units that could reproduce themselves.
- The first living things were simple bacteria. **Nitrifying** bacteria fed on ammonia, producing nitrates which then formed proteins. **Denitrifying** bacteria fed on ammonia, releasing nitrogen to the atmosphere. Other bacteria fed on sulphur compounds.
- Then, around 2200 million years ago, the first green plants developed and photosynthesis began. Photosynthesis used up carbon dioxide and *produced oxygen*.
- The first oxygen reacted fiercely with the elements on Earth's surface, forming oxides. But eventually it started to build up in the atmosphere. It poisoned most of the organisms that fed on other gases. Their descendants, **anaerobic** bacteria, are now found only in places where there is no oxygen (for example, in waterlogged soil or by volcanic vents on the ocean floor).
- Once oxygen began to collect in the atmosphere, the protective layer of ozone started to form.
- Then, around 600 million years ago, when there was enough oxygen and protective ozone, the first simple animals developed.
- Around 1–2 million years ago, our ancestors appeared – over 4 billion years after Earth was formed.

The composition of air has been steady for the last 2000 million years. In Unit 11.1 you'll see how it is maintained that way.

---

1 Make a table to describe the layers of the atmosphere. You should use these headings:
   name height temperature pressure
2 Explain how ozone: **a** is formed **b** protects us.

3 Explain why the atmosphere now contains:
   a less water vapour **b** less carbon dioxide
   c more oxygen   than 4000 million years ago.
4 At first oxygen acted as a *pollutant*. Explain.

10.3 The ocean

The ocean as a giant solution

Water is the most abundant substance on the surface of Earth. Most of it is in the oceans. The Pacific Ocean alone covers almost half the Earth!

Water is an excellent solvent. So the ocean is a giant solution. It contains many different ions, in varying concentrations. The table on the right below shows what's dissolved in it. As you can see, sodium and chloride ions top the list.

Overall, the content of the ocean is a balance between what goes into it and what comes out again. The diagram below shows the various processes involved.

The oceans

- There are five oceans: the Arctic, the Atlantic, the Indian, the Pacific, and the Southern.

- But they are not separate. They are joined together and ocean currents flow between them.

The main substances dissolved in ocean water	Amount (g per kg of sea water)
Chloride ions (Cl^-)	19.2
Sodium ions (Na^+)	10.7
Sulphate ions (SO_4^{2-})	2.7
Magnesium ions (Mg^{2+})	1.4
Calcium ions (Ca^{2+})	0.4
Potassium ions (K^+)	0.38
Hydrogen carbonate ions (HCO_3^-)	0.14
Bromide ions (Br^-)	0.07

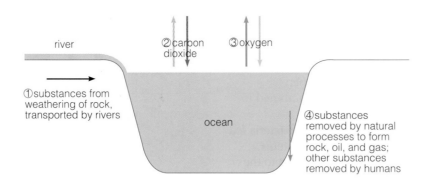

river

②carbon dioxide ③oxygen

①substances from weathering of rock, transported by rivers

ocean

④substances removed by natural processes to form rock, oil, and gas; other substances removed by humans

Let's look at each part of the diagram in turn.

1 Weathering of rock. Rock is broken into fragments, and some of its compounds dissolved, in a process called **weathering**. This eventually turns rock into **soil**, as you'll see on page 269. Rock fragments, soil, and dissolved compounds are carried away by rain water to rivers. Rivers carry them to the ocean.

2 Carbon dioxide. This gas is not that soluble in water. However, the amount taken up by the ocean is greatly boosted by two different processes that go on:

i By photosynthesis. In the top layer of ocean, where there is enough light, the **phytoplankton** live. These are tiny algae and other plants that photosynthesize. They feed on the carbon dioxide in surface water, producing glucose and oxygen. The glucose is turned into starch and other substances.

All ocean life depends on phytoplankton. Fish and other sea animals feed directly on them, or eat other organisms that do.

But photosynthesis does not lock up carbon dioxide for ever. Living things in the ocean, as on land, release some during respiration (page 224). Sea animals also excrete waste material which drops to the ocean floor. This decays, releasing carbon dioxide, which eventually returns to the surface.

Plankton in the ocean. The green ones are phytoplankton. Guess what the green substance is.

ii By chemical reaction. Carbon dioxide dissolves in water in a reversible reaction, forming carbonic acid. The colder the water the more gas dissolves:

$$CO_2(g) + H_2O(l) \rightleftharpoons H_2CO_3(aq)$$
carbonic acid

Carbonic acid is a weak acid. Only some of its molecules break down to form hydrogen ions at any given time:

$$H_2CO_3(aq) \rightleftharpoons HCO_3^-(aq) + H^+(aq)$$
hydrogen carbonate ion   hydrogen ion

The hydrogen ions will react with any basic ions in sea water. For example, they react with carbonate ions like this:

$$H^+(aq) + CO_3^{2-}(aq) \longrightarrow HCO_3^-(aq)$$

As the hydrogen ions get used up, more carbonic acid molecules break down to replace them, maintaining equilibrium. This means that, in turn, more carbon dioxide dissolves. The *overall* reaction is:

$$CO_2(g) + H_2O(l) + CO_3^{2-}(aq) \longrightarrow 2HCO_3^-(aq)$$

This reaction depends on the presence of carbonate ions in surface water. Surface water is continually replaced by water welling up from the deep ocean, carrying carbonate ions, and other basic ions. But that is quite a slow process. So only a limited amount of carbon dioxide can be taken up in this way.

3 Oxygen. Oxygen is slightly soluble in water. Some dissolves from air at the ocean surface. But on balance, the ocean gives up far more than it dissolves, thanks to phytoplankton and photosynthesis. It is estimated that phytoplankton produce around 80% of the oxygen in the atmosphere!

4 Removal of other substances. For example:
- Phytoplankton form tiny protective casings of calcium carbonate. When they die, these fall to the bottom of the ocean to form a sediment. This eventually turns into **chalk**.
- The skeletons and shells of other sea creatures contain calcium carbonate. These collect on the ocean floor as a sediment and eventually form **limestone**.
- Some compounds are removed by humans. Sodium chloride is removed by evaporating sea water (page 234). Magnesium (page 110) and bromine (page 235) are also extracted from it.
- Some solids precipitate from ocean water. For example, calcium carbonate is *very* slightly soluble in water. It will precipitate from warm shallow ocean water, where there isn't enough dissolved carbon dioxide to react with carbonate ions.

More additions to the ocean. A fumarole or sulphide spring, rich in sulphides of iron and copper, spurts up through the ocean floor.

The oceans and global warming

- The ocean dissolves 40–50% of the carbon dioxide we produce by burning fossil fuels.
- If it could dissolve more, we might avoid global warming (page 178).
- One idea is that we could increase uptake by 'farming' phytoplankton.
- But many scientists feel that this kind of interference in nature will just cause other problems.

1 Name: **a** the top two positive ions **b** the top two negative ions present in ocean water.

2 What is *weathering*? How does it affect ocean water?

3 What are *phytoplankton*? Why are they important?

4 Describe two ways in which carbon dioxide is taken up by ocean water.

5 Describe three ways in which substances are removed from ocean water.

10.4 Earth's crust

As you saw on page 155, the crust is Earth's outer layer. It is just a thin skin compared to the inner layers: it varies from about 8 km under the deepest part of the oceans to about 65 km under Mount Everest. So far, no one has managed to dig a hole right through!

Rocks and minerals

Earth's crust is made of rock with a sprinkling of soil, sand, and vegetation on top. Rock in turn is made of **minerals**.
Minerals are inorganic elements or compounds that occur naturally in the Earth and have a regular arrangement of atoms or ions.

Inorganic means they are *not* the remains of living things. The regular arrangement means they are crystalline. For example:

Calcite is calcium carbonate, $CaCO_3$.

Pyrite is iron(IV) sulphide, FeS_2. It is often called fool's gold.

Quartz is silicon dioxide, SiO_2, also known as **silica**.

Some rocks contain just one mineral. Pure limestone rocks contain only calcite. But most are a mixture. For example, granite is a mixture of several minerals including quartz, feldspars, and micas.

Over 90% of the minerals in Earth's crust are **silicates**. These contain silicon and oxygen, plus other atoms depending on the mineral.

> **Some complicated minerals**
>
> ● One type of **feldspar** has the formula $KAlSi_3O_8$.
> ● One type of **mica** has the formula $K_2(MgFe)_6Si_3O_{10}(OH)_2$.
>
> (You don't need to remember these!)

The composition of the crust

If you could break down all the minerals in Earth's crust into their elements, this is what you'd find:

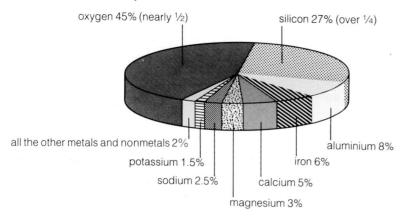

oxygen 45% (nearly ½)

silicon 27% (over ¼)

all the other metals and nonmetals 2%

potassium 1.5%

sodium 2.5%

magnesium 3%

calcium 5%

iron 6%

aluminium 8%

Nearly three-quarters of Earth's crust is made of oxygen and silicon. (Remember all those silicates!) The rest consists mainly of just six metals, with aluminium and iron the most abundant. The remaining 82 elements together make up *only 2%* of the crust.

Most elements are found combined with others – for example in silicates, carbonates, and oxides. But sulphur occurs naturally as the element. So do unreactive metals like gold, platinum, and silver.

Abundance of some metals in Earth's crust / %	
aluminium	8
iron	5.8
manganese	0.1
nickel	0.007
copper	0.006
lead	0.001
tin	0.0002
silver	0.000 008
platinum	0.000 000 5
gold	0.000 000 2

price tends to increase with rarity – but it also depends on how useful the metal is

Rare metals

If a metal makes up less than 0.01% of Earth's crust it is classified as **rare**. All except two of the metals listed on the right are rare. Note how expensive some of them are!

Metal ores

In some places, the mineral deposits in rock contain a high concentration of one metal. So it is worth digging them up to extract the metal. These mineral deposits are called **ores**.
An ore is a mineral deposit that can be mined and refined at a profit.

If a metal is in great demand it may be worth digging up a **low-grade** ore that contains only a small concentration of it. The largest mine in the world is a copper mine at Bingham Canyon (USA). Here copper is mined from rock with a copper content of only 0.65%. Huge volumes of rock are broken up to get a small amount of copper.

The cost of metals	
Metal	Approx cost/kg in 1996
platinum	£8797
gold	£8764
silver	£330
nickel	£4.90
tin	£3.99
copper	£1.21
aluminium	96p
lead	50p
iron	32p

Minerals and wealth

Deposits of important minerals are not spread evenly round Earth. Most countries have limestone and iron ore. But for many minerals, just a few countries own most of the world's supply.

Countries with large deposits of important minerals have the potential to be wealthy. Many developing countries depend on minerals for most of their trade. For example, copper accounts for over 80% of Zambia's exports, and cobalt for over 50% of Zaire's.

Where the biggest deposits are		
Country	Ore	% of world supply
Australia	lead	20
Canada	zinc	15
Chile	copper	27
China	tungsten	40
Guinea	aluminium	26
S. Africa	gold	50
Zaire	cobalt	40

1 a What is a mineral?
   b Is sodium chloride a mineral? Explain.
   c Is coal a mineral? Explain.
2 Name a mineral that contains this element, and give its formula: **a** calcium **b** iron **c** silicon.

3 Why is iron so much cheaper than gold?
4 What do you think a *high-grade* ore is?
5 Name three ores found in the United Kingdom.
6 Which country has the biggest known reserves of:
   a gold? **b** zinc? **c** copper? **d** cobalt?

10.5 Making use of Earth's resources

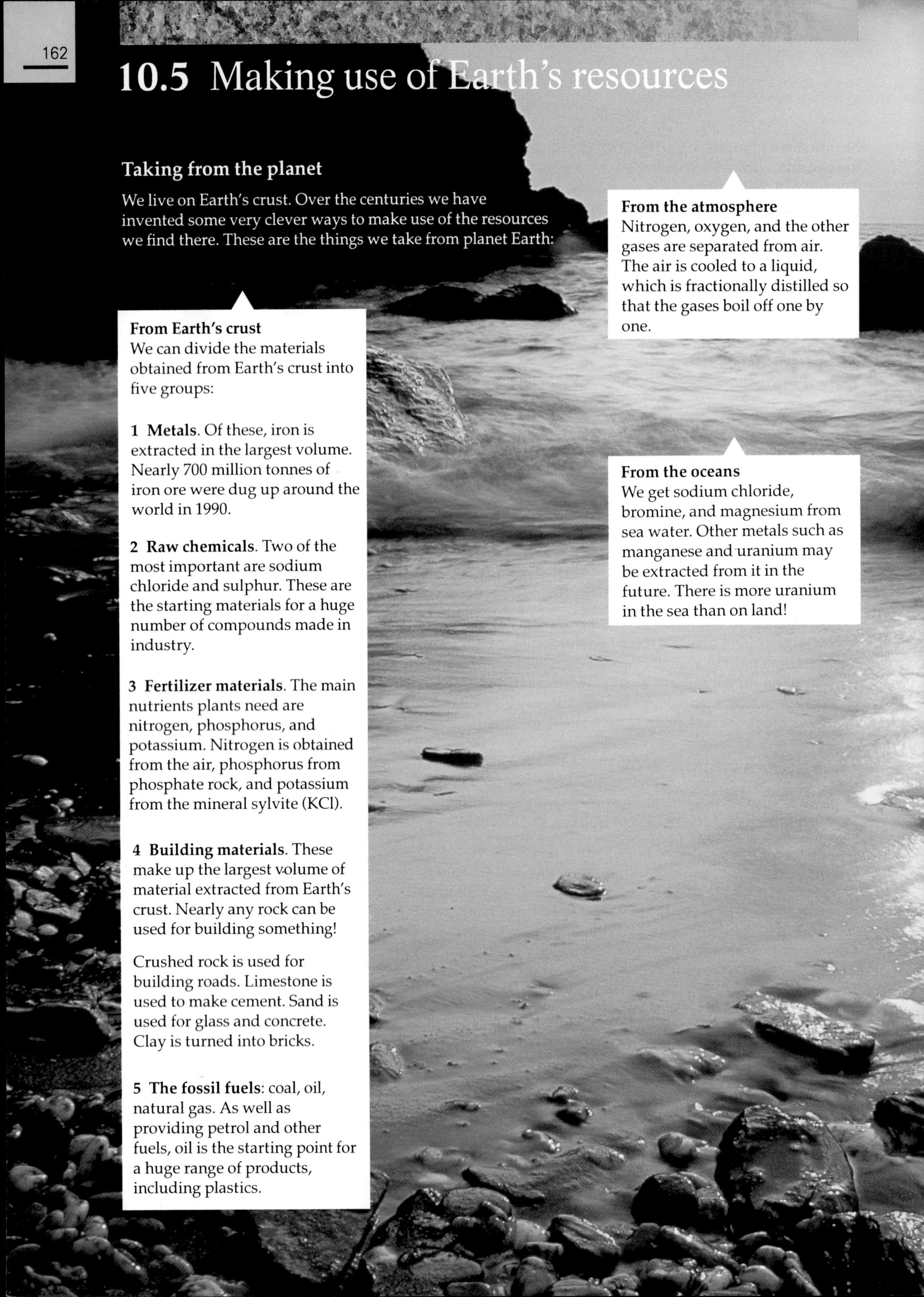

Taking from the planet

We live on Earth's crust. Over the centuries we have invented some very clever ways to make use of the resources we find there. These are the things we take from planet Earth:

From Earth's crust

We can divide the materials obtained from Earth's crust into five groups:

1 Metals. Of these, iron is extracted in the largest volume. Nearly 700 million tonnes of iron ore were dug up around the world in 1990.

2 Raw chemicals. Two of the most important are sodium chloride and sulphur. These are the starting materials for a huge number of compounds made in industry.

3 Fertilizer materials. The main nutrients plants need are nitrogen, phosphorus, and potassium. Nitrogen is obtained from the air, phosphorus from phosphate rock, and potassium from the mineral sylvite (KCl).

4 Building materials. These make up the largest volume of material extracted from Earth's crust. Nearly any rock can be used for building something!

Crushed rock is used for building roads. Limestone is used to make cement. Sand is used for glass and concrete. Clay is turned into bricks.

5 The fossil fuels: coal, oil, natural gas. As well as providing petrol and other fuels, oil is the starting point for a huge range of products, including plastics.

From the atmosphere

Nitrogen, oxygen, and the other gases are separated from air. The air is cooled to a liquid, which is fractionally distilled so that the gases boil off one by one.

From the oceans

We get sodium chloride, bromine, and magnesium from sea water. Other metals such as manganese and uranium may be extracted from it in the future. There is more uranium in the sea than on land!

Are we taking too much?

Earth's resources took billions of years to develop. But we are using them up at an alarming rate. When will they run out? It's hard to predict, for several reasons:

- Much of Earth's crust and most of the oceans are still unexplored. We may find lots of new mineral and oil deposits in the future.
- Some deposits of minerals and fossil fuels are not profitable to develop at present. New technology could change this.
- Replacements may be found for a material, so it will get used up more slowly than expected. For example, glass optic fibres are replacing copper for transmitting electricity.
- Rather than dumping, more waste is being recycled. In the UK about 40% of 'new' steel is in fact recycled steel.

A manganese nodule. These are strewn all over the ocean floor, but they'd cost too much to harvest. And there's another problem – who owns the ocean floor anyway?

This table shows the current predictions for some materials:

Material	When known reserves may run out...
Aluminium	2190
Copper	2026
Iron	2165
Lead	2011
Phosphorus	Not for several centuries
Potassium	Not for several centuries
Tin	2017
Zinc	2010
Oil	2060

Even if these dates are pessimistic, one thing is certain. The materials we take from Earth's crust and oceans are **non-renewable**. We cannot afford to squander them.

Strategic minerals

Some minerals are **strategic**. That means they are essential to a country's economy or its defence, but in relatively short supply. For example, chromium is essential for the alloys used in jet engines, and manganese for high-grade steels.

Governments try to keep on friendly terms with the countries from whom they import strategic minerals. To make sure of a steady supply in the event of war and other disasters, they may also stockpile them. The USA has stockpiles of around 2 million tonnes of manganese and 2 million tonnes of chromium.

A stealth bomber. Its production depends on a supply of metals such as manganese and chromium.

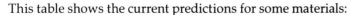

1 Name the five groups of materials extracted from the Earth's crust.
2 Which metal is extracted in the greatest quantity?
3 Give two reasons why it is difficult to predict when mineral reserves will run out.

4 Do you think we'll run out of oxygen? Explain.
5 Explain what the term means, and give an example:
  a non-renewable resource b strategic mineral.
6 Could it ever be a strategic commodity? Why?
  a limestone b oil c uranium d water.

10.6 Example: making use of limestone

In the last Unit you saw the kinds of things we take from Earth's crust and make use of. **Limestone** is a good example.

Limestone is made from the shells and skeletons of dead sea creatures, which accumulated as a thick sediment on the sea floor. Millions of years of heat and pressure turned the sediment into rock. Then, over millions more years, powerful forces raised some sea beds upwards, draining and folding them to form mountains. That explains why some limestone is found inland, miles from the sea!

In this way, bits of dead sea creatures have turned into one of our major raw materials. Every year over 60 million tonnes of limestone are blasted from UK quarries. This is what we use it for:

Chalk

Chalk is another form of calcium carbonate. It's made from the hard casings of tiny sea plants (page 159). Like limestone, it is used to make cement, mortar, and so on.

Chalk is softer than limestone and can be dug out of a quarry with excavators. You don't need to blast it out with explosives.

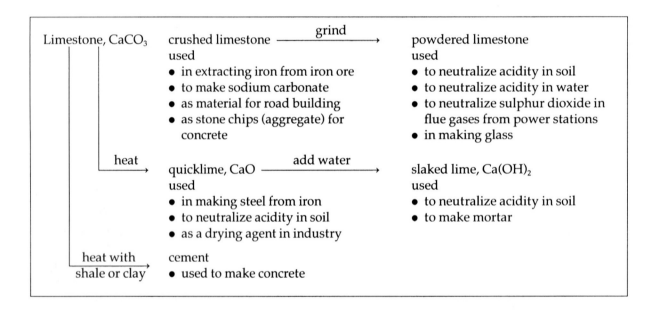

Limestone, CaCO₃

crushed limestone ──── grind ────→ powdered limestone

crushed limestone
used
- in extracting iron from iron ore
- to make sodium carbonate
- as material for road building
- as stone chips (aggregate) for concrete

powdered limestone
used
- to neutralize acidity in soil
- to neutralize acidity in water
- to neutralize sulphur dioxide in flue gases from power stations
- in making glass

heat

quicklime, CaO ──── add water ────→ slaked lime, Ca(OH)₂

quicklime, CaO
used
- in making steel from iron
- to neutralize acidity in soil
- as a drying agent in industry

slaked lime, Ca(OH)₂
used
- to neutralize acidity in soil
- to make mortar

heat with shale or clay

cement
- used to make concrete

The substances made from limestone

Quicklime   When limestone is heated, it breaks down:

$$CaCO_3(s) \rightleftharpoons CaO(s) + CO_2(g)$$

calcium carbonate   calcium oxide   carbon dioxide
(limestone)   (quicklime)

Note that this reaction is reversible. But the carbon dioxide is carried off in a strong current of air before it can react with the quicklime.

This shows a rotary kiln for making quicklime. Limestone is fed in one end, and heated by burning oil, gas, or pulverized coal. Quicklime comes out the other end.

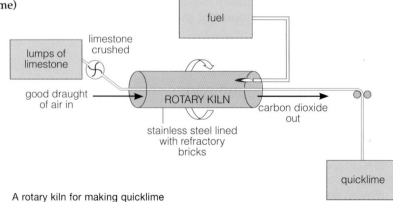

A rotary kiln for making quicklime

Slaked lime When water is added to quicklime the reaction is exothermic. The mixture hisses and steams. Conditions are carefully controlled so that the slaked lime forms as a fine powder:

$$CaO\,(s) \; + \; H_2O\,(l) \longrightarrow Ca(OH)_2\,(s)$$

calcium oxide   calcium hydroxide
(quicklime)   (slaked lime)

Mortar This is used to bind bricks together. It is formed by mixing slaked lime with sand, water, and usually some cement. The exact mix depends on what kind of mortar you want. When the water evaporates the slaked lime absorbs carbon dioxide from the air, forming a hard layer of calcium carbonate.

Cement This grey powder is a mixture of calcium silicates and aluminates. When it is mixed with water it dries into a hard solid. It is made by mixing limestone (or chalk) with shale or clay, heating the mixture very strongly, and grinding up the final lumps of solid with some gypsum (calcium sulphate):

The art of plastering. Like mortar, plaster is a mixture of slaked lime, sand, and cement. It is used for inside walls and dries to a smooth hard surface.

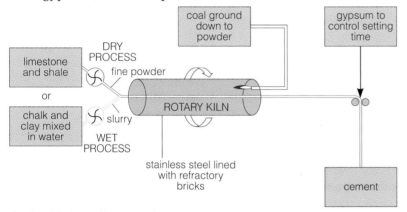
A rotary kiln for making cement

Concrete For this, cement is mixed with **aggregate** (a mixture of limestone chips and sand) and water is added. As the concrete 'sets', the compounds in the cement bond to form a hard mass of **hydrated** crystals – that is, crystals with water molecules bonded inside them.

The setting process is exothermic. If it's too fast, the heat inside a big concrete structure (such as a dam) could be enough to crack the concrete. It's also a nuisance if the concrete you mix on a building site hardens before you're ready to use it. Gypsum is added to cement to slow the setting process down.

Glass It is made by heating limestone, sand, sodium carbonate, and recycled glass in a furnace. You can find out more about this process and the structure of glass on page 58.

Which mix for cement?

Limestone or chalk? Shale or clay? Use whichever is available!

- Silica and clay both contain the same compounds: silica (SiO_2) and alumina (Al_2O_3).
- In the north of England, limestone and shale are most readily available. They are ground into a dry mix.
- In the south it's usually chalk and clay. The clay is first mixed with water and made into a slurry, before the chalk is added. So it's a wet mix.
- A wet mix means you need to burn more fuel to drive the water off.

1 How is: **a** limestone **b** chalk formed?
2 **a** How is quicklime made? Write the equation for the reaction. **b** Why is it important to remove the carbon dioxide? **c** How is this done?
3 How is slaked lime made? Write the equation.
4 Give two uses each for quicklime and slaked lime.
5 **a** What is *cement*? **b** How is it made? **c** Why is gypsum added?
6 What ingredients are mixed to make concrete?
7 List the ingredients for making glass.

10.7 An overview of industrial processes (I)

No matter what you're making from Earth's resources – whether it's cement from limestone, or sodium hydroxide from rock salt, or steel from iron ore – you'll need most or all of the four ingredients below.

1 Key starting material(s)

For example:
- a mineral such as limestone, rock salt, or haematite (an iron ore);
- raw chemicals such as sulphuric acid or ammonia;
- hydrocarbons from oil, for example if you are making plastics.

Saving time and money
- If the materials are big and bulky, you will try to site your plant as close as possible to the source, to cut transport costs and time.
- If they are imported, it makes sense to be close to a port.

2 A source of energy

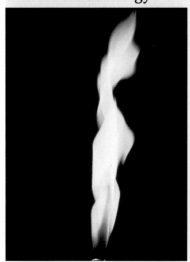

You must *supply* energy …
- for all endothermic reactions including electrolysis (for electrolysis it must be supplied as electricity);
- to compress gases for reactions at high pressure;
- for crystallization, distillation, and other processes for separating and purifying your products.

You must *remove* energy …
- from exothermic reactions, to prevent overheating;
- from compressed gases because they get very hot.

Saving time and money
- If you're smart, you will use heat removed from one part of the plant to supply heat elsewhere, or to generate electricity from steam.
- You'll also think hard about a catalyst: it reduces the amount of energy that's needed for successful reaction to occur.

Water is often used to transfer energy. Pipes filled with steam supply heat. Pipes filled with cold water carry heat away. The pipes are coiled around the reaction tanks.

3 Water

You may need water …
- as a raw material, for example in making sodium hydroxide;
- as a solvent, for example in making fertilizers;
- to supply or remove energy in the form of heat.

You may have to treat the water first to remove impurities, because …
- they may interfere with reactions;
- scale from hard water may block your pipes.

You can get rid of the 'used' water into rivers or sewers, but first you must:
- treat it to remove harmful substances;
- cool it if it's hot, because otherwise it will kill river life.

4 Air

You may need air …
- as a raw material (for example in making sulphuric acid);
- as a coolant: it is cheaper than water;
- for burning fuels.

Air and smoke coming out of your plant may need to be treated:
- to remove harmful gases such as sulphur dioxide;
- to remove dust particles.

The plant

The **plant** is all the equipment and machinery needed for the production process: tanks, pipes, pumps, pressure gauges, furnaces, and so on. It's usually contained in a **factory** or **works**. It needs a great deal of thought and clever design.

- The plant materials must be able to withstand the temperatures and pressures needed for the reactions. For example, the rotary kilns for making quicklime and cement are made of steel and lined with refractory (heat resistant) bricks.
- The plant materials must not react with the chemicals.
- The plant should be designed to use energy as efficiently as possible, to keep costs down.
- Chemicals must not leak out and pollute the air or water. Neither should heat or dust.

The owners of a plant which pollutes the environment could be fined or the plant could even be closed down. Most plants include areas for treating **effluent** (waste liquid) and waste gases before they are released, for example using:

- **limestone.** This will neutralize acidic liquids, or the sulphur dioxide in waste gases from burning coal.
- **charcoal.** This is porous and has a large surface area. It is excellent at adsorbing impurities. So it is used for filtering both air and liquids.
- **scrubbers.** The waste gas is sprayed with water to dissolve harmful compounds before they reach the chimneys.
- **incinerators.** Harmful waste gases such as solvent fumes are burned in an incinerator to give harmless products.
- **ion exchangers.** When waste liquid flows through an ion exchanger, harmful ions (such as mercury or cadmium ions) are replaced by harmless ones. Ion exchangers can be designed to remove any ions you wish. Find out more on page 186.
- **electrostatic precipitators.** In these, an electric current is passed through smoke on its way to the chimneys. The dust particles get charged and cling to electrodes. This method is used at cement works to stop cement dust escaping from kilns. The trapped dust is returned to the cement store.

Who said chemical plants weren't beautiful? The task of constructing them is a challenge.

---

1 Name four things you are likely to need to convert something from Earth's crust into a useful product.
2 Name all the raw materials for making cement.
3 Give three examples of processes in industry that use up energy. (Choose any you wish.)

4 Think of: **a** advantages **b** disadvantages of using water as a carrier for energy.
5 In industry, what is a *plant*?
6 Name two essential properties for a material used to make the container in which reactions occur.

10.8 An overview of industrial processes (II)

Where should the plant be sited?

These are the questions to consider, when choosing a site:

- **Are we close to our key starting material(s)?** This is especially important if they're big and bulky and expensive to transport. Cement works are usually right at the quarry. Steel works are close to ports if they depend on imported iron ore.

- **Have we access to a good transport network?** You need a good road and rail network for bringing in raw materials and sending out the final products to your customers.

- **Are we close to our customers?** You need access to the people who will buy your products.

- **How expensive is energy here?** You may be able to do a deal with a supplier of gas, oil, coal, or electricity.

- **Have we access to sufficient water supplies?**

- **Have we access to other services we need?** For example, sewage works to take the treated effluent from your plant.

- **Will it be easy to get staff?** Not only chemists, chemical engineers, and other technical staff, but also marketing people, administrative staff, accountants, and managers. If you can't find the right people locally you must persuade them to move from elsewhere. That usually means offering an attractive salary.

- **Are there any government grants?** The government offers grants or 'tax holidays' in some areas to attract industry.

- **Does this area have any other advantages?** An area may be suitable because it already has similar or related businesses.

- **Will there be objections from planners or the local people?** People will object if the area is one of great natural beauty, or if they think you'll bring pollution, heavy transport, or noise.

- **Will there be an 'environmental' cost?** Industries involved in mining and quarrying must undertake to restore the landscape once a site is exhausted. You must consider this cost.

Beautiful countryside disfigured by a limestone quarry. These days, companies must present plans for restoring the land when they seek permission to quarry.

Once a gravel pit, this is now an attractive recreational area.

Will you make a profit?

Whether you are making cement or sodium hydroxide or fertilizers or steel, your aim is to make a profit. There are various terms for profit. **Operating profit** is the name given to profit you make before you pay tax:

operating profit (£) =
money in from sales (£) – direct costs (£) – overheads (£)

The more you sell, and the lower the direct costs and overheads, the higher your operating profit will be.

Direct costs   These are the costs directly related to the production process. They include:
- **the cost of the raw materials.**
- **labour costs.** Many processes run 24 hours a day so the workers work in shifts. But many are also highly automated, so not many workers are needed. If your process is highly automated, your labour bill might not be very high.
- **energy.** This is likely to be one of your biggest costs, if not the biggest. You'll want to go for the cheapest source of energy you can find.

Overheads   These are the general costs of running any business. For example:
- the salaries of the managers, secretaries, and other administrative staff.
- the costs of marketing and selling the products.
- the cost of running the canteen, the upkeep of the buildings and grounds, central heating, and so on.
- the cost of leasing the site or equipment or trucks, if your company doesn't own them.

Sales   If your product is in demand, sales will be high. But sales also depend on the economy. For example, in a strong economy a lot of building goes on: roads, bridges, offices, hospitals, and homes. So the demand for limestone products is high. In a recession building work is usually badly hit, and sales fall.

A pep-talk for the marketing team. No sales: no plant!

Good management will help to ensure that your sales are high and costs are low, and that you stay in profit.

---

1 What will be the most important factor in deciding where to site a cement works? Explain why.
2 Why is it important for an industrial plant to be:
   a close to a good road/rail network?
   b within easy reach of customers?
3 Give three concerns you might have if you lived close to a limestone quarry.
4 Is it fair that companies who quarry limestone should have to restore the landscape? Why?
5 What is *operating profit*?
6 Give two examples of overheads, in running a quarrying business.
7 Explain why the limestone quarrying business is healthier when the economy is strong.

10.9 Recycling

We are using up Earth's resources at an ever faster rate. At the same time we are filling up landfill sites with huge amounts of material that could be **recycled**.

Recycling makes environmental sense. Recycling a metal, or glass, or plastic, or paper:
- saves on raw materials
- cuts down the need for mining and extraction, which means less environmental damage, and less waste material from these processes
- needs less energy than producing it in the first place, so less fossil fuel is used up
- uses less water and fewer chemicals than producing it in the first place, and causes less pollution
- saves on landfill sites and other costs of waste disposal.

There are many problems in recycling material from household waste. For a start it has to be collected, sorted, and cleaned. All this uses energy and therefore costs money. But the government has set a target of 25% of household waste to be recycled by the year 2000.

Recycling some common materials

Steel Steel is the world's most recycled metal. About 40% of the 'new' steel produced each year is in fact recycled steel, which is collected as scrap and sent back to the steel plant for remelting.

Much of the scrap steel comes from 'tins', which are steel cans coated with a thin layer of tin. In the UK, over a billion steel cans a year are recycled. Compared with production from iron ore, recycling a tonne of tinplated steel:
- saves 1.5 tonnes of iron ore and 0.5 tonnes of coke
- needs only about $\frac{1}{3}$ the energy
- cuts down on waste particles and fumes by at least $\frac{1}{3}$.

Some steel cans are collected at 'save a can' banks and in house-to-house collections. But most end up at dumps with the rest of the household waste. Here they can be separated by **magnetic extraction**. The waste is passed under a powerful electromagnet. The steel cans cling to the magnet while the rest passes on.

Magnetic extraction can also be carried out after the waste has been incinerated. By this time the tin layer has burned off the steel cans. Otherwise the steel must be **detinned** before going back to the steel plant. This is carried out by electrolysis (page 110), which allows the tin to be recycled too.

In the UK, around 2 million scrap cars a year are also recycled. The crushed cars are fed into shredders which break them up into fist-sized pieces. Then the steel pieces are extracted magnetically. The other materials used in cars — copper, zinc, glass, plastic — would interfere with the recycling process.

In the same way, around 5 million washing machines, cookers, and other appliances are recycled each year.

Steel cans can be sorted from aluminium using a magnet.

Aluminium   4 out of 5 of all drinks cans are made from aluminium. We use about 5.5 billion aluminium drinks cans a year in Britain, and less than a quarter get recycled. So £30 million worth of aluminium a year ends up in landfill sites.

Recycling aluminium takes only $\frac{1}{20}$ of the energy needed to produce it in the first place. At the plant, the cans are shredded and passed by an electromagnet to remove any steel. Next, paint and laquer are burnt off. Then the scrap aluminium is melted and cast into ingots. These are rolled into sheets and turned back into cans.

Glass   The main ingredients in glass are sand, soda ash, and limestone. These are heated together in a furnace to over 1500 °C. Scrap crushed glass is added to the mixture. The scrap is called **cullet**. On average, 20% of the furnace content is cullet, but it could be as much as 75%.

The ingredients for glass are cheap. But recycling is still worthwhile because:
- it takes 20% less energy to recycle glass than to make it
- recycling produces about 20% less pollution
- the glass dumped in landfill sites will never rot away.

Plastics   In 1995, only about 5% of waste plastic was recycled. This is partly because there are around 50 different families of plastics, which makes recycling difficult.

Most plastic bottles are made of polyethene (PE), polyvinyl chloride (PVC), or polyethene terephthalate (PET). A machine with an X-ray sensor can sort PVC bottles from the others by identifying their chlorine content. Then the remaining bottles can be chopped up into bits and dumped into water, where the PET will sink and the PE float. But so far, the best way to separate the bottles is by hand, before they reach the recycling plant.

In future, plastic cartons and containers will be marked with a code to make sorting easier. Scientists are also developing tracers that can be added to plastics and detected by sorting machines.

Over 70 aluminium cans per person per year are thrown into landfill sites in Britain.

The economics of recycling

- To see if recycling saves money, you need to take *all* costs into account, including the costs of collecting and cleaning the waste.

- Materials in short supply are more likely to be recycled. People don't throw their old gold and silver out!

- The more money industry can save by recycling, the more effort it will put in.

- Where recycling is not financially attractive, the government can encourage it by setting targets, making stricter anti-pollution laws, and offering tax incentives.

---

1 What is a *landfill* site? Where is the nearest one to your home?

2 The more material is recycled, the fewer landfill sites are needed. Why is this a good thing?

3 List four other advantages of recycling.

4 Steel has one property that makes recycling easier. Which property is this?

5 What is the big advantage of recycling aluminium?

6 Can you think of anyone who might suffer if a lot more aluminium were recycled?

7 So far, not much plastic is being recycled. Why?

8 Suppose the price of oil shot up. What effect would this have on the recycling of plastic? Explain.

9 What can *you* do to encourage recycling?

Questions on Section 10

1 This table shows how we think the composition of the atmosphere has changed over the last few thousand million years.

Percentage of certain gases in atmosphere

	Millions of years ago					
Gas	4500	4000	3000	2000	1000	Now
hydrogen	5	3	1	0	0	0
carbon dioxide	90	30	10	5	0.03	0.03
nitrogen	5	40	65	75	77	78
oxygen	0	0	0	0	15	21

a Plot a graph of the data, showing all four curves on the same graph. (Use a full sheet of graph paper. Show time on the x axis and percentage of gas on the y axis.)
b i About when did oxygen first appear?
 ii Explain how it was formed.
c i Which gas once present in very large quantities is reduced to only a small quantity today?
 ii Suggest ways in which this gas was removed from the atmosphere.
d Do you think the composition of air is likely to change much over the next million years? Give reasons to support your answer.

2 The bar charts below show the composition of the atmospheres on Venus, Earth, and Mars.

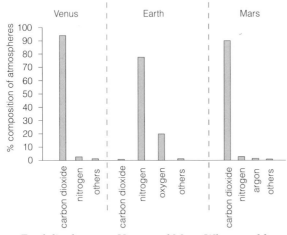

a Earth lies between Venus and Mars. What would you *predict* the composition of its atmosphere to be?
b How does your prediction fit with the actual figures?
c Venus is often called Earth's twin, since it has almost the same size and density.
 i At the start, Earth's atmosphere was similar to that of Venus. What caused it to change?
 ii What conclusions can you draw about Venus?
d The planet Mercury is the smallest planet in our solar system, and closest to the Sun. Mercury has practically no atmosphere. Suggest a reason.

3 Look again at the table of gases in question **1**.
a Name one gas *not* listed in the table which is present in very small quantities in air.
b i Name one gas not listed in the table which helps to protect us from harmful radiation.
 ii Where is this gas found?
 iii Explain how it helps to protect us.
 iv This gas is an _____ of oxygen, O_2. What is the missing word?

4 This table compares the concentration of some ions in river and sea water. For example, 50% of the cations in river water are calcium ions, while only 6% of the cations in sea water are calcium ions.

Concentration (as % of all cations present)

	in river water	in sea water
Ca^{2+}	50	6
Na^+	20	75

Concentration (as % of all anions present)

	in river water	in sea water
HCO_3^-	70	2
Cl^-	20	80

a Which cation is present in a much higher concentration in the sea than in rivers? Why?
b Calcium and hydrogen carbonate ions are present in much lower concentration in sea water than in river water. Why?

5 The water of the Dead Sea contains 230 g of dissolved solids in every litre (dm³) of water. The figure for most other sea water is about 35 g. River water has less than 1 g of dissolved solids in every litre.
a Why is the value for river water so much lower than for sea water?
b Where do the dissolved substances in sea water come from?
c Suggest reasons why Dead Sea water contains more dissolved solids than other sea water does.

6 The **reserves** of a metal are the size of its known ore deposits that are considered worth mining. In 1960 the reserves of aluminium were put at 11 600 million tonnes. They were expected to last 300 years. By 1990 they had increased to 21 700 million tonnes but were expected to last only 194 years.
a Give *two* reasons why a reserve figure is different from the actual amount of the metal left in Earth's crust.
b Give reasons why the figure for aluminium reserves increased between 1960 and 1990.
c Suggest why the prediction for the lifetime of these reserves was *reduced* at the same time.
d It is very difficult to predict the lifetime of known reserves accurately. Why is this?

7 The materials for making glass are soda ash, limestone, and sand. Recycled glass is also added to the mixture.
   a Match each name below to a material above.
      i calcium carbonate
      ii silicon oxide
      iii sodium carbonate
   b Write down the formulae for the three compounds in a.
   c What is another name for recycled glass?
   d Why is it a good idea to add recycled glass to the mixture? Give several reasons.
   e Draw a simple diagram to show the structure of glass. (Hint: use the index!)

8 For bottles, a sensible alternative to recycling is to wash them thoroughly and then refill them.
   This table compares energy costs in the factory for returnable and non-returnable bottles.

Energy needed (MJ)	Returnable	Non-returnable
to make bottle	7.5	4.7
to wash and fill bottle	2.6	2.5

   a What does MJ stand for?
   b More energy is used to *make* a returnable bottle than a non-returnable one. Suggest a reason.
   c In your factory you have to make a choice between non-returnable bottles, or returnable bottles which you will wash and refill.
      i Which is cheaper in terms of energy: to make a returnable bottle and get it back once for refilling, or to make two non-returnable bottles?
      ii What is the minimum number of times you would have to wash and refill a returnable bottle, to make this the cheaper option?
   d So far you have considered the energy costs within the factory. What other energy costs are involved in recycling bottles?

9 Cement is an important building material. A typical cement contains about 60% calcium oxide and 25% silicon oxide.
   a Name the raw material that is the source for:
      i calcium oxide   ii silicon oxide
   b What is the main use of cement?
   c When water is added to cement, an exothermic reaction occurs in which calcium silicate ($CaSiO_3$) is formed.
      i Write a balanced equation for the formation of calcium silicate.
      ii How could you tell a chemical reaction is occurring?
      iii Is the reaction reversible? Explain your answer.

10 Limestone is calcium carbonate, $CaCO_3$. It is quarried on a huge scale.
   a Which elements are present in it?
   b Calculate the percentage of calcium in calcium carbonate. (RAMs: C = 12, O = 16, Ca = 40)
   c Much of the limestone that's quarried is used to make quicklime (CaO) for the steel industry.
      i What is the chemical name for quicklime?
      ii How much quicklime would be obtained from 1000 tonnes of limestone?
   d Powdered limestone is used to improve the water quality in acidified lakes.
      i Suggest how the water might have become acidified in the first place.
      ii How does limestone improve its quality?
      iii Why is the limestone powdered?
   e You live in a limestone area. A new limestone quarry is being proposed for the area. List the advantages and disadvantages to local people.

11 Aluminium is the most abundant metal in Earth's crust, but more and more of it is being recycled.
   a Why is it worth recycling aluminium?
   b A 500 g roll of aluminium foil from a supermarket costs around £2.50. Calculate the cost of a tonne of this material. (1 tonne = 1000 kg)
   c The cost of blocks of aluminium from the factory is about £950 per tonne. Compare this with your answer for b, and explain any difference.

12 Some copper ores are high grade, and some are low grade. This shows how the *average* grade of copper ores mined in the United States has changed over the years 1911–1991.

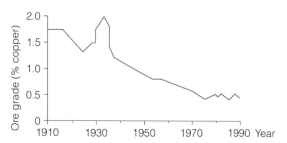

   a What is: i a *high-grade* ore? ii a *low-grade* ore?
   b Which is more profitable to mine?
   c Around which year was the average grade of mined ore highest? Can you suggest a reason?
   d As you can see, the overall trend on the graph is towards low-grade ore.
      i What does this suggest about the copper mines in the United States?
      ii What does it suggest about the demand for copper?

11.1 Another look at air

In Unit 10.2 you saw how Earth's atmosphere developed slowly over several billion years, to give the mixture shown on the right. This mixture seems to have remained more or less unchanged for the last several hundred million years.

How the atmosphere is maintained

The atmosphere is like an enormous gas store. We breathe in oxygen, and use it up when we burn fuel. Plants take in carbon dioxide for photosynthesis. Nitrogen is removed from the air by certain bacteria in soil, and converted to its oxides by lightning. With all that going on, how can the atmosphere's composition remain steady? The answer is that the gases removed from it are in effect **recycled**.

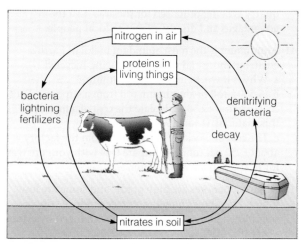

1 Nitrogen circulates between the air, the soil, and living things in the **nitrogen cycle**. You'll learn more about this on page 214.

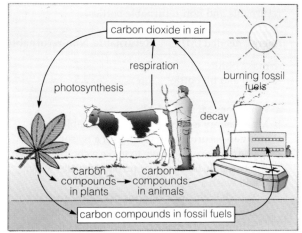

2 Carbon dioxide circulates between the air, the soil, the ocean, and living things in the **carbon cycle**. See page 240 for more.

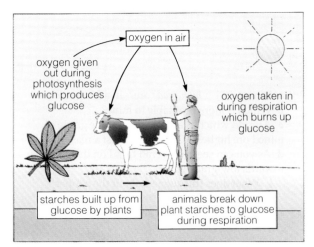

3 The combination of photosynthesis and respiration which recycles carbon dioxide does the same for **oxygen**.

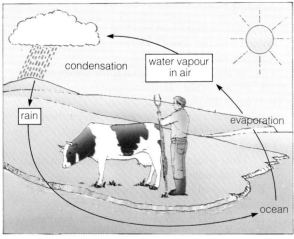

4 And finally, water circulates between the air, the soil, the oceans, and living things in the **water cycle** (page 181).

Measuring the percentage of oxygen in air

The first person to realize that air was a mixture of gases was the French chemist Lavoisier. He experimented by heating mercury in air in a closed vessel. The mercury turned into a red powder, and about one-fifth of the air got used up. (What was the red powder?)

No matter how much mercury Lavoisier started with or how much he heated it, only one-fifth of the air got used up. He tested the remaining gas and found that a candle wouldn't burn in it. He put a mouse in it and the mouse died. Lavoisier concluded that the gas that got used up was also the one that supported life. He named it **oxygen**.

You too can do an experiment along the same lines as Lavoisier's.

The apparatus   A tube of hard glass is connected to two gas syringes A and B. The tube is packed with small pieces of copper wire. One syringe contains 100 cm³ of air. The other is empty:

Lavoisier (1743–1794) in his lab. He met his death on the guillotine during the French Revolution.

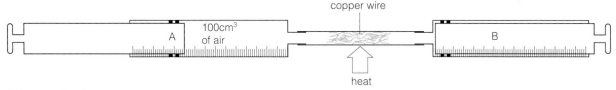

The method   These are the steps:
1 The tube is heated by a Bunsen burner. When A's plunger is pushed in, the air is forced through the tube and into B. The oxygen in it reacts with the hot copper, turning it black. When A is empty, B's plunger is pushed in, forcing the air back to A. This cycle is repeated several times.
2 Heating is stopped after about 3 minutes, and the apparatus allowed to cool. Then all the gas is pushed into one syringe and its volume measured. (It is now less than 100 cm³.)
3 Steps 1 and 2 are repeated until the volume of the gas remains steady. This means that all the oxygen has been used up. The final volume is noted.

The results   Starting volume: 100 cm³. Final volume: 79 cm³.
So the volume of oxygen in 100 cm³ air is 21 cm³.

The percentage of oxygen in air is therefore $\dfrac{21}{100} \times 100 = \textbf{21\%}$.

> **Air dissolved in water**
> - Oxygen is more soluble than nitrogen in water.
> - If you boiled the air out of water, and measured the percentage of oxygen in it, you'd get an answer of 33%.

---

1 Together, what percentage of air do nitrogen and oxygen form?
2 What is the combined percentage of all the other gases in air?
3 About how much more nitrogen is there than oxygen in air, in terms of volume?

4 Which is the most reactive gas in air?
5 What was the red powder Lavoisier obtained during his experiments with mercury and oxygen? Write down its name and formula.
6 Write down the name and formula of the black substance that forms in the experiment above.

11.2 Making use of air

We use oxygen in making steel, and argon to fill light bulbs. But first these gases must be separated from the air. Every year in the UK alone, over 20 million tonnes of air is separated into its component gases by **fractional distillation**.

The fractional distillation of liquid air

The process makes use of the fact that the gases in air have different boiling points. This is how it works:
- first the air is cooled down until it turns into a liquid
- then the liquid air is heated up again. The different gases boil off at different temperatures and are collected one by one.

This diagram shows the air separation plant.

Boiling points of gases in air /°C	
Carbon dioxide (sublimes)	−32
Xenon	−108
Krypton	−153
Oxygen	−183
Argon	−186
Nitrogen	−196
Neon	−246
Helium	−269

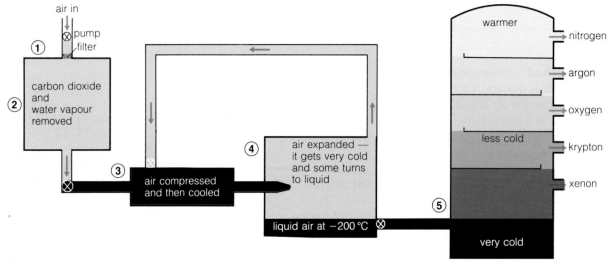

The steps in the process are:
1 Air from outside is pumped into the plant. The filter gets rid of dust.
2 Next the carbon dioxide and water vapour are removed. Otherwise they would freeze later, when the air is cooled, and block the pipes.
3 Next the air is forced into a small space, or **compressed**. That makes it hot. It is then cooled down again by recycling cold air.
4 The cold, compressed air is passed through a jet, into a larger space. It expands, and that makes it very cold.

Steps 3 and 4 are repeated several times, and each time the air gets colder. By the time it reaches −200 °C, all its gases have become liquid, except for neon and helium. These two are removed.

5 The liquid air is pumped into the fractionating column. There it is slowly warmed up. The gases boil off one by one, and are collected in tanks or cylinders. Nitrogen boils off first. Why?

Liquid nitrogen

Making use of oxygen

1 Astronauts carry oxygen with them, and so do deep-sea divers. Aeroplanes also carry their own oxygen supply.
2 In hospitals it is given to patients who can't breathe properly, for example because their lungs are diseased.
3 In steel works it is blown through molten steel to purify it.
4 A mixture of oxygen and **ethyne** (acetylene) is used as the fuel in oxy-acetylene cutting and welding torches. It gives a flame hot enough to melt steel (3150 °C).

To cut steel, the steel is first heated by the flame. Then the cutting torch is switched to supply oxygen only. The oxygen jet reacts with the hot steel to form iron(III) oxide. This reaction is exothermic, and gives out enough heat to melt the steel below it. The force of the jet clears away the molten oxide, exposing the molten steel. So reaction continues until the steel is cut through.

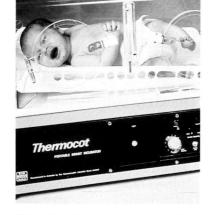

This cot has its own oxygen supply for sick infants.

Making use of nitrogen

1 Liquid nitrogen is very cold (−196 °C) so is used for quick-freezing food in food factories, and freezing liquid in damaged pipes while they're being repaired.
2 It is also used to **shrink-fit** machine parts. One metal part is dipped into liquid nitrogen. It contracts and can then be fitted easily into another part.
3 Because nitrogen is unreactive or **inert**, it is pumped into oil storage tanks to flush out oxygen and reduce the risk of fire. It is also flushed through food packaging to remove oxygen and keep food fresh.

Frozen meat coming out of a liquid nitrogen freezing tunnel

Making use of the noble gases

Their uses rely on the fact that the noble gases are completely inert.
1 They are used for lighting. Argon provides the inert atmosphere in ordinary tungsten light bulbs. Neon is used in advertising signs because it glows red when an electric current is passed through. Krypton and xenon are used in lighthouse lamps.
2 **Arc welding** uses an electric arc to melt the metals being welded. It is often carried out in a 'blanket' of argon to stop oxygen reacting with the metals.
3 Helium is used to fill air ships and balloons, since it is so light.
4 Krypton is used in **lasers**, which produce very intense beams of light. Krypton lasers are used in eye surgery, to prevent bleeding on your retina. The laser is directed to the points where bleeding might occur. The intense light makes blood clot.

Liquid nitrogen is used to freeze liquid in damaged pipes, before repairs.

1 In the separation of air into its gases:
  a why is the air compressed and then expanded?
  b why is argon obtained *before* oxygen?
  c what do you think is the biggest expense? Why?
2 Write down three uses of oxygen gas.
3 Why does a mixture of oxygen and ethyne burn better than a mixture of air and ethyne?
4 What *two* properties of nitrogen make it suitable for keeping food frozen during transportation?
5 Give three uses for noble gases.

11.3 Pollution alert!

For millions of years nature has maintained the atmosphere by means of the cycles shown in Unit 11.1. But for the last 150 years we have increasingly misused it by making it a dump for waste gases.

Pollution from burning coal. In many areas, byelaws permit only smokeless fuels.

The main culprit: the fossil fuels

The burning of fossil fuels – coal, oil, gas, and petrol – is the main source of waste gases in the atmosphere. This is what it produces.

Carbon dioxide This gas is essential to life. Plants take it in for photosynthesis, and living things give it back through respiration. Left to themselves, these processes keep the gas at a constant level.

But every year, we pour an extra 5000 million tonnes of carbon dioxide into the air from burning fossil fuels: from car exhausts, central heating systems, power stations, and factories. At the same time, tropical rainforests which soak it up are being destroyed. The oceans can dissolve some of the extra carbon dioxide, but not all. The rest stays in the atmosphere, upsetting nature's balance.

Carbon dioxide is a **greenhouse gas**. That means it acts as a blanket around the earth, preventing the escape of heat. The more carbon dioxide in the atmosphere, the more heat will be trapped. The result is **global warming**. Over the last hundred years, the average air temperature around the earth has increased by half a degree.

This might not seem much of an increase. But if it keeps on it could lead to melting of the polar ice caps, flooding of low-lying coasts, and unpredictable new weather patterns.

Carbon monoxide When fossil fuels burn in too little air, carbon monoxide (CO) is produced instead of carbon dioxide (CO_2). This gas is colourless, insoluble, and has no smell. It is poisonous even in small concentrations. It reacts with the **haemoglobin** in your blood and stops it carrying oxygen to the body cells.

Sulphur dioxide Coal, oil, and gas all contain sulphur compounds. If these aren't removed they burn to form sulphur dioxide. This gas attacks the lungs, and can affect people with asthma very badly. It also dissolves in rain water to form acid rain (page 229). This damages buildings, trees, and plants, and kills fish and other river life.

Nitrogen oxides These are an indirect result of burning fossil fuels. Inside car engines and power station furnaces, the air gets so hot that nitrogen and oxygen react together, forming oxides of nitrogen. Like sulphur dioxide, these attack the lungs and cause acid rain.

Ozone This is also a by-product of burning fossil fuels. It is formed in heavy traffic in hot weather, when sunlight causes the nitrogen oxides and hydrocarbons from car exhausts to react together. Up in the ozone layer, ozone protects us from the damaging rays of the Sun. But at ground level it can cause sore eyes, noses and throats, headaches, breathing problems and lung damage.

Greenhouse gases

Any gas that traps heat in the atmosphere is a greenhouse gas. At present the main ones are:
- carbon dioxide
- methane
- ozone
- CFCs (see next page)

Who are the chief offenders?

- About 85% of the carbon monoxide in UK air comes from exhausts. (But deaths from this gas are usually caused by faulty gas burners and heaters in badly ventilated rooms.)

- Nearly three-quarters of the sulphur dioxide comes from power stations.

- About 45% of the nitrogen oxides comes from engine exhausts, and about 35% from power stations.

Tiny solid particles   These are mainly particles of carbon (soot) from burning coal in power stations and petrol in motor engines. Exhaust fumes from leaded petrol also contain particles of lead. The carbon particles are coated with harmful hydrocarbons and other compounds, and can get deep into your lungs. They also damage trees and plants and make buildings black and grimy. Lead causes brain damage, particularly to children.

Reducing pollution from fossil fuels

Several things are being done to cut down pollution from fossil fuels:
- The exhausts of new cars are fitted with **catalytic converters** in which harmful gases are converted to harmless ones.
- All new cars must run on lead-free petrol.
- Manufacturers are looking at ways to make car engines more efficient so that they use less petrol, and at alternative fuels.
- Coal is turned into **smokeless fuel** for use in homes.
- New power stations are fitted with plant to stop sulphur dioxide escaping. Gases leaving the power station are passed through limestone, which reacts with the sulphur dioxide to form calcium sulphate. Some of the plant cost is recovered by selling the calcium sulphate to make plaster for the building industry.
- Scientists are looking at ways to make homes and factories more energy efficient, so that we burn less fuel, not more.

But there is a lot more that could be done. For example, if there was cheap and efficient public transport, people would use cars less.

Other harmful substances
Gases that destroy the ozone layer   The main culprits here are chlorine compounds such as chlorofluorocarbons (CFCs). These are used in air conditioning and refrigeration, in aerosols, and as solvents in the electronics industry.

CFCs that leak into the atmosphere can react with ozone (O_3), breaking it down to oxygen (O_2). This causes holes in the ozone layer. It means that more of the sun's harmful radiation reaches Earth, causing skin cancer, eye cataracts, and crop damage.

Waste gases and dust from factories   Many industrial processes produce other harmful gases and dust which can damage lungs and ruin vegetation. Companies can be prosecuted for releasing these into the atmosphere. See page 167 for ways to prevent their escape.

Limestone statues attacked by acid rain. Which one is most sheltered?

International action

Air pollution travels. This means it's everyone's problem!
- The European Union has ordered all power stations in the EU to cut sulphur dioxide emission by 60% by the year 2003.

- In Helsinki in 1989, 81 nations agreed to ban eight chemicals (mainly CFCs) which damage the ozone layer by the year 2000.

- In Rio de Janeiro in 1992, over 160 countries agreed to reduce the emission of greenhouse gases to 1990 levels by the year 2000.

---

1 Explain why carbon dioxide, which is essential to life, can also be a danger.
2 a What is carbon monoxide and how is it formed?
  b Why is carbon monoxide so dangerous?

3 How is limestone used to reduce air pollution?
4 What are CFCs? Why are they harmful?
5 Write a list of the things *you* could do to help cut down the amount of fossil fuel being burned.

11.4 Water and the water cycle

What is water?

Water is a compound of hydrogen and oxygen. Its formula is **H₂O**.
You could make it in the laboratory by burning a jet of hydrogen in
air, as shown on the right. The reaction is fast and may be dangerous:

$$2H_2(g) + O_2(g) \longrightarrow 2H_2O(l)$$
$$\text{hydrogen} + \text{oxygen} \longrightarrow \text{water}$$

The water forms as a gas. It condenses to liquid on an ice-cold tube.

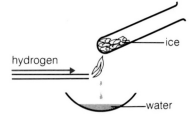

Making water in the lab

Tests for water

If a liquid is water, it will:
1 turn blue when you add white anhydrous copper(II) sulphate
2 turn blue cobalt(II) chloride paper pink.

If a liquid is *pure* water it will boil at 100 °C and freeze at 0 °C, at
normal pressure. But water is such a good solvent that pure water
does not occur naturally. You can produce it by distillation.

Water and living things

As you saw in Unit 10.2, water first appeared on Earth billions of
years ago as steam, in the gases that burst from volcanoes. The steam
cooled, condensed, and formed the oceans.

Life began in water. 3500 million years ago the first simple algae and
bacteria developed in the oceans. From that time on, water has been
essential to life. Every cell in every living thing contains some water.
You are over 70% water. Lettuce is about 96% water!

Your body contains about 35 litres of water altogether. Most is in
your cells, the rest in your blood, saliva, and other body fluids. Every
day you lose about 2.5 litres of water in sweat, urine, and as water
vapour in your breath. If you didn't replace this by eating and
drinking, you'd die in a matter of days.

Water, water everywhere?

- There's over 1.3 billion cubic km
  of water on earth. 97% of it is
  in the oceans.
- The oceans cover over 70% of
  Earth's surface. The Pacific
  alone covers almost half of it.
- But in terms of Earth's overall
  size, the oceans are like a thin
  film of dampness on the
  surface.
- Some regions don't have
  enough water. Many millions
  of lives have been lost through
  the effects of drought.

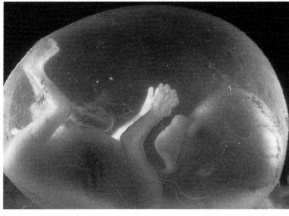

Fourteen weeks after conception you looked like this, floating
in a watery solution.

The water cycle in action

The water cycle

Water circulates between the air, the oceans, and living things through the **water cycle**, which is driven by the Sun:

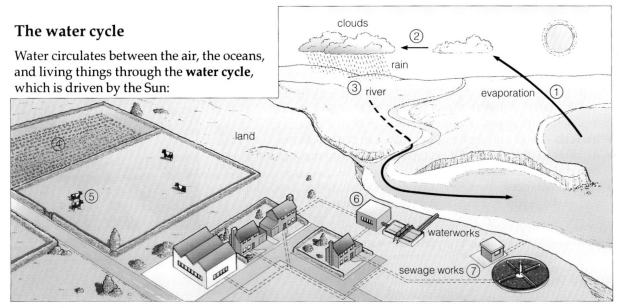

1 Heat from the Sun causes water to evaporate from seas and oceans. The vapour rises, cools, and condenses to form tiny water droplets. The droplets form clouds.
2 The clouds get carried along on air currents. They cool, and the droplets join to form larger drops which fall as rain. Or, if the air is very cold, as hail, sleet, or snow.
3 Some rainwater soaks through the ground, and reappears as springs. Some flows along the ground as streams. The springs and streams feed rivers. The rivers flow to the sea. The main cycle is complete.

So where do living things come into the picture?

4 Plants soak up rainwater through their roots, and use it for photosynthesis and building their cells and fluids. They give out water vapour during respiration.
5 Animals take in water by eating and drinking. They give it out in sweat and urine, and breathe out water vapour during respiration.
6 River water is filtered, cleaned up, and pumped into homes and factories for washing and cooking, or use as a raw material.
7 Waste water from homes and factories is filtered and cleaned up at sewage works, and pumped back into the river.

Stages 6 and 7 are where problems arise. We have dumped all kinds of waste into our water over the years: waste chemicals, raw sewage, fertilizer, organic solvents, and detergents. These can't all be removed at waterworks or sewage works. They harm fish and other river life. They also harm humans. For example, scientists think that certain chemicals in drinking water may be lowering the sperm count in males, which means less chance of having children.

Is the water cycle completely absent here?

1 Is the reaction between hydrogen and oxygen exothermic or endothermic?
2 How would you test a liquid to see if it is:
   a water? b pure water?
3 Ocean water is salty. Use the water cycle to help you explain why river water is *not* salty.
4 Write a list of the ways you personally get involved in the water cycle.

11.5 Our water supply

Where tap water comes from

In Britain, most tap water comes from **rivers** and **reservoirs**. Some is also pumped up through boreholes from natural underground reservoirs called **aquifers**. Water from these sources is never completely pure, especially river water. It may contain:

- **bacteria** – most are harmless, but some can cause disease.
- **dissolved substances** – for example, calcium and magnesium compounds dissolved from rocks.
- **solid substances** – particles of mud, sand, grit, twigs, dead plants, and perhaps tins and rags that people have dumped.

To make the water safe to drink, bacteria must be killed and solid substances removed. This is done at the **waterworks**.

Part of a modern waterworks

At the waterworks

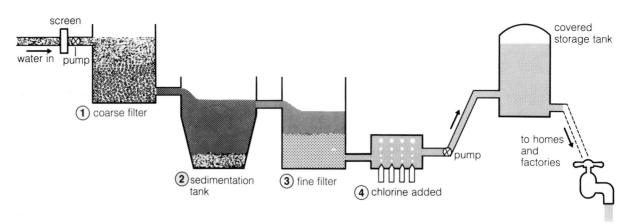

At the waterworks, the water is *usually* pumped through a screen to get rid of the larger bits of rubbish. *What happens next depends on the waterworks, but these are the usual stages:*

1. It goes through a coarse filter which traps larger particles of solid. The filter could be sand or anthracite.
2. In older waterworks it may go to a **sedimentation tank** where chemicals are added to make smaller particles stick together. They sink to the bottom of the tank.
3. Next it goes through a fine filter. This could be sand, or carbon granules with thousands of tiny pores. The carbon removes tastes and smells.
4. Finally chlorine is added to kill bacteria.

Adding fluoride to water   Water naturally contains some fluoride ions. The amount varies from place to place depending on the rocks. Studies have shown that a concentration of 1 mg of fluoride (0.001 g) per litre of water helps to fight tooth decay. So in some areas where the natural concentration is less than this, fluoride is added before the water leaves the waterworks.

Since the action of fluoride takes place on the *surface* of your teeth, fluoride toothpaste may be more effective – especially if you don't drink much tap water!

Fluoride in water

Some people object to the fluoridation of water because they dislike the idea of 'mass medication'. Besides, excess fluoride has been linked to:

- **dental fluorosis**, which is a mottling of the teeth.
- **skeletal fluorosis**, where bones get dense, causing pain and stiffness.
- **fractures**. There appear to be more hip fractures in areas where the concentration of fluoride in water is higher.

But small amounts of fluoride are *good* for you!

Waste water and sewage works

All sorts of things get mixed with tap water: shampoo, toothpaste, detergents, grease, body waste, food, grit, sand, and waste from factories. This mixture goes down the drain, and is called **sewage**. It is piped underground to **sewage works**, where the water in it is cleaned up and fed back to the river. Below is a diagram of the plant:

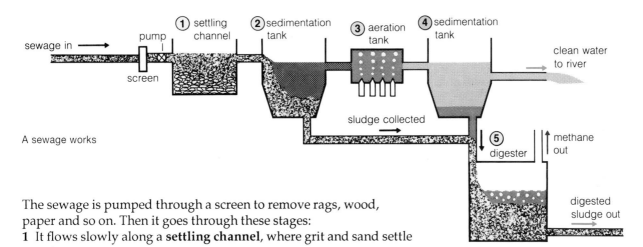

A sewage works

The sewage is pumped through a screen to remove rags, wood, paper and so on. Then it goes through these stages:

1 It flows slowly along a **settling channel**, where grit and sand settle out.
2 Next it passes into a **sedimentation tank**. Here smaller pieces of waste sink slowly to the bottom. This waste is called **sludge**. It is grey and evil-smelling, and contains many harmful things.
3 The water now looks cleaner. It flows into an **aeration tank**, which contains special bacteria growing on sludge. These bacteria feed on harmful things in the water, and make them harmless. For this they need a lot of oxygen, so air is continually pumped through the sludge, from the bottom of the tank. Instead of aeration tanks, some plants use **percolating filters**, where the bacteria live on stones, and the water trickles over them.
4 Next comes another **sedimentation tank**, where any remaining sludge settles out. The water is now safe to put into the river.
5 All the sludge is collected into tanks called **digesters**. Here it is mixed with bacteria which destroy the harmful substances, producing **methane gas**. Methane is a good fuel, so may be used to make electricity for the sewage plant. The digested sludge is burned to ash or sold to farmers as fertilizer.

As you can see, the steps above remove solid waste and many other harmful substances. But they can't remove everything. Some harmful substances in the waste from homes and factories will end up in the river, and perhaps even back in tap water. Not good for fish or humans!

Percolating filters, where bacteria live on the stones

---

1 List ten impurities that you might find in river water.
2 What happens in the sedimentation tanks at waterworks?
3 What is: **a** sewage? **b** sludge?
4 At a sewage plant, describe what happens in:
   a aeration tanks **b** digesters

11.6 Soft and hard water

Scum alert!

As you saw in the last Unit, water is treated at waterworks before being piped to your home.

The treatment removes only the insoluble particles and kills bacteria. So the water still isn't pure. It contains 'natural' compounds dissolved from rocks and soil. It may also contain traces of chemicals dumped from homes, farms, and factories.

If your tap water comes from an area where the rocks contain chalk, limestone, dolomite, or gypsum, it will contain dissolved calcium and magnesium compounds: sulphates and hydrogen carbonates. How can you tell? Soap will give you a clue.

Washday blues. The scum that forms in hard water areas is difficult to wash out of clothes.

Lather or scum?

lather
soap flakes added

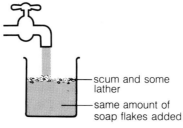
scum and some lather
same amount of soap flakes added

If soap lathers easily, it means the water contains very little calcium and magnesium compounds. It is **soft** water.

But a greyish scum and hardly any lather shows that larger amounts are present. The water is called **hard** water.

The scum forms because the calcium and magnesium compounds react with soap to give an insoluble product. For example:

calcium sulphate + sodium stearate $\longrightarrow$ calcium stearate + sodium sulphate
(soap)   (scum)

How water acquires hardness

Calcium hydrogen carbonate is the main cause of hard water. It forms when rain falls on rocks of **limestone** and **chalk**. These are made of calcium carbonate which is *not* soluble in water. But rainwater contains carbon dioxide dissolved from the air, which makes it acidic. So it reacts with the rocks to form calcium hydrogen carbonate which *is* soluble and ends up in our taps. The reaction is:

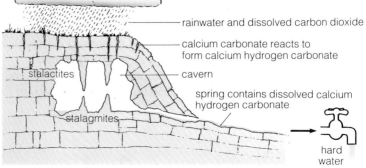

rainwater and dissolved carbon dioxide

calcium carbonate reacts to form calcium hydrogen carbonate

stalactites

cavern

spring contains dissolved calcium hydrogen carbonate

stalagmites

hard water

$$H_2O\,(l) \;+\; CO_2\,(g) \;+\; CaCO_3\,(s) \;\rightleftharpoons\; Ca(HCO_3)_2\,(aq)$$
water   carbon dioxide   calcium carbonate   calcium hydrogen carbonate

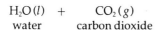

In the same way, rainwater reacts with **dolomite** ($CaCO_3 . MgCO_3$) and **gypsum** ($CaSO_4.2H_2O$) to give the other compounds that make water hard. These reactions are part of the **weathering** process. You can find out more about this in Unit 18.3.

Temporary or permanent?

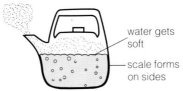

water gets soft

scale forms on sides

kettle of hard water

This water is hard: it contains calcium hydrogen carbonate. On heating, this compound breaks down to form calcium carbonate which is insoluble.

The new compound forms a hard **scale** on the kettle. But the water itself is now soft since the calcium hydrogen carbonate has been removed.

The same reaction occurs inside boilers and pipes in factories and central heating systems. Scale has blocked this pipe almost completely.

Since it can be removed just by boiling, the hardness caused by calcium hydrogen carbonate is called **temporary hardness**. Hardness caused by other compounds is called **permanent hardness** because boiling does not affect it. Even after boiling water with permanent hardness, you'll still get a scum with soap.

Hardness in water: good or bad?

Compare the advantages and disadvantages of hard and soft water. Which kind of water would *you* prefer?

Hard water	Soft water
Advantages	*Advantages*
• The dissolved substances give it a pleasant taste.	• It makes soap lather better and gets clothes cleaner.
• The calcium compounds in it are good for bones and teeth.	• Laundry uses less soap and can be done at lower temperatures.
• There are fewer heart attacks in areas with hard water.	• It means less scum in your bath tub.
• A coating of calcium carbonate inside pipes, boilers, and radiators helps to prevent corrosion.	• It leaves no scale in pipes or boilers.
Disadvantages	*Disadvantages*
• When the coating develops into a scale, it makes pipes, boilers, and radiators less efficient. It must be removed from time to time to prevent blockages.	• Soft water contains more *sodium* ions than hard water does. Sodium is linked to heart disease.
• Hard water uses up more soap than soft water does.	• Soft water dissolves metals such as cadmium and lead from metal pipes. Lead is poisonous. Cadmium has been linked to hypertension. (But it also dissolves copper which your body needs.)
• It leaves a messy scum, and is tougher on laundry.	

1 Hard water wastes soap. Explain why. Use an equation to help you.
2 Chalk is insoluble in water. Why does it dissolve in rainwater? Write an equation for the reaction.

3 What is the difference between *temporary* and *permanent* hardness?
4 Hardness in water costs money. Write down as many reasons as you can to explain why.

11.7 Making hard water soft

Ways to soften hard water

Because of the problems it causes, hard water is often softened for use in factories, laundries, and homes. That means removing the dissolved calcium and magnesium ions. Below are ways to do this.

Boiling This removes temporary hardness, as you saw in the last Unit, by causing calcium carbonate to precipitate. But it uses a lot of fuel which makes it expensive to do on a large scale – unless you need the hot water anyway. (You'll still end up with scale in the boiler!)

Distillation In distillation, the water is boiled and the steam collected, cooled, and condensed. Distilled water is pure water. *All* the dissolved substances have been left behind. Like boiling, it is an expensive option in terms of fuel. But it is essential for some purposes, for example for lab experiments and for making drugs.

It's advisable to use distilled water in steam irons. Why?

Adding washing soda Washing soda is sodium carbonate, Na_2CO_3. It removes both temporary and permanent hardness by precipitating calcium carbonate. For example, it reacts with calcium sulphate like this:

$$Na_2CO_3\,(aq) \quad + \quad CaSO_4\,(aq) \quad \longrightarrow \quad CaCO_3\,(s) \quad + \quad Na_2SO_4\,(aq)$$

| sodium carbonate | calcium sulphate | calcium carbonate | sodium sulphate |

Bath salts contain sodium carbonate to soften the bath water. The sodium sulphate that forms does not affect soap. But the precipitate of calcium carbonate leaves a ring on your bath.

Ion exchange In ion exchange, unwanted ions are removed by replacing them with 'harmless' ions.

A typical ion exchanger is a container full of small beads. These are made of a special plastic called **ion exchange resin**, which has the 'harmless' ions weakly attached to it. They are usually sodium ions, as shown on the right. When hard water flows through, the calcium and magnesium ions in it switch places with the sodium ions and attach themselves to the resin. The sodium ions are carried away:

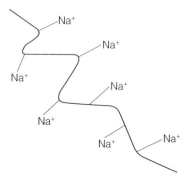

This is only part of a long molecule of resin. (Each bead contains many molecules.)

After a time all the sodium ions have gone so no more hardness can be removed. The resin has to be **regenerated**. A concentrated solution of sodium chloride is poured into it. The sodium ions push the calcium and magnesium ions off the resin and the ion exchanger is ready for use again.

Other ions could be used instead of sodium for the resin. But sodium ions have one big advantage: salt is cheap!

How well do these methods work?

A simple experiment will help you find out. Collect samples of water softened in different ways, and then compare them with distilled water using a lather test. Here's how to do it:

1 Measure a sample of water (say 20 cm³) into a conical flask.
2 Add a little soap solution from a burette, and shake the flask carefully.
3 Add more soap solution, a little at a time, until a lather lasting at least half a minute is obtained.
4 Note the volume of soap solution used.

In the experiment in the photograph, water from the same tap was used for A, B, and C. For A it was left untreated. For B it was boiled. For C it was passed through an ion exchanger. D is distilled water. Here are the results:

Water sample	Volume of soap solution used (cm³)
A (untreated tap water)	10
B (boiled)	7
C (ion exchanger)	0.5
D (distilled)	0.5

Is the ion exchanger an effective way of removing hardness? What about boiling? Which type(s) of hardness did A contain?

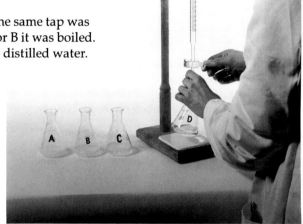

Comparing hardness in water samples, using the lather test

A little more about ion exchangers

- Ion exchangers for water softening can be any size, from small ones designed for home use to large ones for factories.
- A typical ion exchanger for the home has two tanks. One contains the resin beads. The other contains a concentrated solution of sodium chloride (brine). A meter records how much water has gone through the resin. The brine is then automatically flushed through at an appropriate stage.
- Ion exchangers aren't used only for water softening. For example, they can be designed to remove nitrate ions, or to recover valuable metal ions from waste solution. Both anions and cations can be removed.
- They aren't always made of plastic either. Some are made of natural or synthetic compounds called **zeolites**. These are silicates in which some silicon atoms are replaced by aluminium atoms. They have an open porous structure.

---

1 Which method of water softening would you expect to be: **a** cheapest? **b** most expensive?
2 Why is sodium carbonate used in bath salts?
3 **a** Explain how an ion exchanger works.
  b Suggest another ion to use instead of sodium.

4 Explain how the experiment above shows that:
  a the tap water contains both types of hardness
  b most of the hardness in it is permanent
  c ion exchange is just as effective as distillation in removing hardness.

1 Copy and complete the following paragraph:
Air is a .... of different gases. 99% of it consists of the two elements .... and .... One of these, ..., is needed for respiration, which is the process by which living things obtain the .... they need. The two elements above can be .... from liquid air by ...., because they have different ....
Much of the .... obtained is used to make nitric acid and fertilizers. Some of the remaining 1% of air consists of two compounds, ... and ....
One of these is important because it is taken in by plants, in the presence of ...., to form ....
The rest of the air is made up of elements called the .... These are all members of Group ... of the Periodic Table.

2 Oxygen and nitrogen, the two main gases in air, are both slightly soluble in water. Using the apparatus below, a sample of water was boiled until $100\,cm^3$ of dissolved air had been collected.

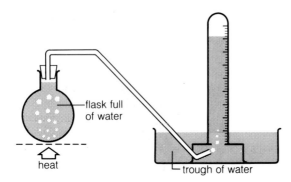

- flask full of water
- heat
- trough of water

This air was then passed over heated copper. Its final volume was $67\,cm^3$.
a Is air more soluble or less soluble in water, as the temperature rises? How can you tell?
b The copper reacts with the oxygen in the dissolved air. Write an equation for the reaction.
c Draw a diagram of the apparatus used for passing air over heated copper.
d What volume of oxygen was present in $100\,cm^3$ of dissolved air?
e Calculate the approximate percentages of oxygen and nitrogen in dissolved air.
f What are the percentages of these gases in atmospheric air?
g Explain why the answers are different, for parts **e** and **f**.
h Which gas is more soluble in water, nitrogen or oxygen?
i What is the biological importance of air dissolved in water?

3 Air is a *mixture* of different gases.
a Which gas makes up about 78% of the air?
b Only one gas in the mixture will allow things to burn in it. Which gas is this?
c How are the gases in the mixture separated from each other, in industry?
d Which noble gas is present in the greatest amount in air?
e Which gas containing sulphur is a major cause of air pollution?
f Name two other gases that contribute to air pollution.
g Name one substance which is not a gas but which also pollutes the air.

4 In the catalytic converters fitted to new cars, carbon monoxide and oxides of nitrogen in the exhaust gas are converted to other substances.
a Explain why we want to remove these gases.
b What is meant by a *catalytic* reaction?
c In one catalytic reaction, nitrogen monoxide (NO) reacts with carbon monoxide to form nitrogen and carbon dioxide. Write an equation for this.
d Name one air pollution problem that is *not* solved by the use of catalytic converters.

5 One cause of hard water is the reaction between calcium carbonate, carbon dioxide, and water.
a Write the equation for this reaction. (Note: it is a reversible reaction.)
b Name the soluble compound formed.
c i Which ion causes the hardness?
ii Why does this ion affect the ability of soap to form a good lather?
d i How can the equilibrium reaction be reversed?
ii Why does this remove hardness from water?
iii What is this type of hardness called?
e You can also remove hardness from water using washing soda. Write an equation for this.

6 In an ion exchange column for softening water, calcium and magnesium ions are replaced by sodium ions. The total charge on the ion exchange resin must always be zero.
a Write the formulae for the three metal ions.
b How many sodium ions will be exchanged for:
i one calcium ion? **ii** one magnesium ion?
c After a time, the ion exchanger will no longer remove hardness. Why?
d The resin is *regenerated* using a concentrated solution of sodium chloride. Explain this.
e Calculate the maximum mass of calcium ion that could be removed by $100\,g$ of sodium chloride.
(Metal ions have the same mass as their atoms. The RAMs are: Na $=23$, Cl $=35.5$, Ca $=40$.)

	Sample		
Test	A	B	C
1 Shaken with soap solution	poor lather	good lather	poor lather
2 Boiled first and then shaken with soap solution	good lather	good lather	poor lather
3 Some bath salts added, shaken with soap solution after filtering	good lather	good lather	good lather

7 Some samples of water were tested in the laboratory. The results are shown in the table above.
a Only one of the samples could have been pure water. Which one? Explain your answer.
b The other two samples were both from hard water areas.
   i Which contained temporary hardness?
   ii Which contained permanent hardness?
Explain how you were able to tell them apart.
c Name one substance that could cause the hardness in sample C.

8 A distilled water
  B tap water
  C calcium hydroxide solution
  D sodium chloride solution
  E unknown solution
Ten drops of soap solution were added to 10 cm³ of each of the above liquids. After shaking for ½ minute, these heights of lather were obtained:

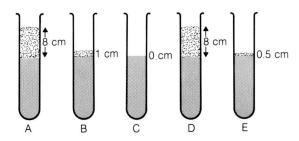

a Does the tap water come from a hard water or soft water area? Explain your answer.
b Does sodium chloride form a precipitate with the soap solution? Explain your answer.
c Does calcium hydroxide cause hardness? How can you tell from the experiment?
d What can you say about liquid E, the unknown solution?
e Another sample of solution E was boiled for 5 minutes, and then the test repeated. This time a lather height of 8 cm was obtained. Explain this result.
f What result would you expect for a further sample F, which is water from an ion exchanger?
g What results would be obtained, using 10 drops of a soapless detergent in place of the soap solution?

9 a Name the two elements from which water is made.
b Name a solid which is very soluble in water.
c Name a metal which reacts quickly with cold water.
d Describe an experiment you could carry out in the laboratory to find out whether tap water contains any dissolved solids.
e Describe a method of completely removing dissolved solids from tap water.

10 25 cm³ samples of water from four different areas were tested with soap solution, to see how much soap solution was needed for a lather that lasted at least half a minute. The experiment was repeated a second time using samples that had been boiled, and then a third time using samples that had been passed through an ion exchanger. The results are shown in the table below.

	Volume of soap solution/cm³		
Sample	Untreated	Boiled	Passed through ion exchanger
A	14	1.9	1.9
B	16	16	1.8
C	25	20	1.9
D	1.8	1.8	1.8

a Which of the samples is the hardest water? Why do you think so?
b Which sample behaves like distilled water? Explain your choice.
c Decide whether the hardness is temporary, permanent, or both, in:
   i sample A
   ii sample B
   iii sample C
d Name a chemical which could be responsible for the hardness in:
   i sample A
   ii sample B
e Write an equation for the reaction in which the temporary hardness is removed, in samples A and C.
f Explain how an ion exchanger removes hardness from water.

12.1 Metals and nonmetals

There are 105 different elements. Of these, 84 are **metals** and 21 are **nonmetals**.
This means that over three-quarters of the elements are metals.

The properties of metals

Metals *usually* have these properties:

1 They are **strong** under tension and compression. That means they can withstand stretching and crushing without breaking.
2 They are **malleable**. That means they can be hammered and bent into shape without breaking.
3 They are **ductile**: they can be drawn out to make wires.
4 They are **sonorous**: they make a ringing noise when you strike them.
5 They are shiny when polished.
6 They are good conductors of electricity and heat.
7 They have high melting and boiling points. (They are all solid at room temperature, except mercury.)
8 They have high densities. That means they feel 'heavy'.
9 They react with oxygen to form oxides. For example, magnesium burns in air to form magnesium oxide. Metal oxides are **bases**, which means they react with acids to form salts.
10 When metals form ions, the ions are positive. For example, in the reaction between magnesium and oxygen, magnesium ions (Mg^{2+}) and oxide ions (O^{2-}) are formed, as shown on page 45.

The last two properties above are called **chemical properties**, because they are about chemical changes in the metals. The other properties are **physical properties**.

Some of the metals

Aluminium, Al
Calcium, Ca
Copper, Cu
Gold, Au
Iron, Fe
Lead, Pb
Magnesium, Mg
Potassium, K
Silver, Ag
Sodium, Na
Tin, Sn
Zinc, Zn

What is density?

The density of a substance is a measure of how 'heavy' it is.

$$\text{Density} = \frac{\text{mass (in grams)}}{\text{volume (in cm}^3)}$$

Compare these:

1 cm^3 of iron
mass = 7.86 g
density = 7.86 g/cm^3

1 cm^3 of lead
mass = 11.34 g
density = 11.34 g/cm^3

Think of two reasons why metals are used to make drums . . .

. . . and three reasons why they are used for saucepans.

All metals are different

The properties on the last page are typical of metals. But not all metals have *all* of these properties. For example:

Iron is a typical metal. It is used for gates like these because it is both malleable and strong. It is used for anchors because of its high density. It melts at 1530 °C. But unlike most other metals, it is **magnetic**.

Sodium is quite different. It is so soft that it can be cut with a knife, and it melts at only 98 °C. It is so light that it floats on water, but it reacts immediately with the water, forming a solution. No good for gates …

Gold melts at 1064 °C. Unlike most other metals it does not form an oxide – it is very unreactive. But it is malleable and ductile, and looks attractive. So it is used for making jewellery.

No two metals have exactly the same properties. You can find out more about the differences between them on the next four pages.

Comparing metals with nonmetals

Only 21 of the elements are nonmetals. Nonmetals are quite different from metals. They *usually* have these properties.

1 They are not strong, or malleable, or ductile, or sonorous. In fact, when solid nonmetals are hammered, they break up – they are **brittle**.
2 They have lower melting and boiling points than metals. (One of them is a liquid and eleven are gases, at room temperature.)
3 They are poor conductors of electricity. Graphite (carbon) is the only exception. They are also poor conductors of heat.
4 They have low densities.
5 Like metals, most of them react with oxygen to form oxides:

sulphur + oxygen $\longrightarrow$ sulphur dioxide

But unlike metal oxides, these oxides are not bases. Many of them dissolve in water to give *acidic* solutions.
6 When they form ions, the ions are negative. Hydrogen is an exception – it forms the ion H^+.

Some of the nonmetals
Bromine, Br
Carbon, C
Chlorine, Cl
Helium, He
Hydrogen, H
Iodine, I
Nitrogen, N
Oxygen, O
Sulphur, S

1 Make two lists, showing twenty *metals* and fifteen *nonmetals*. Give their symbols too.
2 Try to think of a metal that is not malleable at room temperature.
3 Suggest reasons why:
   a silver is used for jewellery;
   b copper is used for electrical wiring.

4 For some uses, a highly sonorous metal is needed. Try to think of two examples.
5 Try to think of *two* reasons why:
   a mercury is used in thermometers;
   b aluminium is used for beer cans.
6 Look at the properties of nonmetals, above. Which are *physical* properties? Which are *chemical*?

12.2 Metals and reactivity (I)

On the last page you saw that all metals are different.
The next few pages compare the way some metals react, to see how different they are.

1 The reaction between metals and oxygen

Look at the way sodium reacts with oxygen:

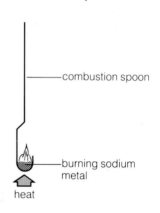

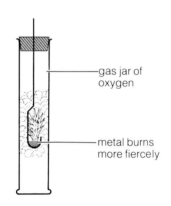

Sodium is stored under oil. Why?

A small piece of sodium is put in a combustion spoon and heated over a Bunsen flame. It melts quickly, and catches fire.

Then the spoon is plunged into a jar of oxygen. The metal burns even more fiercely, with a bright yellow flame.

The steps above can be repeated for other metals. This table shows what happens:

Metal	Behaviour	Order of reactivity	Product
Sodium	Catches fire with only a little heating. Burns fiercely with a bright yellow flame	most reactive	Sodium peroxide, Na_2O_2, a pale yellow powder
Magnesium	Catches fire easily. Burns with a blinding white flame		Magnesium oxide, MgO, a white powder
Iron	Does not burn, but the hot metal glows brightly in oxygen, and gives off yellow sparks		Iron oxide, Fe_3O_4, a black powder
Copper	Does not burn, but the hot metal becomes coated with a black substance		Copper oxide, CuO, a black powder
Gold	No reaction, no matter how much the metal is heated	least reactive	——

If a reaction takes place, the product is an oxide.
Sodium reacts the most vigorously with oxygen. It is the **most reactive** of the five metals. Gold does not react at all – it is the least reactive of them. The arrow on the table shows the **order of reactivity**.

2 The reaction between metals and water

Metals also show differences in the way they react with water. For example:

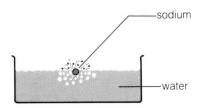

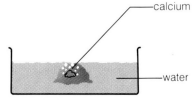

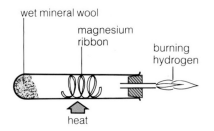

Sodium reacts violently with cold water, whizzing over the surface. Hydrogen gas and a clear solution of sodium hydroxide are formed.

The reaction between calcium and cold water is slower. Hydrogen bubbles off, and a cloudy solution of calcium hydroxide forms.

Magnesium reacts very slowly with cold water, but vigorously when heated in steam: it glows brightly. Hydrogen and solid magnesium oxide are formed.

This table shows the results for other metals too:

Metal	Reaction	Order of reactivity	Products
Potassium	Very violent with cold water. Catches fire	most reactive	Hydrogen and a solution of potassium hydroxide, KOH
Sodium	Violent with cold water		Hydrogen and a solution of sodium hydroxide, NaOH
Calcium	Less violent with cold water		Hydrogen and calcium hydroxide, $Ca(OH)_2$, which is only slightly soluble
Magnesium	Very slow with cold water, but vigorous with steam		Hydrogen and solid magnesium oxide, MgO
Zinc	Quite slow with steam		Hydrogen and solid zinc oxide, ZnO
Iron	Slow with steam		Hydrogen and solid iron oxide, Fe_3O_4
Copper Gold	No reaction	least reactive	——

Notice that the first three metals in the list produce hydroxides. The others produce oxides, if they react at all.
Now compare this table with the one on the opposite page. Is sodium more reactive than iron each time? Is iron more reactive than copper each time?

1 Describe how magnesium and iron each react with oxygen. Write balanced equations for the reactions.
2 Which is more reactive, copper or iron?
3 Which is more reactive, sodium or zinc?

4 What gas is always produced if a metal reacts with water?
5 Describe how magnesium reacts with steam. Write a balanced equation for the reaction.

12.3 Metals and reactivity (II)

3 The reaction of metals with dilute hydrochloric acid

Some metals react with dilute acid, some don't. When they do, they drive out or **displace** hydrogen from the acid. Hydrochloric acid has the formula HCl (*aq*). Compare these results:

Metal	Reaction with hydrochloric acid	Order of reactivity	Products
Magnesium	Vigorous	most reactive	Hydrogen and a solution of magnesium chloride, $MgCl_2$
Zinc	Quite slow		Hydrogen and a solution of zinc chloride, $ZnCl_2$
Iron	Slow		Hydrogen and a solution of iron(II) chloride, $FeCl_2$
Lead	Slow, and only if the acid is concentrated		Hydrogen and a solution of lead(II) chloride, $PbCl_2$
Copper Gold	No reaction, even with concentrated acid	least reactive	

Now compare this table with the last two tables. Is iron always more reactive than copper? Is magnesium always more reactive than iron?

4 Competition between metals for oxygen

The reactions with oxygen, water, and hydrochloric acid show that iron is more reactive than copper. Now look at this experiment.

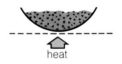
heat

This is a mixture of powdered iron and copper(II) oxide. On heating, the reaction starts.

The mixture glows, even after the Bunsen is removed. Iron(II) oxide and copper are formed.

Here iron and copper are competing for oxygen. Iron wins:

$$Fe (s) + CuO (s) \longrightarrow FeO (s) + Cu (s)$$
iron + copper(II) oxide $\longrightarrow$ iron(II) oxide + copper

By taking away the oxygen from copper, iron is acting as a **reducing agent** (page 90). Other metals behave in the same way when heated with the oxides of less reactive metals.

When a metal is heated with the oxide of a less reactive metal, it will remove the oxygen from it. The reaction is exothermic.

A competition reaction for repairing railway lines. Aluminium and iron(III) oxide are heated together to give molten iron, which runs into gaps between the rails. This is called the Thermit process.

5 Displacement of one metal by another

— iron nail

— blue solution of copper(II) sulphate

— coating of copper on nail
— pale green solution

This time, an iron nail is placed in copper(II) sulphate solution.

Soon copper appears on the nail. The solution turns green.

Here iron and copper are competing to be the compound in solution. Once again iron wins. It drives out or **displaces** copper from the copper(II) sulphate solution, just as it drove it from its oxide. Green iron(II) sulphate is formed:

$$Fe\,(s)\;+\;CuSO_4\,(aq)\;\longrightarrow\;FeSO_4\,(aq)\;+\;Cu\,(s)$$

iron + copper(II) sulphate $\longrightarrow$ iron(II) sulphate + copper
(blue) (green)

Other metals displace less reactive metals in the same way.
A metal will always displace a less reactive metal from solutions of its compounds.

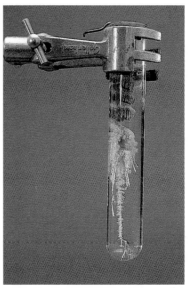

When copper wire is placed in silver nitrate solution, the solution turns blue and crystals of silver form on the wire. Which metal is more reactive?

6 Competition between metals and carbon

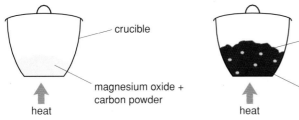

— crucible

— magnesium oxide + carbon powder

heat

— beads of molten lead

— lead(II) oxide + carbon powder

heat

Magnesium oxide is mixed with powdered carbon and heated in a crucible. No reaction!

When lead(II) oxide is used instead, silvery beads of molten lead appear.

In competition against magnesium for oxygen, carbon loses. But against lead, a less reactive metal, it wins. It **reduces** the lead(II) oxide to lead. In this way it proves itself more reactive than lead:

$$C\,(s)\;+\;PbO\,(s)\;\longrightarrow\;CO\,(g)\;+\;Pb\,(s)$$

carbon + lead(II) oxide $\longrightarrow$ carbon monoxide + lead

Carbon is more reactive than some metals. It will reduce their oxides to the metals.

Redox again!

- All the reactions shown in this unit are redox reactions.
- Use OIL RIG to check:
  Oxidation **I**s **L**oss of electrons
  Reduction **I**s **G**ain of electrons.
- In reactions 4, 5 and 6, metal ions gain electrons and become atoms. So they are reduced. The reducing agent is another metal, or carbon.

---

1 Describe how iron and lead react with hydrochloric acid. Write balanced equations for the reactions.
2 Write a rule for the reaction of a metal with:
   a oxides of other metals
   b solutions of compounds of other metals.
3 Will copper react with lead(II) oxide? Explain why.

4 What would you *see* when zinc is added to copper(II) sulphate solution? (Zinc sulphate is colourless.)
5 Explain how the Thermit process works.
6 Tin does not react with iron(II) oxide. But it reduces lead(II) oxide to lead. Arrange tin, iron, and lead in order of decreasing reactivity.

12.4 The reactivity series

What is the reactivity series?

In the last two Units we compared the reactions of different metals. You saw how some were always more reactive than others. In fact we can list them in order of reactivity. The list is called the **reactivity series**. Here it is:

Potassium, K	most reactive	
Sodium, Na		
Calcium, Ca		above this line,
Magnesium, Mg		metal oxides can't be
Aluminium, Al		reduced by carbon
Zinc, Zn	increasing	metals above this line
Iron, Fe	reactivity	react with acids,
Lead, Pb		displacing **hydrogen**
Copper, Cu		
Silver, Ag		
Gold, Au	least reactive	

A metal's position in the reactivity series will give you clues about its uses. Only unreactive metals are used to make coins.

Useful things to remember about the reactivity series

1 The more reactive the metal, the more it 'likes' to form compounds. So only copper, silver and gold are ever found as *elements* in Earth's crust. The other metals are *always* found as compounds.

2 When a metal reacts, it gives up electrons to form ions. **The more reactive the metal, the more easily it gives up electrons**.

3 The more reactive the metal, the more **stable** its compounds. *Stable* means difficult to break down. For example when you heat sodium nitrate you get sodium nitrite:

$$2NaNO_3 (s) \longrightarrow 2NaNO_2 (s) + O_2 (g)$$

But copper(II) nitrate breaks down further, to the oxide, giving off nitrogen dioxide:

$$2Cu(NO_3)_2 (s) \longrightarrow 2CuO (s) + 4NO_2 (g) + O_2 (g)$$

The table on the right shows the pattern for the nitrates and hydroxides.

4 The more reactive the metal, the more difficult it is to **extract** from its compounds (since the compounds are stable). For the most reactive metals you'll need the toughest method of extraction: electrolysis.

5 The less reactive metals have been known and used since ancient times, because they are easiest to extract. But aluminium was scarcely used before 1886, when it was first extracted by electrolysis. Look at the table on the right.

6 If you stand two metals in an electrolyte and join them up with a copper wire, you'll get a current! Electrons flow from the more reactive metal to the less reactive one. For more, see next page.

	Heating the nitrate gives ...	Heating the hydroxide gives ...
K	nitrite	no change
Na	nitrite	no change
Ca	oxide	oxide
Mg	oxide	oxide
Al	oxide	oxide
Zn	oxide	oxide
Fe	oxide	oxide
Pb	oxide	oxide
Cu	oxide	oxide
Ag	metal	metal

When it began to be widely used ...

Aluminium	1890
Zinc	1500
Iron	1400 BC
Lead	2000 BC
Copper	4000 BC

• Compare this list with the reactivity series. What do you notice?

The reactivity series and cells

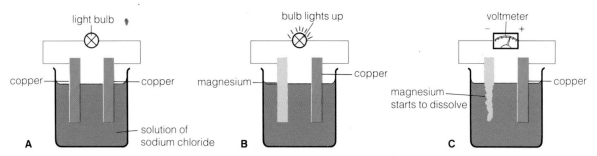

A

Two copper strips are wired up to a light bulb and placed in a solution of sodium chloride. The solution is an **electrolyte** – it *can* conduct electricity. But nothing happens.

B

If one copper strip is replaced by a *magnesium* strip, the bulb lights up. Electricity is being produced. Electrons are flowing through the wires even though there is no battery.

C

Here the bulb is replaced by a **voltmeter**. This measures the 'push' or **voltage** that makes the electrons flow. It is 2.7 volts. The needle shows the direction of the electron flow.

Where do the electrons come from? The answer is this: Magnesium can give up electrons more readily than copper. So magnesium atoms give up electrons and go into solution as ions. The electrons flow along the wire to the copper strip. This arrangement is called a **cell**. The magnesium strip is the **negative pole** of the cell. The copper strip is the **positive pole**.

A cell consists of two different metals and an electrolyte. In the cell, chemical energy produces electricity. The more reactive metal becomes the negative pole from which electrons flow.

The experiment can be repeated with other metals. As long as the strips are made of different metals, electrons will flow. But the voltage changes with the metals, as this table shows.

Metal strips	Volts
Copper and magnesium	2.70
Copper and iron	0.78
Lead and zinc	0.64
Lead and iron	0.32

The correct name for a torch battery is a 'dry cell'. It contains two different metals and an electrolyte. The electrolyte is a paste rather than a liquid, because a liquid would leak.

Of these metals, copper and magnesium are furthest apart in the reactivity series. They give the highest voltage. Lead and iron are closest. Look at the voltage they give.
The further apart the metals are in the reactivity series, the higher the voltage of the cell.
You can find out more about cells on page 296.

1 Why is sodium never found uncombined in nature?
2 Which will break down more easily on heating, magnesium nitrate or silver nitrate? Explain.
3 Why is magnesium more reactive than copper?
4 Explain why the bulb lights in experiment B above.
5 Why doesn't the bulb light in experiment A?

6 Will the bulb in B light if a sugar solution is used instead? Explain your answer.
7 Which pair of metals gives the highest voltage? Iron, zinc; copper, iron; silver, magnesium. Why?
8 In question 7, which metal in each pair becomes the negative pole?

Questions on Section 12

1 Read the following passage about the physical properties of metals.

Elements are divided into metals and nonmetals. All metals are <u>electrical conductors</u>. Many of them have a high <u>density</u> and they are usually <u>ductile</u> and <u>malleable</u>. All these properties influence the way the metals are used. Some metals are <u>sonorous</u> and this leads to special uses for them.

a Explain the meaning of the words underlined.

b Copper is ductile. How is this property useful in everyday life?

c Aluminium is hammered and bent to make large structures for use in ships and aeroplanes. What property is important in the shaping of this metal?

d Name one metal that has a *low* density.

e Some metals are cast into bells. What property must the chosen metals have?

f Add the correct word: *Metals are good conductors of* .... *and electricity.*

g Name one other physical property of metals and give two examples of how this property is useful.

2 a Write a short passage, like that in question 1, about the physical properties of nonmetals.

b Give one way in which the *chemical* properties of nonmetals and metals differ.

c Name *two* nonmetals which each show a metallic property, and describe the property.

3 From this list, choose elements to match the descriptions below:

aluminium   iron   argon   chlorine
iodine   gold   sulphur   carbon
bromine   copper potassium hydrogen

A a black solid which conducts electricity and cannot be melted in the school laboratory

B a yellow solid which burns with a blue flame

C a grey-black solid which forms a purple vapour

D a dense pale green gas which kills bacteria

E a pinkish brown solid which conducts electricity but does not react with dilute hydrochloric acid

F a red-brown liquid which reacts with iron to form a solid product

G a gas which does not react with anything and is used as the inert gas in tungsten light bulbs

H a solid which is attracted to a magnet and forms coloured compounds

I a fairly reactive solid which appears to be inactive because of a surface oxide layer

J a solid which reacts violently with water and burns with a lilac flame

K a flammable gas which is lighter than air and burns to form water

L a coloured solid which is very dense and chemically unreactive.

4 This shows metals in order of reactivity:

sodium (most reactive)
calcium
magnesium
zinc
iron
lead
copper
silver (least reactive)

a Which element is stored in oil?

b Which elements will react with *cold* water?

c Choose one metal that will not react with cold water but will react with steam. Draw a diagram of suitable apparatus to demonstrate this reaction. (You must show how the steam is produced.)

d Name the gas given off in **b** and **c**.

e Name another reagent that reacts with metals to give the same gas.

f Which of the metals will *not* react with oxygen when heated?

g How does iron react when heated in oxygen?

h How would you expect: **i** lead **ii** calcium to react when heated in oxygen? (Hint: look at the table on page 192 and the full reactivity series on page 196.)

5 Look again at the list of metals in question 4. Because zinc is more reactive than iron, it will remove the oxygen from iron(III) oxide, on heating.

a Write a word equation for the reaction.

b Decide whether these chemicals will react together, when heated:

  i magnesium + lead(II) oxide
  ii copper + lead(II) oxide
  iii magnesium + copper(II) oxide
  iv iron + magnesium oxide

c For those that react:

  i describe what you would *see*
  ii write a word equation
  iii write a balanced equation.

d What name is given to reactions of this type?

6 When magnesium powder is added to copper(II) sulphate solution, solid copper forms. This shows that magnesium is more reactive than copper.

a Write a word equation for the reaction.

b Use the list of metals in question 4 to decide whether these will react together:

  i iron + copper(II) sulphate solution
  ii silver + calcium nitrate solution
  iii zinc + lead(II) nitrate solution

c For those that react:

  i describe what you would *see*
  ii write a word equation
  iii write a balanced equation.

d What name is given to reactions of this type?

7 This table shows the densities of some metals:

Metal	Density (g/cm³)
aluminium	2.7
calcium	1.6
copper	8.9
gold	19.3
iron	7.9
lead	11.4
magnesium	1.7
sodium	0.97

a Arrange the metals in order of increasing density.
  i What is meant by *density*?
  ii Which metal is least dense?
  iii Which one is most dense?
  iv A block of metal has a volume of 20 cm³ and a mass of 158 g. Which metal is it?
b Arrange the metals in order of reactivity.
  i What is the density of the most reactive metal?
  ii What is the density of the least reactive metal?
  iii Does there appear to be a relationship between density and reactivity? If yes, what?
c Using low-density metals for vehicles saves money in terms of fuel costs and road and rail repairs. Explain why.
d Which of the low-density metals above is the most suitable for vehicles? Why? Give three reasons.
e What would first need to be done to this metal to make it strong enough for use in vehicles?
f Suggest some substances you could use for part **e**, and explain your choice.

8 Strips of copper foil and magnesium ribbon were cleaned with sandpaper and then connected as shown below. The bulb lit up.

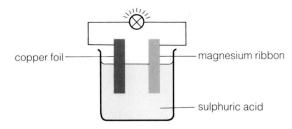

a Why were the metals cleaned with sandpaper?
b Name the electrolyte used.
c Explain why the bulb lit up.
d Which is the more reactive of the two metals?
e Which releases electrons into the circuit?
f In this arrangement, energy is being changed from one form to another. Explain.
g What is this type of arrangement called?
h Give two good reasons why this particular arangement could not be used commercially as a battery.

9 Chromium is a typical transition metal. It reacts with hydrochloric acid in a similar way to iron. It forms a green ion with the formula Cr^{3+}.
a i What would you see when chromium is added to dilute hydrochloric acid?
  ii Write a balanced equation for the reaction.
b i What would you see if powdered chromium was added to copper(II) sulphate solution?
  ii Write a balanced equation for this reaction.
c Chromium can be obtained from its oxide by heating with zinc metal.
  i Which is more reactive, chromium or zinc?
  ii Write a balanced equation for the reaction.
d Chromium can also be obtained at the cathode by electrolysis of a solution containing the Cr^{3+} ion. Write an ionic equation for the cathode reaction.
e Another transition metal, titanium, is extracted from its chloride by heating with sodium. What can you say about its reactivity?

10 Nickel is below iron but above lead in the reactivity series. Use the series to predict the reaction of nickel:
a on heating in air
b with cold water
c with concentrated hydrochloric acid
d on heating with copper(II) oxide
e when it is placed in a solution of lead nitrate.

11 Use the reactivity series to explain why:
a calcium is not used for household articles
b copper is used for water pipes
c powdered magnesium can be used in fireworks
d gold is an excellent material for filling teeth.

12 Write equations for the reactions that occur when:
a magnesium reacts with steam
b zinc reacts with dilute hydrochloric acid
c aluminium is heated with iron(III) oxide
d copper wire is placed in silver nitrate solution.

13

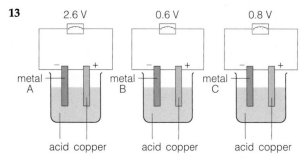

Look at the three cells above.
a How can you tell that the three unknown metals are all more reactive than copper?
b Place the metals in order, most reactive first.
c What voltage would be obtained if the cell was constructed using: **i** A and B? **ii** B and C?
d For each of the cells in **c** state which metal is the negative terminal.

13.1 Finding and mining metal ores

Mineral exploration

Exploring for minerals is detective work. It can cost hundreds of thousands of pounds, and the results are usually disappointing. That's why most exploration is carried out by big multinational companies who can stand the losses. But when a company does get lucky, it hopes for enough profit to make up for the disappointments.

Exploration is a long, slow process. It is carried out in a systematic way, and the company may decide to give up at one of several points. These are the main stages:

Reminder from Unit 10.3

- Rocks are a mixture of one or more minerals.
- A mineral is an inorganic crystalline element or compound that occurs naturally in Earth's crust.
- An ore is a mineral deposit that can be mined and refined at a profit.

choose another region

1 A region is chosen   Perhaps because it's similar to other regions where ores have been found, or has been mined at some time in the past.

continue? — no

yes

2 The preliminary study   This is desk work. The company studies geological maps, satellite images, and any other data it can get hold of.

continue? — no

yes

3 The survey   Samples of soil and sediment from streams are collected, and rocks measured for radioactivity and other properties, at regular points over the region. The information is analysed for clues about ores.

continue? — no

yes

4 Exploratory drilling   Costs really start to mount up here.
- First, the company must enter into legal and financial agreements with the owner of the land. (It could be the government of a foreign country.)
- Then it must move in heavy drilling equipment. In a remote region this can be very expensive.
- Plugs are drilled from the rocks and the minerals analysed.

continue? — no

yes

5 The feasibility study   By this stage the company has a clear picture of the mining potential of the region. Now it must decide whether mining will be **profitable**. It must take into account:
- the likely future prices for the metal, based on current prices and uses
- the quality of the ore
- any particular difficulties in mining or extracting it
- the availability of roads, railways, and ports for transporting it
- the cost of providing housing and other amenities for mine workers and their families
- possible political developments that may affect the mine's future
- government requirements in terms of taxation, pollution control, restoration of the land after mining, and so on
- the needs and objections of the local people

continue? — no

yes

6 The mine is constructed   Once it is completed, mining will begin.

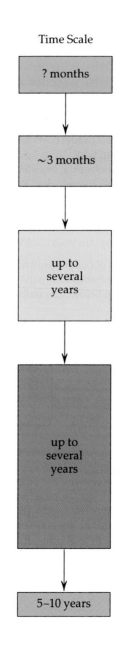

Time Scale

? months

~3 months

up to several years

up to several years

5–10 years

Environmental problems related to mining

Without mining we'd have no cars or computers or washing machines. But mining can also bring big environmental problems.

- Huge amounts of rock are dug up to get small amounts of ore. For example, 1000 tonnes of rock may produce just 5 tonnes of copper. This leaves huge scars on the landscape (if it's opencast mining) or huge holes underground.
- Holes from underground mining cause land to sink or **subside**. This is happening around the salt beds in Cheshire. The result is severe damage to buildings.
- Unwanted rock material gets heaped up in **tips**. These are unsightly. They can be unstable and therefore dangerous.
- Waste material gets washed into streams and rivers. The **sediment** that builds up chokes rivers and alters their routes.
- Everything for miles around may get covered with **dust**.
- Poisonous compounds (for example of lead, cadmium, and arsenic) are found in many ores. These can get washed into the soil and streams from mining processes. They kill fish and plant life and can end up in your food.

Governments are getting ever tougher with mining companies about damage to the environment. Sadly, in developing countries where much mining takes place, laws may be less strict.

Social and economic problems related to mining

- Native people may be forced off land to make way for mining, as happened to aboriginals in Australia in the 1960s and 70s.
- In many developing countries, governments can't afford to set up mines. They rely on deals with foreign mining companies. These companies have all the expertise, which gives them a lot of power. There's a risk that poor countries may be exploited.
- When mining is carried out in a developing country by a multinational company, the profits usually leave the country.
- Mining does create jobs. But local people may not have the skills that are needed – for example, to look after complex equipment or analyse data. Most of the best jobs may go to outsiders.
- Mining *may* bring other industries to an area. For example, factories to make metal goods. But very often this doesn't happen. The ore may even be sent abroad for processing.
- When a mine closes down – as it will eventually – local people may be left with nothing to show but a ruined environment. For example, when the tin mines on the Jos Plateau in Nigeria closed in the 1980s they left a disaster area.

When is a mineral worth mining?		
Metal	% in Earth's crust	Minimum grade of ore extracted (% of metal in ore)
Aluminium	8	25
Iron	5.8	25
Zinc	0.008	2.5
Copper	0.006	0.5
Tin	0.0002	0.2
Silver	0.000008	0.01
Gold	0.0000002	0.0008

The largest quarried hole in the world: Bingham Canyon copper mine in Utah is 800 metres deep. It is an opencast mine.

1 What does *mineral exploration* mean?
2 Why is it usually carried out by multinational companies?
3 **a** What is a *feasibility study*? **b** At what point will a mining company carry it out? **c** Why not sooner?
4 List six factors a mining company will take into account during a feasibility study.

5 **a** As the table above shows, it may be profitable to mine rock containing as little as 0.01% silver. How many tonnes of this rock will give 1 tonne of silver?
b What environmental problem(s) can mining huge quantities of rock cause?
c Name: **i** three *other* environmental problems
ii three economic problems related to mining.

13.2 Extracting metals from their ores

What is extraction?

Bauxite, an ore of aluminium. It is mainly aluminium oxide, Al_2O_3.

Haematite, an ore of iron. It is mainly iron(III) oxide, Fe_2O_3.

Zinc blende, an ore of zinc. It is mainly zinc sulphide, ZnS.

Chalcopyrite, an ore of copper. It is mainly copper(II) sulphide, CuS.

Above are four important ores obtained by mining. But mining is just the first step. The ore must then be decomposed to give the metal. This is called **extraction**. There are two stages.

1 Preparation The unwanted rock material in ore is called **gangue**. Ore is treated to remove as much gangue as possible. For example:
- It is crushed. Then it is dropped into water, where the fragments containing metal sink faster. Or jets of air are blown at it, when the lighter waste material gets carried to one side.
- A method called **froth flotation** is used with sulphide ores (like the last two above). The ore is powdered and made into a slurry with water. Then 'frothing' chemicals are added. Sulphides are attracted to these chemicals. When air is blown through the slurry, froth rises to the top of the tank carrying the metal sulphides with it. They are skimmed off and dried. The gangue sinks.

> **Getting rid of the gangue**
>
> Ore is usually separated from gangue at the mine.
> - Why do you think this is? (Think of the costs of transporting bulky materials.)
> - How does it add to environmental problems at mines?

2 The extraction process This depends on how reactive the metal is. The more reactive it is, the more stable its compounds are, and the more difficult and expensive to break down. Look at the table:

Metal	Method of extraction from ore		
Potassium Sodium Calcium Magnesium Aluminium	Electrolysis		
Zinc Iron Lead	Heating with carbon or carbon monoxide	method of extraction more powerful	method of extraction more expensive
Copper	Roasting in air		
Silver Gold	Occur naturally as elements		

metals more reactive   ores more difficult to decompose

Examples of the different methods of extraction

1 **Electrolysis.** This is used for extracting aluminium from aluminium oxide or **alumina**, obtained from bauxite. The alumina must be melted first (see page 207):

$$2Al_2O_3\,(l) \longrightarrow 4Al\,(l) + 3O_2\,(g)$$

alumina $\longrightarrow$ aluminium + oxygen

2 **Heating with carbon monoxide.** This is used for extracting iron from haematite in the blast furnace (page 208):

$$Fe_2O_3\,(s) + 3CO\,(g) \longrightarrow 2Fe\,(l) + 3CO_2\,(g)$$

iron(III) oxide + carbon monoxide $\longrightarrow$ iron + carbon dioxide

3 **Heating with carbon.** The zinc ore zinc blende is first converted into zinc oxide. Then this is heated in a furnace with coke:

$$ZnO\,(s) + C\,(s) \longrightarrow Zn\,(s) + CO\,(g)$$

zinc oxide + carbon $\longrightarrow$ zinc + carbon monoxide

4 **Roasting in air.** This is used for extracting copper from chalcopyrite:

$$CuS\,(s) + O_2\,(g) \longrightarrow Cu\,(l) + SO_2\,(g)$$

copper(II) sulphide + oxygen $\longrightarrow$ copper + sulphur dioxide

Extraction means reduction!

Note that during the extraction of a metal, *the ore is reduced*. This is obvious in examples 1–3 above, where oxygen is removed from the metal oxide. It is also obvious in example 4 if you think about the wider definition of reduction: *a gain in electrons*. The copper ion in copper sulphide gains electrons. It is reduced to copper:

$$Cu^{2+} + 2e^- \longrightarrow Cu$$

Titanium: an unusual reduction

Not all ores are reduced by the methods shown in the table. It's a case of finding the cheapest method that works and is safe! For example, titanium occurs as the ore **rutile**, which is titanium dioxide. Rutile is first converted into titanium choride, $TiCl_4$. This is then reduced by heating with sodium or magnesium:

$$TiCl_4\,(l) + 4Na\,(l) \longrightarrow Ti\,(s) + 4NaCl\,(l)$$

titanium(IV) chloride + sodium $\longrightarrow$ titanium + sodium chloride

Sodium and magnesium are used because they are more reactive than titanium. The heating is carried out in an inert atmosphere of argon. Why?

Pollution alert!

Extraction produces many substances that *could* cause pollution. For example:

- sulphur dioxide from the reduction of sulphide ores. It causes acid rain.
- harmful metal dust.
- poisonous compounds (for example of lead, cadmium, and arsenic) which occur in ores such as chalcopyrite.

Care must be taken in designing the plant to make sure they can't escape!

Useful by-products of extraction

- Sulphur dioxide produced during the extraction of sulphide ores is turned into sulphuric acid.
- Many metal ores contain tiny amounts of gold, silver, and other precious metals. These are recovered and sold.

---

1 What does *extraction* mean?
2 Before extraction, the gangue is removed. Why?
3 During extraction the ore is *reduced*. Explain.

4 Why is iron *not* extracted by electrolysis?
5 In the extraction of titanium: **a** which is more reactive, sodium or titanium? **b** why is argon used?

13.3 Making use of metals

Pure metals and alloys

The way a metal is used depends on its **properties**:

Pure aluminium can be rolled into very thin sheets, which are quite strong but easily cut. So it is used for milk bottle tops and cooking foil.

Pure lead is soft, and bends easily without being heated. It also resists corrosion. So it is used to seal off brickwork around chimneys.

Pure copper is easily drawn into wires, and is an excellent conductor of electricity. So it is used for electrical wiring around the home.

Sometimes a metal is most useful when it is pure. For example, copper is not nearly such a good conductor when it contains impurities.
But many metals are more useful when they are *not* pure. Iron is the most widely-used metal of all, and it is almost never used pure:

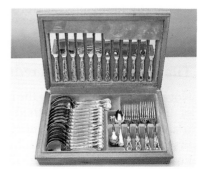

Pure iron is no good for building things, because it is too soft and stretches easily, as you can see in the photo above. Besides, it rusts easily too.

But when a little carbon (0.5%) is mixed with it, the result is **mild steel**. This is hard and strong. It is used for buildings, bridges, ships and car bodies.

When nickel and chromium are mixed with iron, the result is **stainless steel**. This is hard and rustproof. It is used for car parts, kitchen sinks, and cutlery.

You can see that the properties of the iron have been changed by mixing other substances with it.
The properties of any metal can be changed by mixing other substances with it. The mixtures are called alloys.
The added substances are usually metals, but sometimes nonmetals like carbon or silicon. An alloy is usually made by melting the main metal and then dissolving the other substances in it. (See page 54!)
Turning a metal into an alloy increases its range of uses.

Uses of pure metals

This table summarizes some uses of pure metals:

Metal	Uses	Properties that make it suitable
Sodium	A coolant in nuclear reactors Extraction of titanium	Conducts heat well. Melts at only 98 °C, so the hot metal will flow along pipes. Is more reactive than titanium and melts easily.
Aluminium	Overhead electricity cables (with a steel core for strength) CDs and CD-ROMs	A good conductor of electricity (not as good as copper, but cheaper and much lighter); resists corrosion. Provides a cheap reflective coating.
Zinc	Coating iron, to give **galvanized** iron	Protects the iron from rusting.
Tin	Coating steel cans or 'tins'	Unreactive and non-toxic. Protects the steel from rusting.
Nickel	Electroplating steel	Resists corrosion, sticks well to steel, shiny and attractive to look at.
Titanium	Teeth implants and replacement hip joints Tailpipes for aircraft	Light, strong, resists corrosion, non-toxic, and ductile so can be easily shaped. Light, strong, resists corrosion, ductile.

Uses of alloys

There are thousands of different alloys. Here are just a few!

Alloy	Made from	Special properties	Uses
Cupronickel	75% copper 25% nickel	Hard-wearing, attractive silver colour	'Silver' coins
Stainless steel	70% iron 20% chromium 10% nickel	Does not rust	Car parts, kitchen sinks, cutlery, tanks and pipes in chemical factories
Manganese steel (Hadfield steel)	85% iron 13.8% manganese 1.2% carbon	Very hard	Springs
A titanium alloy	92.5% titanium 5% aluminium 2.5% tin	High strength at high temperatures	Jet engine components
Brass	70% copper 30% zinc	Harder than copper, does not corrode	Musical instruments
Bronze	95% copper 5% tin	Harder than brass, does not corrode, sonorous	Statues, ornaments, church bells
Solder	70% tin 30% lead	Low melting point	Joining wires and pipes

1 Why is iron more useful when it is mixed with a little carbon?
2 What are *alloys*? How are they made?
3 Explain why tin is used to coat food tins.

4 Name an alloy that:
  a has a low melting point b never rusts.
5 Which metals are used to make:
  a stainless steel? b brass? c bronze?

13.4 More about aluminium

From rocks to rockets

Aluminium is the most abundant metal in Earth's crust.
Its main ore is **bauxite**, which is aluminium oxide mixed with impurities like sand and iron oxide. The impurities make it reddish brown.

These are the steps in obtaining aluminium:

1 When a company discovers bauxite, it does a feasibility study to see if mining will be profitable. If yes, the mine is constructed and mining begins.

2 Bauxite usually lies near the surface, so it is easy to dig up. This is a bauxite mine in Jamaica. Everything gets coated with red-brown bauxite dust.

3 From the mine, the ore is taken to a bauxite plant, where it is treated to remove the impurities. The result is white **aluminium oxide**, or **alumina**.

4 The alumina is taken to another plant for electrolysis. Much of the Jamaican alumina is shipped to plants like this one, in Canada or the USA ...

5 ... where electricity is cheaper. There it is electrolysed to give aluminium. The metal is made into sheets and blocks, and sold to other industries.

6 It is used to make beer cans, cooking foil, saucepans, racing bikes, TV aerials, aeroplanes, ships, and so on. It is usually made into an alloy first.

A closer look at the electrolysis

In step 5 above you saw that aluminium is obtained from alumina by electrolysis. This is an expensive method of extraction. (Why?)
The electrolysis is carried out in a huge steel tank. The tank is lined with graphite, which acts as the cathode. Huge blocks of graphite hang in the middle of the tank, and act as anodes.

Pure alumina melts at 2045 °C. It would be expensive, and dangerous, to keep the tank at that temperature. Instead, the alumina is dissolved in molten **cryolite** for the electrolysis. (Cryolite is another aluminium compound, with a much lower melting point.)

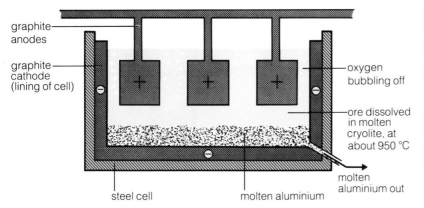

The steel tanks in which electrolysis takes place

When the alumina dissolves, its aluminium ions and oxide ions become free to move.

At the cathode   The aluminium ions receive electrons:

$$4Al^{3+} + 12e^- \longrightarrow 4Al$$

The aluminium atoms collect together, and drop to the bottom of the cell as molten metal. This is run off at intervals.

At the anodes   The oxygen ions give up electrons:

$$6O^{2-} \longrightarrow 3O_2 + 12e^-$$

Oxygen gas bubbles off. But unfortunately it attacks the graphite anodes and eats them away, so they must be replaced from time to time.

The overall equation

The overall equation for the extraction of aluminium is:

$$2Al_2O_3\,(l) \longrightarrow 4Al\,(l) + 3O_2\,(g)$$

The alumina is *reduced*.
Oxygen is removed from it.

Some properties of aluminium

1 Aluminium is a bluish-silver, shiny metal.
2 Unlike most metals, it has a low density – it is 'light'.
3 It is a good conductor of heat and electricity.
4 It is malleable and ductile.
5 It is non-toxic.
6 It is not very strong when pure, but it can be made stronger by mixing it with other metals to form alloys (page 204).

These properties lead to the wide range of uses for aluminium, given in step 6 on the opposite page.

This underground train is made of aluminium strengthened with small amounts of other metals.

---

1 Copy and complete: The chief ore of aluminium is called . . . . It is first purified to give . . . . or . . . . , which has the formula . . . . Then this is . . . . to give aluminium.
2 Draw the cell for the electrolysis of aluminium.

3 Why do the aluminium ions move to the cathode?
4 What happens at the cathode?
5 Why must the anodes be replaced from time to time?
6 List six uses of aluminium. For each, say what properties of the metal make it suitable.

13.5 More about iron

The extraction of iron

Iron is the second most abundant metal in Earth's crust. To extract it, three substances are needed:

1 **Iron ore**. The chief ore of iron is called **haematite**. It is mainly iron oxide, Fe_2O_3, mixed with sand.
2 **Limestone**. This is mainly calcium carbonate, $CaCO_3$.
3 **Coke**. This is made from coal, and is almost pure carbon.

These three substances are mixed together to give a mixture called **charge**. The charge is heated in a tall oven called a **blast furnace**. Several reactions take place, and finally liquid iron is produced.

In the blast furnace A blast furnace is like a giant chimney, at least 30 m tall. It is made of steel, and lined with fireproof bricks. The charge is added through the top. Hot air is blasted through the bottom, making the charge glow white-hot. These reactions are:

1 The coke reacts with oxygen in the air, giving **carbon dioxide**:

$$C\ (s) + O_2\ (g) \longrightarrow CO_2\ (g)$$

2 The limestone decomposes to **calcium oxide** and **carbon dioxide**:

$$CaCO_3\ (s) \longrightarrow CaO\ (s) + CO_2\ (g)$$

3 The carbon dioxide reacts with more coke, giving **carbon monoxide**:

$$C\ (s) + CO_2\ (g) \longrightarrow 2CO\ (g)$$

4 This reacts with iron oxide in the ore, giving liquid **iron** which trickles to the bottom of the furnace:

$$Fe_2O_3\ (s) + 3CO\ (g) \longrightarrow 2Fe\ (l) + 3CO_2\ (g)$$

5 The calcium oxide from step 2 (a basic oxide) reacts with silica (an acidic oxide) in the sand of the ore, to form **calcium silicate** or **slag**:

$$CaO\ (s) + SiO_2\ (s) \longrightarrow CaSiO_3\ (l)$$

The slag runs down the furnace and floats on the iron.

The slag and iron are drained from the bottom of the furnace. When the slag solidifies it is sold, mostly for road building. Only *some* of the iron is left to solidify, in moulds. The rest is taken away while still hot, and turned into steel.

A stockpile of iron ore

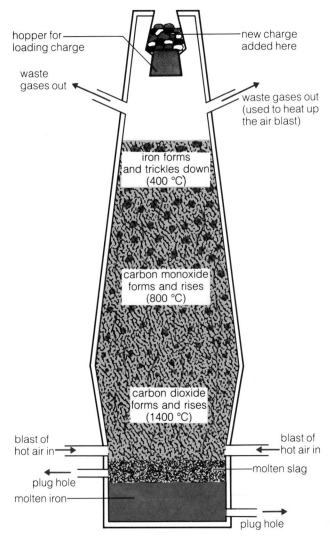

hopper for loading charge
new charge added here
waste gases out
waste gases out (used to heat up the air blast)
iron forms and trickles down (400 °C)
carbon monoxide forms and rises (800 °C)
carbon dioxide forms and rises (1400 °C)
blast of hot air in
blast of hot air in
plug hole
molten slag
molten iron
plug hole

Cast iron

The iron from the blast furnace usually contains a lot of carbon (up to 4%) as well as other impurities. You saw that some of the iron is allowed to solidify in moulds or **casts**. It is called **cast iron**. The carbon makes it very hard, but also brittle – it snaps under strain. So these days it is only used for things like gas cylinders, manhole covers, railings, and storage tanks, which are not likely to get bent during use.

The famous bridge at Ironbridge is made of cast iron. Opened in 1781, it was the first iron bridge in the world. Modern bridges are made of steel.

Steel

Most of the iron from the blast furnace is turned into alloys called **steels**. This is how they are made:

1 **First, unwanted impurities are removed from the iron.** This is done in an **oxygen furnace**. The molten metal is poured into the furnace. A lot of scrap iron is added too. Then some calcium oxide is added, and a jet of oxygen turned on. The calcium oxide reacts with the acidic impurities, forming a slag that can be skimmed off. Oxygen reacts with the others, and they burn away. For some steels, *all* the impurities are removed. But many steels are just iron plus a small amount of carbon - enough to make the metal hard, but not brittle. So the carbon content has to be checked continually. When it is correct, the oxygen is turned off.
2 **Then, other elements may be added.** As you saw on page 204, different elements affect iron in different ways. The added elements are carefully measured out, to give steels of exactly the required properties.

Molten iron being poured into an oxygen furnace

There are thousands of different steels. Below are just three of them:

Name	Contains			Special property	Uses
Mild steel	99.5% Fe,	0.5% C		Hard but easily worked	Buildings, car bodies, machinery
Hard steel	99% Fe,	1% C		Very hard	Blades for cutting tools
Duriron	84% Fe,	1% C,	15% Si	Not affected by acid	Tanks and pipes in chemical factories

1 Name the raw materials for extracting iron.
2 Write an equation for the reaction that gives iron.
3 The calcium carbonate in the blast furnace helps to purify the iron. Explain how, with equations.
4 Name a waste gas from the furnace.

5 The slag and waste gas are both useful. How?
6 What is cast iron? Why is it brittle?
7 Explain how mild steel is made.
8 What makes *hard steel* harder than mild steel?
9 Explain why most iron is turned into steel.

13.6 Corrosion

When a metal is attacked by air, water or other substances in its surroundings, the metal is said to **corrode**:

Damp air quickly attacks potassium, turning it into a pool of potassium hydroxide solution. (That is why potassium is stored under oil!)

Iron also corrodes in damp air. So do most steels. The result is **rust** and the process is called **rusting**. It happens especially quickly near sea water.

But gold is unreactive and never corrodes. This gold mask of King Tutankhamun was buried in his tomb for over 3000 years, and still looks as good as new.

In general, the more reactive a metal is, the more readily it corrodes.

The corrosion of iron and steel

The corrosion of iron and steel is called **rusting**. The iron is **oxidized** to give a compound with the formula $Fe_2O_3.H_2O$. The reaction needs both air and water, as these tests show.

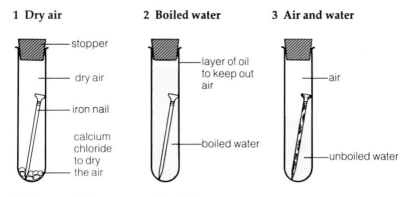

1 Dry air

- stopper
- dry air
- iron nail
- calcium chloride to dry the air

2 Boiled water

- layer of oil to keep out air
- boiled water

3 Air and water

- air
- unboiled water

Nails 1 and 2 do not rust. Nail 3 does rust.

How to stop rust. When iron is made into stainless steel, it does *not* rust. But stainless steel is too expensive to use in large amounts. So other methods of rust prevention are needed.

Below are some of the methods. They mostly involve *coating the metal with something, to keep out air and water*:

1 **Paint.** Steel bridges and railings are usually painted. Paints that contain lead or zinc are mostly used, because these are especially good at preventing rust. For example, 'red lead' paints contain an oxide of lead, Pb_3O_4.
2 **Grease.** Tools, and machine parts, are coated with grease or oil.

Steel girders for building bridges are first sprayed with special paint.

3 **Plastic.** Steel is coated with plastic for use in garden chairs, bicycle baskets, and dish racks. Plastic is cheap and can be made to look attractive.

4 **Galvanizing.** Iron for sheds and dustbins is usually coated with zinc. It is called **galvanized iron**.

5 **Tin plating.** Baked beans come in 'tins' which are made from steel coated on both sides with a fine layer of tin. Tin is used because it is unreactive and non-toxic. It is deposited on the steel by electrolysis, in a process called **tin plating**.

6 **Chromium plating.** Chromium is used to coat steel with a shiny protective layer, for example on car bumpers. Like tin, the chromium is deposited by electrolysis.

7 **Sacrificial protection.** Magnesium is more reactive than iron. When a bar of magnesium is attached to the side of a steel ship, or oil rig, or underwater pipe, it corrodes instead of the steel. When it is nearly eaten away it can be replaced by a fresh bar. This is called **sacrificial protection**, because the magnesium is sacrificed to protect the steel. Zinc can be used in the same way.

'Tins' are steel cans plated with tin.

Does aluminium corrode?

Aluminium is more reactive than iron, so you might expect it to corrode faster in damp air. In fact, clean aluminium starts corroding immediately, but the reaction quickly stops:

rust flakes

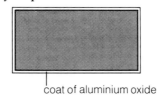

coat of aluminium oxide

When iron corrodes, rust forms in tiny flakes. Damp air can get past the flakes, to attack the metal below. In time, it rusts all the way through.

But when aluminium reacts with air, an even coat of aluminium oxide forms. This seals the metal surface and protects it from further attack.

The layer of aluminium oxide can be made thicker by electrolysis, to give even more protection. This process is called **anodizing**. The aluminium is used as the anode of a cell in which dilute sulphuric acid is electrolysed. Oxygen forms at the anode and reacts with the aluminium, so the layer of oxide grows. (See page 111).

Anodized aluminium is used for cookers, fridges, cooking utensils, saucepans, window frames, wall panels, and so on. The oxide layer can easily be dyed to a bright colour.

Anodized aluminium is used for door and window frames.

1 What is *corrosion*?
2 What two substances cause rusting?
3 Steel that is tin-plated does not rust. Why?
4 In one method of rust prevention, the steel is not coated with anything. Which method is this?

5 Explain why magnesium can prevent the rusting of steel. Why would zinc do instead?
6 Why doesn't aluminium corrode right through?
7 How can a layer of aluminium oxide be made thicker? What is the process called?

Questions on Section 13

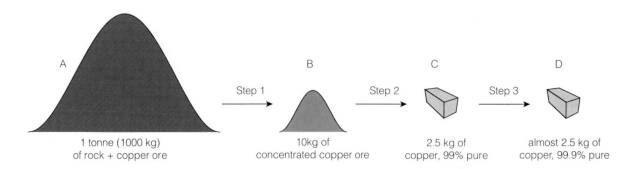

A		B		C		D

1 tonne (1000 kg) of rock + copper ore — Step 1 → 10kg of concentrated copper ore — Step 2 → 2.5 kg of copper, 99% pure — Step 3 → almost 2.5 kg of copper, 99.9% pure

1 The diagram above shows stages in obtaining copper from a low-grade ore. This ore contains copper(II) sulphide, CuS. It may also contain small amounts of silver, gold, platinum, iron, zinc, cadmium, and arsenic.
   a What is an *ore*?
   b What is a *low-grade* ore?
   c What name is given to the waste rock in an ore?
   d i What process is used in step 1 to concentrate the sulphide ore? Explain how it works.
   ii The waste material from this process is in the form of mud. Explain why.
   e i What process is carried out in step 2, to extract the copper from the ore?
   ii Write an equation for the reaction.
   iii What is this type of reaction called?
   iv The copper is 99% pure. Suggest some impurities it may contain.
   f i What process is carried out at step 3 to purify the metal? (Hint: try page 110.)
   ii What will the main cost in this process be?
   iii As well as pure copper, this process may produce other valuable substances. Explain.
   g List all the environmental problems that may arise in going from A to D.

2 Read this passage.

In 1981 a copper mine opened at Ok Tedi in Papua New Guinea. Gold is found with the copper at this mine. It produces 200 000 tonnes of copper a year and 350 000 ounces of gold. The joint owners are BHP, an Australian company (52%), Metall Mining Corporation, a Canadian company (18%), and the government of Papua New Guinea (30%).

Before the mine opened, around 35 000 villagers lived in small clusters of bush huts around the area. They lived by fishing, hunting, and growing crops in their food gardens. There were no schools or health care, and the average life span was not much more than 30. Many people died from malaria.

Since then, the mining company has spent around £170 million on building roads and other infrastructure. Schools and health centres have been opened and water tanks and shower blocks built. The average life span has risen to around 50. Around £10 million a year is paid to the villages in wages and lease payments. Now people buy their food in stores and cook on cookers in their kitchens instead of outside on open fires. Some even fly to Port Moresby, a 2 hour flight away, to go shopping.

Every day around 80 000 tonnes of waste are washed from the mine into the Ok Tedi river. This has killed fish and flooded food gardens. The rain forest is dying as tree roots get submerged. The villages nearest the mine suffer from clouds of dust and fumes. The men, who in the past would have hunted all day, now sit idly smoking and chewing betelnut. No one fishes or swims any more.

In 1994, landowners living downstream from the mine lodged a writ in an Australian court claiming nearly £2 billion compensation from the mining company for damage to health and the environment. But in 1995 the Papua New Guinea government made it illegal for its people to claim for damages. Arguments about the case continue.

a Write a list of all the benefits the mine has brought to the local people. Divide them into social and economic benefits.
b Now write a list of the problems it has brought. Divide them into suitable groups.
c What problems are likely to arise in the future when the mine closes down? Suggest as many as you can.
d Does the mine benefit the rest of Papua New Guinea? Explain.
e Why did the mining company spend so much on infrastructure?
f Who gains most of the profits from the mine?
g Why might the government of Papua New Guinea have decided to make claims for damages illegal?
h On balance, based on the information above, do you feel the local people have a right to compensation?

3 List 10 goods and 10 services we would not enjoy if there was no mining.

4 Some information about the extraction of four different metals is shown below.

Metal	Formula of main ore	Method of extraction
Iron	Fe_2O_3	Heating with carbon
Aluminium	$Al_2O_3.2H_2O$	Electrolysis
Copper	Cu_2S	Roasting the ore
Sodium	$NaCl$	Electrolysis

a Give the chemical name of each ore.
b Arrange the four metals in order of reactivity.
c How are the more reactive metals extracted?
d i How is the least reactive metal extracted?
   ii Why can't this method be used for the more reactive metals?
e Aluminium is a lot more expensive than iron even though both ores are relatively cheap. Why is this?
f Which of the methods would you use to extract:
   i potassium? ii lead? iii magnesium?
(Hint: look at the reactivity series on page 196.)
g Gold is a metal found native in the Earth's crust. Explain what *native* means.
h Where should gold go, in your list for b?
i Name another metal which occurs native.

5 The metals zinc and lead are obtained from ores which are compounds containing the metal and sulphur in the molar ratio 1:1.
a Name the compounds in these ores.
b Write down their formulae.

In the extraction of the metal, the compounds are roasted in air to obtain the oxide of the metal. Then the oxide is heated with coke to obtain the metal and carbon monoxide.
c i Write equations for the roasting of the ores.
   ii What type of reaction is this?
d i Write equations for the reactions with coke.
   ii Which substances are reduced?
e Which gives the greater mass of metal, 1000 g of lead ore or 1000 g of zinc ore? Explain your answer.

6 Explain why the following metals are suitable for the given uses. (There should be more than one reason in each case.)
a Aluminium for window frames.
b Iron for bridges.
c Copper for electrical wiring.
d Lead for roofing.
e Zinc for coating steel.
f Titanium for replacement hip joints.
g Platinum for jewellery.
h Tin for coating food cans.
i Chromium for coating car bumpers.

7 Many metals are more useful when mixed with other elements than when they are pure.
a What name is given to the mixtures?
b What metals are found in these mixtures?
brass solder stainless steel manganese steel
c Describe the useful properties of each mixture.

8 Bauxite is the hydrated oxide of a certain metal. The metal is extracted from it by electrolysis.
a Which metal is extracted from bauxite?
b The compound cryolite is also needed for the extraction. Why is this?
c What are the electrodes made of?
d i At which electrode is the metal obtained?
   ii Write an equation for the reaction that takes place at this electrode.
e i What product is released at the other electrode?
   ii This product reacts with the electrode itself. What problem does that cause?
f Give three uses of the metal obtained.
g To improve its resistance to corrosion, the metal is often anodized. How is this carried out? What happens to the surface of the metal?

9 a Draw a diagram of the blast furnace.
   Show clearly on your diagram:
   i   where air is 'blasted' into the furnace
   ii where the molten iron is removed
   iii where the second liquid is removed.
b i Name the three raw materials added at the top.
   ii What is the purpose of each material?
c i Name the second liquid that is removed from the bottom of the furnace.
   ii When it solidifies, does it have any uses? If so, name one.
d i Name a waste gas from the top of the furnace.
   ii Does this gas have a use? If so, what?
e Write an equation for the reaction that produces the iron.
f Most of the iron that is obtained from the blast furnace is used to make steel. What element, other than iron, is present in most steel?

10 *aluminium gold iron tin magnesium*
   mild steel calcium stainless steel
a In the above list of metals and alloys, only four are resistant to corrosion. Which are they?
b Explain why each is resistant to corrosion.
c Which of the other metals or alloys will corrode most quickly? Explain your answer.
d Which metal in the list is the most widely used?
e Give another name for the corrosion of this metal and write an equation for the reaction.
f List five methods used to prevent this reaction and explain how each method works.

14.1 Nitrogen

The nitrogen cycle

Nitrogen is a colourless, odourless, unreactive gas – and you couldn't survive without it. It continually circulates between the air, the soil, and living things in a set of processes called the **nitrogen cycle**.

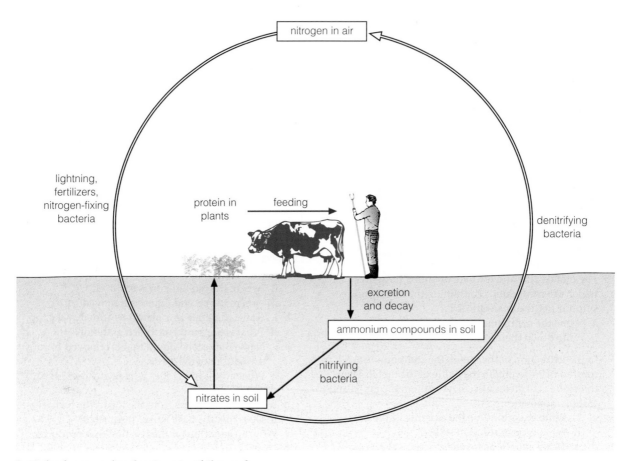

Let's look more closely at parts of the cycle:

1 The conversion of nitrogen from air to nitrates in the soil
This happens in three different ways:

In the heat of a **lightning flash**, nitrogen and oxygen react to form nitrogen oxides. These dissolve to form acid rain, which reacts with compounds in soil to make nitrates.

Some **bacteria** can also convert nitrogen to nitrates. They are **nitrogen-fixing**. They live in soil, or in swellings called **root nodules** on the roots of pea, bean, and clover plants.

Humans turn nitrogen into **fertilizers** in factories. These are compounds such as ammonium nitrate and ammonium phosphate. Farmers spread them on soil.

2 From nitrates to proteins   Plants take in nitrates through their roots and convert them to proteins. (But pea, bean, and clover plants don't need to do this since they can fix nitrogen from the air directly.)

3 From plants to animals   Animals eat the plants, and in turn are eaten by other animals. The proteins are broken down into amino acids during digestion, and built up again to form new proteins. In this way proteins are passed along the food chain. You need them to build up your body tissues.

4 Excretion and decay   Animals excrete ammonium compounds. Bacteria also feed on the remains of dead plants and animals, producing ammonium compounds. These are converted to nitrates by **nitrifying bacteria**.

5 From nitrates back to nitrogen   Bacteria called **denitrifying bacteria** complete the cycle. They live in heavy, wet soils. They break down nitrates, releasing nitrogen back into the air.

three shared pairs of electrons

N   N

The bonding in nitrogen. Since three pairs of electrons are shared, the bond is a triple bond. It can be as shown as $N \equiv N$.

The properties of nitrogen

1 It is a colourless gas with no smell.
2 It is only slightly soluble in water.
3 It is very unreactive compared with oxygen.
4 However, it will react with hydrogen to form **ammonia**, with the help of high pressure, moderate temperature, and a catalyst. The reaction is reversible:

$$N_2(g) + 3H_2(g) \rightleftharpoons 2NH_3(g)$$

This reaction is the first step in making nitrogen fertilizers. You can find out more about it on page 218.
5 It will also combine with oxygen at high temperatures to form oxides of nitrogen. As you saw opposite, this happens naturally during thunderstorms. It also happens inside car engines and power station furnaces. The result is air pollution and acid rain (page 178).

Uses of nitrogen

Its main use is in the manufacture of ammonia, nitric acid, and fertilizers. But some nitrogen is separated from air by fractional distillation (page 176) and used:
- to flush out food packaging and keep food fresh (since it is inert)
- to quick-freeze food, freeze liquid in damaged pipes, and shrink-fit machine parts (since liquid nitrogen is very cold).

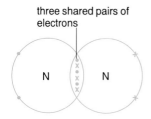

Exhaust gases contain oxides of nitrogen as well as many other unwelcome pollutants.

1 What is the *nitrogen cycle*?
2 List three ways in which nitrogen from the air is made available to plants.
3 Why does the nitrogen in air not get used up?

4 a Draw a molecule of nitrogen.
  b The bond in it is called a *triple* bond. Why?
  c Is the bond weak or strong? Explain your answer.
5 List five nitrogen compounds and their formulae.

14.2 Hydrogen

Hydrogen is the lightest of all the elements. It is so light that there is none in the air: it has escaped into the outer atmosphere. But overall, it is the most common element in the Universe. (See Unit 10.1.)

How hydrogen is made in industry

It is made from North Sea gas, which is mainly methane.
The North Sea gas is mixed with steam and passed over a catalyst:

$$CH_4\ (g) + H_2O\ (g) \overset{catalyst}{\rightleftharpoons} CO\ (g) + 3H_2\ (g)$$

This is a reversible reaction. A high temperature and pressure are needed to give a good yield. Then the carbon monoxide is converted to carbon dioxide in another reversible reaction:

$$CO\ (g) + H_2O\ (g) \overset{catalyst}{\rightleftharpoons} CO_2\ (g) + H_2\ (g)$$

The carbon dioxide is removed by 'scrubbing' the gases with an alkali. This leaves hydrogen.

The properties of hydrogen

1 It is the lightest of all gases. It is about 20 times lighter than air.
2 It is colourless and has no smell.
3 It is almost insoluble in water.
4 It combines with oxygen to form water. A mixture of the two gases will explode when lit. So take care! The reaction is:

$$2H_2\ (g) + O_2\ (g) \longrightarrow 2H_2O\ (l)$$

The reaction gives out so much energy that it is used to fuel space rockets where the reaction takes place in **fuel cells**. (See page 297.)
5 Hydrogen acts as a reducing agent, by removing oxygen. For example, copper(II) oxide is reduced to copper, by heating it in a stream of hydrogen. The hydrogen is oxidized to water:

$$CuO\ (s) + H_2\ (g) \longrightarrow Cu\ (s) + H_2O\ (l)$$

Uses of hydrogen

- It is used to 'harden' vegetable oils to make margarine (page 261).
- It is used as a fuel for space rockets.

A promising fuel for the future?

Hydrogen may become an important fuel for cars and homes in the future, as we run out of oil and gas. It has two big advantages:
- Its reaction with oxygen produces just water. No pollution!
- It is a 'renewable' resource: it can be made by electrolysing acidified water (page 109). As cheaper sources of electricity for electrolysis are developed, this may become an attractive option.

Sunshine, thanks to hydrogen

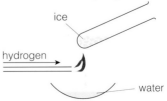

A jet of hydrogen burns in air or oxygen with a blue flame. It can be condensed to water on a cold surface.

---

1 Describe, with equations, how hydrogen is made:
  a in industry   b in the lab (Try page 292!)
2 How does hydrogen react with copper(II) oxide?

3 a Explain why hydrogen can be used as a fuel.
  b Give two advantages it has as a fuel.
  c What disadvantages does it have?

14.3 Ammonia

Ammonia is a gas with the formula NH_3.

In industry It is made in industry by reacting nitrogen with hydrogen, as you'll see on the next page.

In the lab It is made in the lab by heating any ammonium compound with a strong base. The base drives out or *displaces* ammonia from the compound. For example:

$$2NH_4Cl\,(s) + Ca(OH)_2\,(s) \longrightarrow CaCl_2\,(s) + 2H_2O\,(l) + 2NH_3\,(g)$$

ammonium   calcium   calcium   ammonia
chloride   hydroxide   chloride

This reaction can be used as a test to identify an ammonium compound. If a compound gives off ammonia when heated with a base it must be an ammonium compound.

The properties of ammonia

1 It is a colourless gas with a strong, choking smell.
2 It is less dense than air.
3 It is easily liquefied by cooling to −33 °C or by compressing. So it is easy to transport in tanks and cylinders.
4 It reacts with hydrogen chloride gas to form a white smoke. The smoke consists of tiny particles of solid ammonium chloride:

$$NH_3\,(g) + HCl\,(g) \longrightarrow NH_4Cl\,(s)$$

This reaction can be used to test whether a gas is ammonia.
5 It is very soluble in water. (It shows the fountain effect.)
6 The solution turns red litmus blue – it is **alkaline**. That means it contains hydroxide ions. Some of the ammonia has reacted with water to form ammonium ions and hydroxide ions:

$$NH_3\,(aq) + H_2O\,(l) \rightleftharpoons NH_4{}^+\,(aq) + OH^-\,(aq)$$

As only some of its molecules form ions, ammonia is a **weak** alkali.
7 Since ammonia solution is alkaline, it reacts with acids to form salts. For example, with nitric acid it forms ammonium nitrate. Ammonium nitrate is an important fertilizer.

$$NH_3\,(aq) + HNO_3\,(aq) \longrightarrow NH_4NO_3\,(aq)$$

8 It can be used to test for copper(II) compounds:
(i) If a solution contains a copper(II) compound, it will react with ammonia solution to give a blue precipitate of copper(II) hydroxide.
(ii) When more ammonia solution is added, the precipitate will dissolve, giving a deep blue solution.
The blue solution proves that a copper(II) compound was present. The deep blue colour is due to ions with the formula $[Cu(NH_3)_4]^{2+}$.

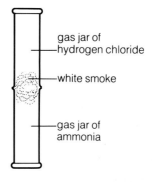

gas jar of hydrogen chloride

white smoke

gas jar of ammonia

What is the chemical name for the white smoke?

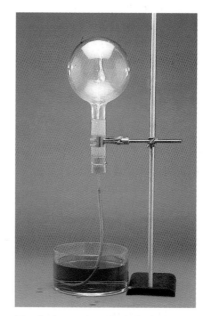

The flask contains ammonia. It dissolves in the first drops of water that reach the top of the tube, so more water rushes up to fill the vacuum. Any soluble gas will show this effect.

---

1 Write down three physical properties of ammonia.
2 What is the boiling point of ammonia?
3 A solution of ammonia is alkaline. Explain why.
4 Ammonia solution is a weak alkali. Why?

5 How would you make *pure* ammonium sulphate, starting with ammonia solution? Write an equation for the reaction.
6 Describe a test for ammonia gas. (See page 293.)

14.4 Ammonia and nitric acid in industry

Making ammonia in industry

In industry, ammonia is made from nitrogen and hydrogen in a process called the **Haber process**. First you must get the two gases:

- **hydrogen** is made from methane (North Sea gas) and steam:

$$CH_4(g) + 2H_2O(g) \xrightarrow{\text{catalyst}} CO_2(g) + 4H_2(g)$$

- **nitrogen** is obtained by burning hydrogen in air to remove oxygen. When the water vapour that forms in the reaction is cooled and condensed, the remaining gas is mainly nitrogen.

As you saw on page 134, the reaction between nitrogen and hydrogen is **reversible**. At a certain point, ammonia breaks down at the same rate as it forms. So conditions must be chosen to give the highest possible yield for the lowest cost. These conditions are:
- high pressure
- ammonia removed, so that more will form
- moderate heat and a catalyst, to help the reaction reach equilibrium quickly.

These are the steps in the Haber process:

1 The two gases are mixed, and the mixture is **scrubbed** to get rid of impurities.
2 It is **compressed** to a pressure of about 200 atmospheres.
3 Then it goes to the **converter**. This is a round tank containing beds of hot iron. The iron is a catalyst for the reaction:

$$N_2(g) + 3H_2(g) \rightleftharpoons 2NH_3(g)$$

4 A mixture of all three gases leaves the converter. It is cooled until the ammonia condenses. Then the nitrogen and hydrogen are pumped back to the converter for another chance to react.
5 The ammonia is run into tanks and stored as a liquid under pressure.

Part of the ICI ammonia plant at Billingham in Cleveland

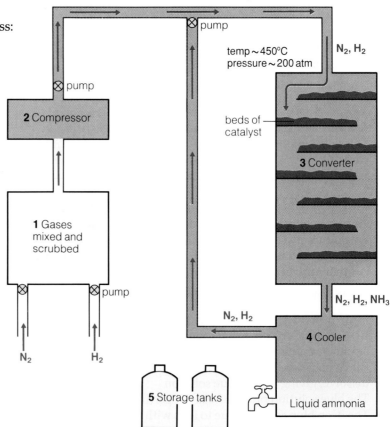

The Haber process

Uses of ammonia

- It is used to make fertilizers such as ammonium nitrate and ammonium sulphate.
- It is used to make household cleaners, dyes, explosives, and nylon.
- A lot is used to make nitric acid.

Making nitric acid in industry

A lot of the ammonia from the Haber process is used to make **nitric acid**. The raw materials for nitric acid are **ammonia**, **air** and **water**. This flow chart shows the stages in the process:

Part of the ICI nitric plant at Billingham. It is built close to ICI's ammonia plant. Can you explain why?

```
┌─────────────────────────────────────────────┐
│              Ammonia, NH₃                    │
└─────────────────────────────────────────────┘
                    │   1 Mixed with air
                    │   2 Passed over a heated catalyst
                    │     (a gauze of platinum/rhodium)
                    ↓
┌─────────────────────────────────────────────┐
│           Nitrogen monoxide, NO             │
│  4NH₃(g) + 5O₂(g) ⟶ 4NO(g) + 6H₂O(g)        │
└─────────────────────────────────────────────┘
                    │   3 Cooled
                    │   4 More air added
                    ↓
┌─────────────────────────────────────────────┐
│           Nitrogen dioxide, NO₂             │
│      2NO(g) + O₂(g) ⟶ 2NO₂(g)               │
└─────────────────────────────────────────────┘
                    │   5 More air added
                    │   6 Mixed with water
                    ↓
┌─────────────────────────────────────────────┐
│              Nitric acid, HNO₃              │
│  4NO₂(g) + O₂(g) + 2H₂O(l) ⟶ 4HNO₃(aq)      │
└─────────────────────────────────────────────┘
```

The **Ammonia, NH₃** box leads to:

1 Mixed with **air**
2 Passed over a heated catalyst (a gauze of platinum/rhodium)

Nitrogen monoxide, NO
$$4NH_3(g) + 5O_2(g) \longrightarrow 4NO(g) + 6H_2O(g)$$

3 Cooled
4 More air added

Nitrogen dioxide, NO₂
$$2NO(g) + O_2(g) \longrightarrow 2NO_2(g)$$

5 More air added
6 Mixed with **water**

Nitric acid, HNO₃
$$4NO_2(g) + O_2(g) + 2H_2O(l) \longrightarrow 4HNO_3(aq)$$

The overall result is that ammonia is oxidized to nitric acid. Chemical engineers must make sure that no nitrogen monoxide or nitrogen dioxide can escape from the plant, since these gases cause acid rain. Besides, if any nitric acid goes down the drain, it will end up in the river, killing fish and other river life.

Uses of nitric acid

- Most is used to make **fertilizers**.
- Some is used to make explosives such as trinitrotoluene (TNT).
- Some is used in making nylon and terylene.
- Some is used in making drugs.

---

1 Ammonia is made from nitrogen and hydrogen.
  a How are the nitrogen and hydrogen obtained?
  b What is the process for making ammonia called?
  c What catalyst is used? What does it do?
  d Write an equation for the reaction.
  e The reaction is *reversible*. What does that mean?
  f What steps are taken to improve the yield of ammonia? Explain each step.

2 Look at the diagram on the opposite page. Why are the catalyst beds arranged that way?
3 a Name the raw materials for making nitric acid.
  b Write word equations for the three reactions that take place during its manufacture.
  c The result of the reactions is that ammonia is *oxidized* to nitric acid. What does that mean?
  d What is the main use of nitric acid?

14.5 Fertilizers

Plants need carbon dioxide, light, and water for photosynthesis. But they also need:

- nitrogen to make proteins for growth and root development
- potassium to promote growth
- phosphorus to help leaves develop.

They also need smaller amounts of many other elements, including calcium, iron, and sulphur. They obtain these things from compounds in the soil. But when the same crops are grown in the same soil year after year, the supplies of compounds in the soil get exhausted, and crops suffer. That's where fertilizers come in.

Fertilizers

Fertilizers are added to the soil to replace the elements used up by plants, and to help crops grow. Animal manure is a natural (organic) fertilizer. Artificial fertilizers are compounds such as:
ammonium nitrate, NH_4NO_3
ammonium phosphate, $(NH_4)_3PO_4$
potassium chloride, KCl

In fact potassium chloride occurs naturally as the mineral **sylvite**. To turn it into fertilizer, the rock is simply crushed, ground, and washed. But most fertilizers are made by neutralization reactions. For example, ammonium nitrate is made by neutralizing nitric acid with ammonia:

$$NH_3(aq) \; + \; HNO_3(aq) \; \longrightarrow \; NH_4NO_3(aq)$$

Choosing a fertilizer

There are two main types of fertilizer on sale:
- **straight N** fertilizer for farmers who want only nitrogen, for example for growing grass for cattle feed. This is usually ammonium nitrate.
- **NPK compound fertilizers**, which are usually a mixture of ammonium nitrate, ammonium phosphate, and potassium chloride. Different mixtures are made up to suit different needs. Farmers can have their soil tested to see which mixture would suit best.

Fertilizers are big business

Millions of tonnes of fertilizers are produced every year. Without them, world food production would probably be halved. Fertilizer factories are usually huge, and consist of several plants side by side. Ammonia is made in one plant, nitric acid in another, and so on. Each plant is carefully controlled so that it is making just the right amount of a substance at the right time.

The flow plan for a fertilizer factory is shown on the next page. Plant safety and pollution control are major concerns. Can you explain why?

Nitrogen, potassium, and phosphorus working wonders

Cow manure contains urea, a natural (organic) fertilizer with the formula $CO(NH_2)_2$. But farmers usually treat the grass to artificial fertilizers rich in nitrogen.

A bag of NPK compound fertilizer. It is 20% nitrogen and 8% phosphorus, by mass. What percentage is potassium?

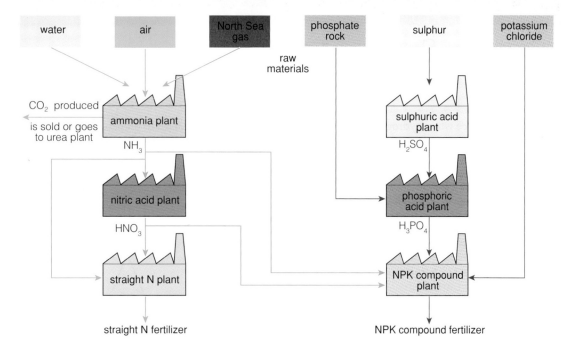

water   air   North Sea gas   phosphate rock   sulphur   potassium chloride

raw materials

CO_2 produced is sold or goes to urea plant

ammonia plant
NH_3

sulphuric acid plant
H_2SO_4

nitric acid plant
HNO_3

phosphoric acid plant
H_3PO_4

straight N plant

NPK compound plant

straight N fertilizer

NPK compound fertilizer

Flow plan for a fertilizer factory

The problem with fertilizers

Fertilizers allow farmers to grow a bigger crop in the same soil, year after year. That sounds good, but there are problems.

- Nitrates are very soluble. Rain washes or **leaches** them out of the soil and into rivers.
- In the rivers, the nitrates promote the growth of tiny water plants called **algae**. These form a scum on the water. When they die, bacteria feed on the remains. In the process, they use up the oxygen dissolved in the water. Fish and other river life die from oxygen starvation, and the river becomes choked and lifeless. This is called **eutrophication**.
- Some algae are poisonous to fish and humans. People swimming through these algae or swallowing the water may get rashes, eye irritation, muscle pains, vomiting, and diarrhoea.
- Excess nitrate in the river may get into drinking water. It increases the risk of **blue-baby syndrome**. What happens is this: bacteria in a feeding bottle or a baby's body convert the nitrate to nitrite. This gets taken up instead of oxygen, by the haemoglobin in blood. The baby turns blue and can die.

Note that potassium from fertilizers also gets leached into rivers – but it does not cause a problem. Phosphates *could* cause a problem – but they are less soluble than nitrates so little leaching occurs.

The answer is for farmers to use fertilizers sparingly, and apply them very carefully – and not when they expect rain!

Monitoring water quality

- Throughout the UK, the water quality in rivers, lakes, and reservoirs is regularly monitored. Because of the possible health risks, one thing that's measured is nitrate concentration.
- It is very difficult to remove excess nitrate at waterworks. If there's a nitrate problem in drinking water, it is usually easier to take water from a 'cleaner' source instead.

Fertilizers are not the only problem

- Some of the nitrate and phosphate in rivers is due to fertilizer run-off from farms.
- But some is waste from factories.
- Household waste from sewage works also contributes. (Detergents and shampoos contain phosphates, for example.)

---

1 Plants need nitrogen, phosphorus, and potassium. Explain why.

2 What are *fertilizers*? Why are they needed?

3 What is: **a** a straight N fertilizer?
b an NPK compound fertilizer?

4 **a** What is *eutrophication*?
b What causes it?
c Suggest steps you could take to help a eutrophic river recover.

5 Describe one health problem related to fertilizers.

Questions on Section 14

1 The reaction that takes place when hydrogen is passed over iron(II) oxide is:

FeO (s) + H₂ (g) ⟶ Fe (s) + H₂O (g)

a Complete the sentences below.
In the reaction the iron(II) oxide is .... to iron and the hydrogen is .... to water. In the reaction, hydrogen is acting as a ....
b Draw a diagram of the apparatus that could be used to demonstrate this reaction in the laboratory.
c Suggest why carbon monoxide and not hydrogen is used in industry, to extract iron from iron oxide.
d Give one use of hydrogen, in industry.

2 This paragraph is about the element nitrogen. Rewrite it, choosing the correct item from each pair in brackets.

Air is a (mixture/compound) which contains ($\frac{1}{5}$/$\frac{4}{5}$) nitrogen. The symbol for nitrogen is (N/N₂) and the gas is made of (molecules/atoms) represented by the formula (N/N₂). It is therefore a (monatomic/diatomic) gas. Nitrogen is needed by plants to make (proteins/sugars). Most plants are unable to take nitrogen directly from the (water/air), so some 'fixing' of the gas is required. Some is fixed by (rain/lightning) and some is fixed by (leaves/bacteria), but most is fixed artificially by making (pollutants/fertilizers). For these, the nitrogen is first reacted with (hydrogen/oxygen) to give the (gas/solid) ammonia. This is turned into (sulphuric/nitric) acid, which is then used to make fertilizers such as ammonium nitrate.

3 The diagram below represents the **nitrogen cycle**. Say where the following words should fit in the diagram. You will need to use some words twice.

nitrogen   air   ammonia
nitrates   proteins   bacteria

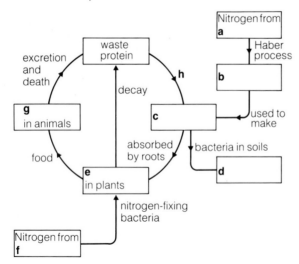

4 This apparatus can be used to prepare *safely* a solution of ammonia in water:

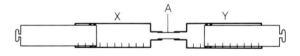

Ammonia gas is given off when the test tube is heated.

a Name two compounds which could be heated together in the test tube.
b Write an equation for the reaction.
c Why would it be dangerous to dip the glass tube into the water, *without* using the filter funnel? (Hint: think about the fountain experiment.)
d If a few drops of litmus solution were added to the water in the trough, what colour change would be seen during the experiment?
e If the water was replaced by a dilute solution of nitric acid, what salt would be formed in the trough?

5 The diagram shows apparatus for an experiment to make ammonia. X and Y are two gas syringes, connected to a combustion tube A. At the start, syringe X contained 75 cm³ of hydrogen and syringe Y contained 25 cm³ of nitrogen.

a Copy the diagram, labelling the gases.
b How would you make the gases mix?
c The reaction needs a catalyst. Why?
d Name a suitable catalyst.
e Where should the catalyst be placed? Add the catalyst to your diagram.
f Where should the apparatus be heated? Show this on your diagram.
g How would you show that some ammonia had been obtained, at the end of the experiment?

6 Look at question 2. Make up a similar question about the properties of ammonia. Use brackets to provide alternatives from which another pupil could choose.

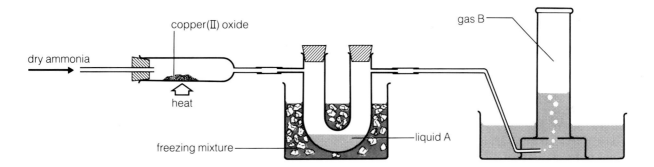

7 Using the above apparatus, dry ammonia is passed over heated copper(II) oxide. The gases given off are passed through a cooled U-tube. A liquid (A) forms in the U-tube and a colourless gas (B) collects in the gas jar.
  a The copper(II) oxide is reduced to copper. What would you see as the gas passes over the heated copper oxide?
  b Why is the U-tube surrounded by a freezing mixture?
  c The liquid A is found to turn blue cobalt chloride paper pink and to have a boiling point of 100 °C. Identify liquid A.
  d Is gas B soluble or insoluble in water?
  e Identify gas B. (Hint: look at the other chemicals in this reaction.)
  f Write a word equation for the reaction.
  g The copper oxide is *reduced* by the ammonia. Explain what this means.
  h How will the mass of the heated tube and contents change, during the reaction?

8 The manufacture of ammonia and nitric acid are both very important industrial processes.

  A AMMONIA
  a Name the raw materials used.
  b Which two gases react together?
  c Why are the two gases scrubbed?
  d Why is the mixture passed over iron?
  e What happens to the *unreacted* nitrogen and hydrogen?
  f Why is the manufactured ammonia stored at high pressure?

  B NITRIC ACID
  a Name the raw materials used.
  b Which chemicals react together to form nitric acid?
  c What would happen if the gauze containing platinum and rhodium was removed?
  d Why must the chemical plant be constantly checked for leaks?

9 Write equations for the chemical reactions in question 8.

10 Write a paragraph explaining why ammonia and nitric acid are very important chemicals.

11 Ammonium compounds and nitrates are of great importance as fertilizers.
  a Why do these compounds help plant growth?
  b Name one *natural* fertilizer.
  c Name two compounds containing nitrogen which are manufactured for use as fertilizers. Write the chemical formulae for these compounds.
  d Name two elements other than nitrogen which plants need, and explain their importance to the plants.
  e Why are some fertilizers not suitable for quick-growing vegetables like lettuce?
  f Some fertilizers are acidic. What is usually added to soils to correct the level of acidity?
  g Land which is intensively farmed needs regular applications of fertilizer. Explain why.
  h Fertilizers obviously have advantages. But many people are worried about the increasing use of fertilizers, especially nitrates, by farmers. Can you suggest why?

12 Ammonia gas is bubbled into copper(II) sulphate solution. At first a blue precipitate forms. This then turns to a deep blue solution.
  a What is the blue precipitate? (Hint: remember that ammonia solution contains hydroxide ions.)
  b Explain why the colour deepens as more ammonia gas is added.
  c This reaction of ammonia is a useful one. Explain why.

13 Nitric acid acts both as an oxidizing agent and as an acid. Say which way it is behaving when it:
  a is neutralized by sodium hydroxide.
  b reacts with a carbonate giving carbon dioxide gas.
  c reacts with copper releasing nitrogen dioxide gas.
  d reacts with copper oxide to form copper nitrate.
  e reacts with dry sawdust, causing it to burst into flames.
  f reacts with magnesium, giving a dark brown gas.
  g turns litmus solution red.

15.1 Oxygen

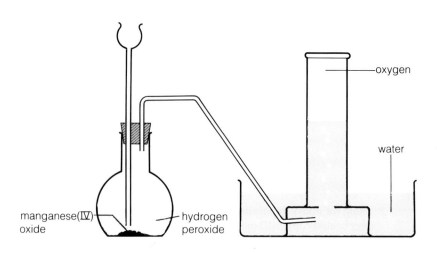

Making oxygen in the lab

Hydrogen peroxide is a colourless liquid, with the formula H_2O_2. It decomposes to water and oxygen:

$$2H_2O_2\,(aq) \longrightarrow 2H_2O\,(l) + O_2\,(g)$$

This reaction is used to prepare oxygen in the lab. But it is very slow, so black manganese(IV) oxide is added as a catalyst.

Oxygen is only slightly soluble in water. How can you tell, from the apparatus?

The properties of oxygen

1 It is a clear, colourless gas, with no smell.
2 It is only slightly soluble in water.
3 It is very reactive. It reacts with a great many substances to produce **oxides**, and the reactions usually give out a lot of energy. For us, its two most important reactions are **respiration** and the **combustion of fuels**. These are very similar, as you will see.

Respiration   This is the process that keeps us alive. During respiration, oxygen reacts with glucose in our bodies. The reaction produces carbon dioxide, water, and the energy we need:

$$C_6H_{12}O_6\,(aq) + 6O_2\,(g) \longrightarrow 6CO_2\,(g) + 6H_2O\,(l) + energy$$
$$glucose + oxygen \longrightarrow carbon\,dioxide + water + energy$$

The glucose comes from digested food, and the oxygen from air:

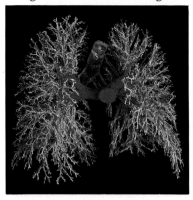

When we breathe in, air travels to our lungs. It passes along tiny tubes to the lungs' surface. The oxygen diffuses through the surface, into the blood.

The blood carries it to the millions of cells in the body, along with glucose from digested food. Respiration takes place in each cell.

The energy from respiration keeps our heart and muscles working. It also keeps us warm. Without it, no body reactions could go on. We would die.

The carbon dioxide and water from respiration pass from the cells back into the blood. The blood carries them to the lungs, and we breathe them out.

Respiration goes on in the cells of *all* living things. Fish use the oxygen dissolved in water, which they take in through their gills. Plants use oxygen from the air, and take it in through tiny holes in their leaves.

Combustion of fuels Fuels are substances we burn to get energy – usually in the form of heat. The burning needs oxygen:

North Sea gas is a fuel. It is mainly **methane**. It is pumped into homes from gas wells in the North Sea.

In the pipes of gas cookers, fires and boilers, the methane is mixed with air. When the mixture is lit, the methane . . .

. . . reacts with the oxygen in the air, giving out energy as heat and light. The heat is used to cook, and heat homes and water.

The equation for the reaction is:

methane + oxygen $\longrightarrow$ carbon dioxide + water + energy
$$CH_4(g) + 2O_2(g) \longrightarrow CO_2(g) + 2H_2O(l) + energy$$

Compare this with the equation for respiration. What do you notice?

Petrol, oil, coal and wood are also used as fuels. Each is a mixture of compounds containing carbon and hydrogen. On burning in plenty of oxygen, they all produce carbon dioxide, water and energy.

If there is only a limited amount of oxygen, **carbon monoxide** (CO) is produced instead of carbon dioxide, when fuels burn. This is a problem, because carbon monoxide is a deadly poisonous gas.

Test for oxygen Things burn much faster in pure oxygen than in air. The reason is that the oxygen in air is **diluted** by nitrogen and other gases. This gives us a way to test a gas, to see if it is oxygen:

1 A wooden splint is lit. Then the flame is blown out. The splint keeps on glowing, because the wood is reacting with oxygen.
2 The glowing splint is plunged into the unknown gas.
3 If the gas is oxygen, the splint immediately bursts into flame.

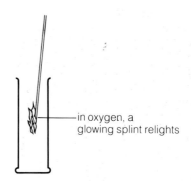

— in oxygen, a glowing splint relights

The test for oxygen

---

1 Draw a labelled diagram, showing how you would make oxygen from hydrogen peroxide.
2 This method produces 'damp' oxygen. Why? Can you suggest a way to dry it?
3 What is respiration? Where does it take place?

4 How do fish obtain oxygen for respiration?
5 What is a fuel? Give four examples.
6 Write down the equations for respiration and the burning of methane. In what ways are they alike?
7 How would you test a gas to see if it was oxygen?

15.2 Oxides

On page 224 you saw that oxygen reacts with many substances to form **oxides**. There are different types of oxide, as you will see below.

Basic oxides

Look at the way these metals react with oxygen:

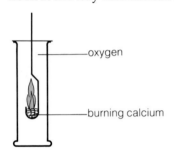

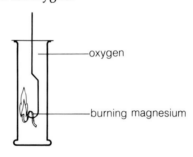

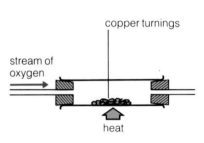

Some grains of calcium are lit over a Bunsen flame, then plunged into a jar of oxygen. They burn brightly, leaving a white solid called **calcium oxide**:

$$2Ca\,(s) + O_2\,(g) \longrightarrow 2CaO\,(s)$$

Magnesium ribbon is lit over a Bunsen flame, and plunged into a jar of oxygen. It burns with a brilliant white flame, leaving a white ash called **magnesium oxide**:

$$2Mg\,(s) + O_2\,(g) \longrightarrow 2MgO\,(s)$$

Copper is too unreactive to catch fire in oxygen. But when it is heated in a stream of the gas, its surface turns black. The black substance is **copper(II) oxide**:

$$2Cu\,(s) + O_2\,(g) \longrightarrow 2CuO\,(s)$$

The way each metal reacts depends on its **reactivity**.
The last reaction above produces copper(II) oxide, which is insoluble in water. But it does dissolve in dilute acid:

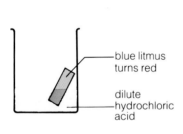

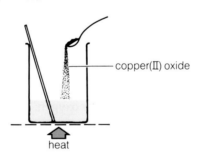

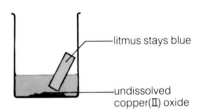

This is dilute hydrochloric acid. It turns blue litmus paper red, as all acids do.

Copper(II) oxide dissolves in it, when it is warmed. But after a time, no more will dissolve.

The resulting liquid has no effect on blue litmus. So the oxide has **neutralized** the acid.

Copper(II) oxide is called a **base**, or **basic oxide**, since it can neutralize acid. The products are a salt and water:

$$\text{base} \quad + \quad \text{acid} \quad \longrightarrow \quad \text{salt} \quad + \quad \text{water}$$
$$CuO\,(s) + 2HCl\,(aq) \longrightarrow CuCl_2\,(aq) + H_2O\,(l)$$

Calcium oxide and magnesium oxide behave in the same way – they too can neutralize acid, so they are basic oxides.
In general, metals react with oxygen to form basic oxides.

Acidic oxides

Now look at the way these nonmetals react with oxygen:

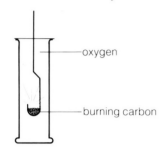

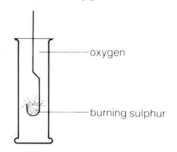

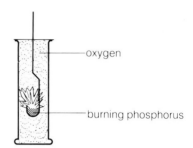

Carbon powder is heated over a Bunsen, until it is red-hot. It is then plunged into a jar of oxygen. It glows bright red and the gas **carbon dioxide** is formed:

$$C\,(s) + O_2\,(g) \longrightarrow CO_2\,(g)$$

Sulphur catches fire over a Bunsen, and burns with a blue flame. In pure oxygen it burns even more brightly. The product is **sulphur dioxide**, a gas:

$$S\,(s) + O_2\,(g) \longrightarrow SO_2\,(g)$$

Phosphorus bursts into flame in air or oxygen, without being heated. (That is why it is stored under water.) A white solid called **phosphorus pentoxide** is formed:

$$P_4\,(s) + 5O_2\,(g) \longrightarrow P_4O_{10}\,(s)$$

The first reaction above produces carbon dioxide, which is slightly soluble in water:

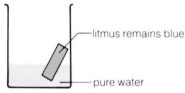

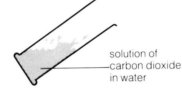

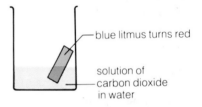

This is pure water. It is neutral. That means it has no effect on red *or* blue litmus paper.

When it is poured into the jar of carbon dioxide, and shaken well, it dissolves some of the gas.

The solution turns blue litmus paper red, so it is **acidic**. It is called **carbonic acid**.

Carbon dioxide is called an **acidic oxide**, because it dissolves to form an acid. Sulphur dioxide and phosphorus pentoxide also dissolve in water to form acids, so they too are acidic oxides.
In general, nonmetals react with oxygen to form acidic oxides.

Neutral oxides

Nitrogen forms the acidic oxide **nitrogen dioxide**, NO_2. But it also forms an oxide which is not acidic, called **dinitrogen oxide**, N_2O. Dinitrogen oxide is quite soluble in water, but the solution is neutral – it has no effect on litmus. So dinitrogen oxide is a **neutral oxide**. Another example of a neutral oxide is carbon monoxide, CO.

> These oxides are unusual because they react with acids *and* bases:
> Aluminium oxide Al_2O_3
> Zinc oxide   ZnO
> Lead(II) oxide   PbO
> They are called **amphoteric** oxides.

---

1 How could you show that magnesium oxide is a base?
2 Copy and complete: Metals usually form .... oxides while nonmetals form .... oxides.
3 Why is phosphorus stored under water?
4 What colour change would you see, on adding litmus solution to a solution of phosphorus pentoxide?
5 Name two nonmetal oxides that are *neutral*.

15.3 Sulphur and sulphur dioxide

Sulphur

Sulphur is the sixteenth most abundant element in Earth's crust.
- It occurs as the element in huge underground sulphur beds in Poland, Mexico, and the US.
- It is also found combined with metals in many metal ores. For example in **pyrite** (FeS_2) and **galena** (PbS).
- It is found in natural gas as hydrogen sulphide (H_2S), and in crude oil as organic sulphur compounds.

Extracting the sulphur

From sulphur beds   Most sulphur is obtained from the underground beds, in an almost pure state, by pumping down superheated water to melt it and then pumping the molten sulphur to the surface. (It melts at 115 °C.)

From oil and gas   An increasing amount of sulphur is recovered from oil and natural gas. Oil companies are forced to remove sulphur compounds before the oil or gas is used. (Otherwise, when it is burned as fuel, sulphur dioxide will form, and cause acid rain.) Sulphur is then extracted from the sulphur compounds.

Natural gas contains sulphur in the form of hydrogen sulphide, and the sulphur compounds in oil are converted into hydrogen sulphide. Some of the hydrogen sulphide is then burned in air to form sulphur dioxide:

$$2H_2S\,(g) \quad + \quad 3O_2\,(g) \longrightarrow \quad 2SO_2\,(g) \quad + \quad 2H_2O\,(l)$$
$$\text{hydrogen sulphide} \qquad\qquad\qquad \text{sulphur dioxide}$$

The sulphur dioxide is reacted with the remaining hydrogen sulphide to give sulphur:

$$SO_2\,(g) \;+\; 2H_2S\,(g) \xrightarrow[\substack{\text{iron(III) oxide}\\\text{catalyst}}]{300\,°C} 3S\,(g) \;+\; 2H_2O\,(l)$$

The properties of sulphur

1. It is a brittle yellow solid.
2. It is made up of crown-shaped molecules, each with eight atoms.
3. It can take two different crystalline forms, as you can see on the right. Different forms existing in the same state are called **allotropes**.
4. Because it is molecular, it has quite a low melting point (115 °C). It melts easily in a Bunsen flame.
5. Like other nonmetals, it does not conduct electricity.
6. Like most nonmetals, it is insoluble in water.
7. It reacts with metals to form sulphides. With iron it forms iron(II) sulphide:

$$Fe\,(s) \;+\; S\,(s) \longrightarrow FeS\,(s)$$

8. It burns in oxygen to form sulphur dioxide:

$$S\,(s) \;+\; O_2\,(g) \longrightarrow SO_2\,(g)$$

A molecule of sulphur

This is a crystal of **rhombic** sulphur, the allotrope that is stable at room temperature.

If you heat rhombic sulphur to above 96°C, the molecules rearrange themselves to form needle-shaped crystals of **monoclinic** sulphur.

Uses of sulphur

- Around 90% of it is used to make sulphuric acid.
- Some is added to rubber, for example for car tyres, to toughen it. This is called **vulcanization** of rubber.
- Some is used to make drugs, pesticides, matches, and paper.
- Some is added to cement to make **sulphur concrete**. Unlike ordinary concrete, this is not attacked by acid. So it is used for walls and floors in plants where acid is used.

Sulphur dioxide

Sulphur dioxide is formed when sulphur burns in air. Its formula is SO_2. It has these properties:

1 It is a colourless gas, with a strong, choking smell.
2 It is heavier than air.
3 It is soluble in water. The solution is acidic because the gas reacts with water to form **sulphurous acid**, H_2SO_3:

$$H_2O\,(l) + SO_2\,(g) \longrightarrow H_2SO_3\,(aq)$$

Sulphur dioxide is therefore an acidic oxide. The acid easily decomposes again to sulphur dioxide and water.
4 It reacts with hydrogen sulphide, as you saw on the opposite page, to give sulphur and water.
5 It acts as a bleach when it is damp or in solution. Some coloured things lose colour when they lose oxygen – that is, when they are **reduced**. Sulphur dioxide bleaches them by reducing them.
6 In the same way it reduces acidified potassium dichromate solution. This contains the orange dichromate ion, $Cr_2O_7^{2-}$. Sulphur dioxide reduces it to the green chromium ion, Cr^{3+}:

$$3SO_2\,(g) + 2H^+\,(aq) + Cr_2O_7^{2-}\,(aq) \longrightarrow 3SO_4^{2-}\,(aq) + H_2O\,(l) + 2Cr^{3+}\,(aq)$$
$$\text{from the acid} \qquad \text{orange} \qquad\qquad\qquad\qquad\qquad\qquad \text{green}$$

As you'll see on page 293, this is used as a test for the gas.
7 When it escapes into the air from engine exhausts and factory chimneys, it causes air pollution. It attacks the breathing system in humans and other animals. It dissolves in rain to give acid rain. Acid rain damages buildings, metalwork, and plants.

Death by sulphur dioxide. These trees were killed by acid rain.

INGREDIENTS
(greatest first):
Oatflakes, Sultanas, Raisins, Toasted Wheatflakes, Wheatflakes, Barley Flakes, Sweetened Dried Banana Chips (min. 3.3%) (Banana, Coconut Oil, Sugar, Flavourings), Dried Coconut (min. 2.9%), Dried Apricot (min. 2.1%), Roasted Hazelnuts (min. 2.1%), Flaked Brazil Nuts (min. 1.8%), Dried Dates (min. 1.8%), Sunflower Seeds (min. 1.4%), Dried Pineapple (min. 0.7%), Preservative (Sulphur dioxide).

Sulphur dioxide (sometimes shown on food labels as E220) is used in preserving dried fruit.

Uses of sulphur dioxide

- It is used to bleach wool, silk, and wood pulp for making paper.
- It is used as a sterilizing agent in making soft drinks and jam, and in drying fruit. It stops the growth of bacteria and moulds.
- Its main use is in the manufacture of sulphuric acid (page 230).

---

1 Name three sources of sulphur in Earth's crust.
2 a Some natural gas contains up to 15% hydrogen sulphide. Why is it important to remove this?
  b How is sulphur obtained from H_2S?
3 Sulphur burns in oxygen to form sulphur dioxide. Write an equation for the reaction.

4 Sulphur dioxide dissolves in water to form an acid.
  a What is the acid called?
  b Write an equation for the reaction.
5 Sulphur dioxide is heavier than air. How will this contribute to air pollution?
6 Explain how sulphur dioxide bleaches things.

15.4 Sulphuric acid

How sulphuric acid is made

More sulphuric acid is made each year than any other chemical. Most of it is made by the **Contact process**. The raw materials are:
- sulphur, air, and water... or
- sulphur dioxide, air, and water. The sulphur dioxide is produced during the extraction of metals such as lead and zinc from their sulphide ores (page 203).

Starting with sulphur, the steps in the Contact process are:

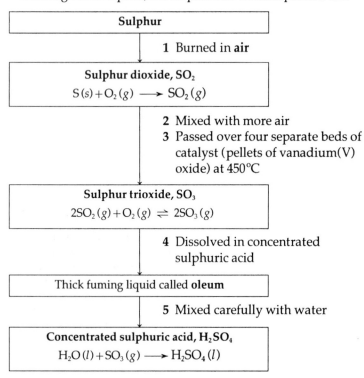

```
┌─────────────────────────────────────┐
│              Sulphur                 │
└─────────────────────────────────────┘
              │  1 Burned in air
              ▼
┌─────────────────────────────────────┐
│      Sulphur dioxide, SO₂           │
│   S(s)+O₂(g) ⟶ SO₂(g)               │
└─────────────────────────────────────┘
              │  2 Mixed with more air
              │  3 Passed over four separate beds of
              │    catalyst (pellets of vanadium(V)
              │    oxide) at 450°C
              ▼
┌─────────────────────────────────────┐
│      Sulphur trioxide, SO₃          │
│   2SO₂(g)+O₂(g) ⇌ 2SO₃(g)           │
└─────────────────────────────────────┘
              │  4 Dissolved in concentrated
              │    sulphuric acid
              ▼
┌─────────────────────────────────────┐
│   Thick fuming liquid called oleum   │
└─────────────────────────────────────┘
              │  5 Mixed carefully with water
              ▼
┌─────────────────────────────────────┐
│  Concentrated sulphuric acid, H₂SO₄ │
│   H₂O(l)+SO₃(g) ⟶ H₂SO₄(l)          │
└─────────────────────────────────────┘
```

Step 1: Burned in **air**

$$S(s) + O_2(g) \longrightarrow SO_2(g)$$

Step 2: Mixed with more air
Step 3: Passed over four separate beds of catalyst (pellets of vanadium(V) oxide) at 450°C

$$2SO_2(g) + O_2(g) \rightleftharpoons 2SO_3(g)$$

Step 4: Dissolved in concentrated sulphuric acid

Step 5: Mixed carefully with water

$$H_2O(l) + SO_3(g) \longrightarrow H_2SO_4(l)$$

Part of the ICI sulphuric acid plant at Billingham. Since it is such a dangerous chemical, safety is a major concern.

This warning sign is displayed on all containers containing sulphuric acid. Do you understand the message?

Things to note about the Contact process

- The reaction between sulphur dioxide and oxygen in step 3 is reversible (page 132). Sulphur trioxide continually breaks down again to sulphur dioxide and oxygen. So the mixture is passed over several beds of catalyst to let the gases react again.
- The sulphur trioxide is removed between the last two beds of catalyst (step 4) in order to increase the yield.
- The catalyst will not work below 400°C. It works better as the temperature rises. But at higher temperatures the yield of sulphur trioxide falls since the reaction in step 3 is exothermic. So 450°C is chosen as a compromise.
- To keep the temperature down to 450°C, heat must be removed from the catalyst beds. This is done using cold water pipes. The water is converted to steam and used to generate electricity for the plant, or sold to nearby factories for heating.
- The sulphur trioxide is dissolved in concentrated acid rather than water, in step 4. If it is dissolved in water, a thick mist of acid forms. This would be a serious pollution hazard.

Pressure and the Contact process

- In theory, increasing the pressure should increase the yield of sulphur trioxide in step 3. (Why?)
- But the increase in yield is so small that it is not usually worth the extra expense.
- So in practice the reaction is usually carried out at atmospheric pressure.

Uses of sulphuric acid

Nearly every industry uses some sulphuric acid: it is the cheapest acid to buy. Its main uses are in the manufacture of:
- fertilizers
- paints, pigments, and dyestuffs
- fibres and plastics
- soaps and detergents.

It is also used as the acid in car batteries.

The properties of sulphuric acid

The concentrated acid It is a colourless oily liquid. It is a **dehydrating agent**: it can remove water. It will dehydrate sugar, paper, and wood. These are all made of carbon, hydrogen, and oxygen. The acid removes the hydrogen and oxygen as water, leaving carbon behind. For example:

$$C_6H_{12}O_6\,(s) \xrightarrow{\text{conc}} 6C\,(s) + 6H_2O\,(l)$$
$$\text{sugar} \quad H_2SO_4\,(l) \quad \text{carbon} \quad \text{water}$$

It will dehydrate flesh in the same way, so it's very dangerous! It will also dehydrate blue copper(II) sulphate crystals, removing the water of crystallization to give the **anhydrous** salt:

$$CuSO_4.5H_2O\,(s) \xrightarrow{\text{conc}} CuSO_4\,(s)$$
$$\text{blue} \quad H_2SO_4\,(l) \quad \text{white}$$

The dilute acid It has typical acid properties:

1 It turns blue litmus red.
2 It reacts with metals to give hydrogen, and salts called **sulphates**.
3 It reacts with metal oxides and hydroxides to give sulphates and water.
4 It reacts with carbonates to give sulphates, water, and carbon dioxide.

The dilute acid is made by carefully adding concentrated acid to water. Never do it the other way round, because so much heat is produced that the acid could splash out and burn you.

Concentrated sulphuric acid was added to two teaspoons of sugar – and this is the result. It dehydrated the sugar, leaving carbon behind. Think what it could do to flesh!

Sulphuric acid for drying gases

- Since concentrated sulphuric acid is a dehydrating agent, it can also be used to dry gases. It will remove the water from them.
- For example, it is used to dry hydrogen chloride when this gas is made in the lab. (Why might this gas in particular be damp?)
- But you would *not* use it to dry ammonia. Why?

The information on page 132 about reversible reactions will help you answer these questions.

1 For making sulphuric acid, name: **a** the process
  b the raw materials **c** the catalyst.
2 The reaction between sulphur dioxide and oxygen is *reversible*. What does that mean?
3 **a** Is the *breakdown* of sulphur trioxide to sulphur dioxide endothermic or exothermic?
  b Does *more* sulphur trioxide break down, or less, as the temperature rises? Why?

4 **a** Why is a catalyst needed?
  b At 500 °C, the catalyst makes sulphur trioxide form even faster. Why isn't this temperature used?
5 Explain how this helps to increase the yield of sulphur trioxide:
  a Several beds of catalyst are used.
  b The sulphur trioxide is removed by dissolving it.
6 Identify two *oxidation* reactions in the manufacture of sulphuric acid.
7 Write down three uses of sulphuric acid.

15.5 Chlorine

How chlorine is made

Chlorine is very reactive so it's never found as the free element in Earth's crust. It occurs mainly as **sodium chloride** or rock salt.

In industry   Chlorine is made in industry by the electrolysis of molten sodium chloride (page 110) or brine (page 112). Brine is a concentrated solution of sodium chloride in water.

In the lab   Chlorine is made in the lab by oxidizing concentrated hydrochloric acid. The oxidizing agent is manganese(IV) oxide. The apparatus is shown on the right. It must be set up in a fume cupboard. Why?

The acid is dripped onto the manganese(IV) oxide while the mixture is heated gently. Look at how the gas is collected. Is it heavier or lighter than air?

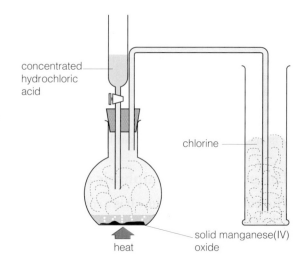

The reaction is:
$$2HCl(aq) + [O] \longrightarrow H_2O(l) + Cl_2(g)$$
from the
oxidizing agent

The properties of chlorine

1 It is a greenish-yellow poisonous gas, with a choking smell.
2 It is heavier than air.
3 It is soluble in water. The solution is called **chlorine water**.
   It is acidic because chlorine reacts with water to form *two* acids:

$$Cl_2(g) + H_2O(l) \longrightarrow \quad HCl(aq) \quad + \quad HOCl(aq)$$
$$\text{hydrochloric acid} \quad \text{hypochlorous acid}$$

The hypochlorous acid slowly decomposes, giving off oxygen:

$$2HOCl(aq) \longrightarrow 2HCl(aq) + O_2(g)$$

4 Chlorine water acts as a bleach. This is because the hypochlorous acid can lose its oxygen to other substances – it **oxidizes** them. Some coloured substances lose their colour when oxidized.

Chlorine is used in swimming pools to kill bacteria . . .

. . . and was used in World War One to kill people. This man and his horses were prepared.

5 Like other bleaches, chlorine water also acts as a sterilizing agent.
   It kills bacteria and other germs.
6 Hydrogen burns in chlorine to form hydrogen chloride.
   The reaction can be explosive:

$$H_2(g) + Cl_2(g) \longrightarrow 2HCl(g)$$

7 Chlorine also combines with most metals, forming metal chlorides.
   For example, when it is passed over heated aluminium the
   aluminium glows white hot and turns into aluminium chloride:

$$2Al(s) + 3Cl_2(g) \longrightarrow 2AlCl_3(s)$$

   Look what's happening in terms of ions:

$$2Al + 3Cl_2 \longrightarrow 2Al^{3+} + 6Cl^-$$

   Aluminium has lost electrons. It has been oxidized by chlorine.
   Chlorine has gained electrons. It has been reduced.
8 If chlorine is bubbled into iron(II) chloride solution, iron(III)
   chloride is formed:

$$2FeCl_2(aq) + Cl_2(g) \longrightarrow 2FeCl_3(aq)$$

   Look again at what's happening in terms of ions:

$$2Fe^{2+} + Cl_2 \longrightarrow 2Fe^{3+} + 2Cl^-$$

   The iron has lost electrons. It has been oxidized by chlorine. The
   chlorine has gained electrons. It has been reduced.
9 Chlorine is more reactive than the halogens below it in Group 7.
   So it will displace them from solutions of their compounds.
   **A more reactive halogen will always displace a less reactive
   halogen from solutions of its compounds.**
   So colourless potassium iodide solution turns red-orange when
   chlorine is bubbled through it, because iodine is displaced:

$$2KI\,(aq) + Cl_2(g) \longrightarrow 2KCl\,(aq) + I_2\,(aq)$$
   colourless   orange

   The same happens with potassium bromide – see page 93.

Uses of chlorine

It is used:
- to make the plastic polyvinyl chloride (PVC)
- to make hydrochloric acid (page 234)
- to make solvents such as tetrachloroethane (for dry cleaning)
- in making bleaches, disinfectants, and insecticides
- to sterilize drinking water and water in swimming pools.

Chlorine reacts violently with aluminium.

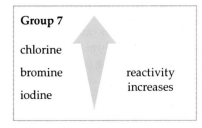

Group 7

chlorine

bromine   reactivity
                                 increases
iodine

---

1 Write an equation to show how chlorine is made in the
   lab. How is the gas collected?
2 Explain why a solution of chlorine in water:
   a is acidic   b is able to bleach things.
3 Write an equation for the reaction between chlorine
   and sodium. Explain why it is a redox reaction.

4 a Iodine in solution is red-brown. Potassium iodide
      is colourless. What would you *see* when chlorine is
      bubbled through potassium iodide solution?
   b Write an equation for the reaction.
   c What is being oxidized in this reaction?
   d What is being reduced?

15.6 Some compounds of chlorine

Hydrogen chloride

Hydrogen chloride is made in industry by burning hydrogen in chlorine. The gas has these properties:

1 It is heavier than air.
2 It has a choking smell, and it irritates the eyes and lungs.
3 It dissolves very easily in water, to form **hydrochloric acid**. Like all acids, this contains hydrogen ions:

$$HCl\,(aq) \longrightarrow H^+\,(aq) + Cl^-\,(aq)$$

4 Hydrogen chloride reacts with ammonia to form white smoke, made of tiny particles of solid ammonium chloride (page 217):

$$NH_3\,(g) + HCl\,(g) \longrightarrow NH_4Cl\,(s)$$

This reaction is used to test for ammonia, or hydrogen chloride.

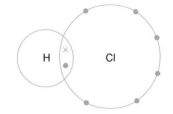

Hydrogen chloride is a covalent compound. This shows the bonding.

Hydrochloric acid

Hydrochloric acid is made in industry by dissolving hydrogen chloride in water. It is a typical acid. It reacts:

- with metals to give hydrogen and salts called **chlorides**.
- with metal oxides and hydroxides to form chlorides and water.
- with carbonates to form chlorides, water, and carbon dioxide.

Sodium chloride

Sodium chloride (NaCl) is the most important chloride of all.

- It occurs naturally in sea water, and as **rock salt** in salt beds.
- The simplest way to obtain it is to evaporate sea water. This method has been used for centuries in hot dry coastal places. But most salt is obtained by mining underground salt beds like those in Cheshire, left behind by the evaporation of ancient seas.
- It is the starting point for many important chemicals. For example, the electrolysis of brine gives sodium hydroxide, chlorine, and, hydrogen, and is the basis of the chlor-alkali industry (page 113).
- It has been valued since earliest times for improving the flavour of food, and as a preservative. Roman soldiers were paid in salt!
- It is used to melt ice on the roads in winter. But the drawback is that car bodies and lamp posts rust faster when there's salt around. It is also damaging to trees.
- Sodium ions are essential for body fluids. But excess salt can cause high blood pressure.

To make hydrogen chloride, add concentrated sulphuric acid to sodium chloride. To dissolve the gas safely in water, use a filter funnel. Why is the funnel safe? Why is a glass tube not safe?

Collecting salt obtained by evaporation, from salt pans in Namibia

---

1 Write equations to show how hydrochloric acid reacts with: **a** sodium hydroxide **b** zinc.
2 How would you test for hydrogen chloride gas?

3 Give four reasons why sodium chloride is such an important chemical.
4 What is the disadvantage of using salt to melt ice?

15.7 Bromine and iodine

Like chlorine, bromine and iodine belong to group 7 of the periodic table. They have similar properties to chlorine, but are less reactive.

1 Like chlorine, when they react with metals they gain electrons to form negative ions. (Cl^-, Br^-, and I^- are called **halide ions**.)
2 They combine directly with hydrogen to form the colourless gases hydrogen bromide (HBr) and hydrogen iodide (HI).
3 Like hydrogen chloride, these gases dissolve very easily in water to form strong acids: **hydrobromic acid** and **hydroiodic acid**. They are so soluble that you must dissolve them with care.
4 As you saw on page 233, **a more reactive halogen will displace a less reactive one from solutions of its compounds**. So bromine reacts with potassium iodide solution like this:

$$2KI\,(aq) + Br_2\,(aq) \longrightarrow 2KBr\,(aq) + I_2\,(aq)$$

Extraction of bromine from sea water

Bromine and iodine are both obtained from sea water. The process depends on the fact that chlorine is more reactive and will displace them from their compounds. This is how bromine is extracted:

1 Chlorine is bubbled into sea water, displacing bromine:

$$Cl_2\,(g) + 2Br^-\,(aq) \longrightarrow 2Cl^-\,(aq) + Br_2\,(aq)$$

2 Bromine vapour is carried away on a stream of air.

3 The stream of air is mixed with sulphur dioxide gas and a fine mist of water. This removes the bromine from it. The sulphur dioxide reduces the bromine to hydrobromic acid:

$$Br_2\,(g) + 2H_2O\,(l) + SO_2\,(g) \longrightarrow 2HBr\,(aq) + H_2SO_4\,(aq)$$

4 The mixture passes down a tall tower, while steam and chlorine pass up. The hydrobromic acid is converted back into bromine:

$$2HBr\,(aq) + Cl_2\,(g) \longrightarrow Br_2\,(l) + 2HCl\,(aq)$$

It ends up as an impure liquid at the bottom of the tower, under an aqueous layer. It is distilled to purify it.

Uses of bromine, iodine, and their compounds

- Bromine is used in making fuel additives, dyes, and pesticides.
- Iodine is used in making dyes, printing inks, animal feeds, and pharmaceuticals. It is also used as an antiseptic.
- Like silver chloride, silver bromide and silver iodide are used to coat photographic film and paper. (See page 94 for more.)

Group 7

chlorine

bromine

iodine

- Like other elements, halogens react so that their atoms can achieve a full outer shell of electrons.
- A halogen atom needs just one more electron to achieve a full outer shell.
- Chlorine is more reactive than bromine and iodine because its atoms are smaller. The outer shell is nearer the nucleus so can attract an extra electron more readily.

See page 35 for more.

chlorine

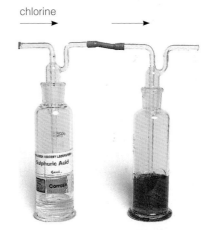

Dried chlorine being bubbled into a solution of potassium bromide. What reaction takes place? Compare it with the method for extracting bromine from sea water.

1 Why is bromine less reactive than chlorine?
2 What are *halide ions*?
3 How would you expect bromine to react with a solution of sodium chloride? Explain.
4 Name the acids formed by: **a** hydrogen chloride **b** hydrogen bromide **c** hydrogen iodide.
5 Describe the steps in the extraction of bromine from sea water.

Questions on Section 15

1 Look at each description below in turn. Say whether it fits oxygen, or sulphur, or chlorine.
   a Quite soluble in water
   b Solid at room temperature
   c Reacts with metals to form oxides
   d Exists in more than one solid form
   e When damp, removes the colour from dyes
   f Burns in air with a blue flame
   g Reacts with hydrogen to form water
   h A poisonous gas
   i Is added to rubber to make it tough and strong
   j Relights a glowing splint
   k Is colourless
   l Reacts with other elements to form chlorides
   m Forms a gaseous oxide which causes acid rain when burnt.

2 Oxygen reacts with many different elements.
   a Copy and complete the following table to show the results of combustion experiments carried out using jars of oxygen gas.

Element	What you would see in the gas gar
potassium	
calcium	
carbon	
phosphorus	
magnesium	
copper	
sulphur	

   b Which element does not need heating?
   c Which element does not catch fire?
   d In the case of potassium, the product is soluble in water. Which of the other products are soluble?
   e Which products give acidic aqueous solutions?
   f Which of the elements form basic oxides?

3 The elements sulphur and oxygen are in the same group of the periodic table, so they have similar properties. But there are also some differences between them. Use the information on pages 224–229 to answer these questions.
   a Oxygen is a nonmetal. Is sulphur a nonmetal?
   b Do the elements look alike at room temperature? Explain your answer.
   c Both the elements are molecular. What type of bonding do their molecules contain?
   d In what way are their molecules different?
   e Sulphur combines with hydrogen to form the gas hydrogen sulphide, H_2S. Name one way in which this differs from the compound formed between oxygen and hydrogen.

4 The following diagrams represent molecules of different substances that contain sulphur.
   a Write the chemical formula for each substance, then name the substance.

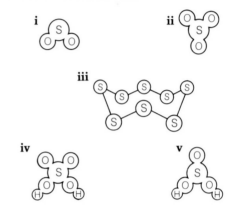

   b How would you convert:
   substance iii into substance i?
   substance i into substance ii?
   substance i into substance v?

5 Below is a flow chart for the Contact process.

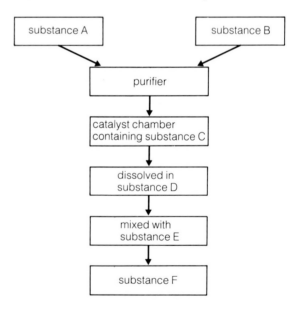

   a Name substances A, B, C, D, E, and F.
   b Why is a catalyst used?
   c Write a chemical equation for the reaction that takes place on the catalyst.
   d Why is the production of substance F very important? Give three reasons.
   e Copy out the flow chart, writing in the full names of the different substances.

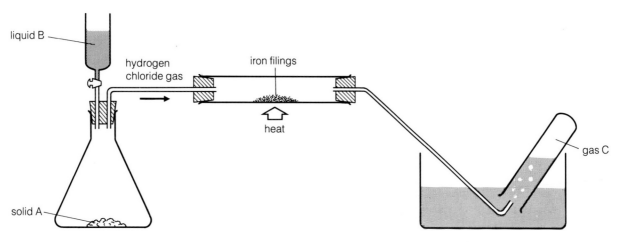

6 The following table shows the properties of certain oxides.

| | | When added to water: | |
Oxide	State at 20 °C	Is energy given out?	pH of solution
Magnesium oxide	solid	no	above 7
Calcium oxide	solid	yes	above 7
Copper(II) oxide	solid	no	—
Sulphur dioxide	gas	yes	below 7
Carbon dioxide	gas	no	below 7

a Why is no pH given for copper(II) oxide?
b Which compounds give a rise in temperature when added to water? Why?
c Which two compounds contain only nonmetals?
d From the information in the table, what conclusions can you draw about:
  i the pH of solutions of oxides?
  ii the state of oxides at room temperature (20 °C)?

7 Sulphuric acid can act as a dehydrating agent and an oxidizing agent, as well as an acid. Decide how it is behaving in each of the following cases:
a It turns blue litmus red.
b It reacts with magnesium to give hydrogen.
c It can be used to dry hydrogen chloride gas.
d It neutralizes sodium hydroxide to form a salt.
e It reacts with copper, releasing sulphur dioxide.
f It turns sugar black.
g It turns blue copper sulphate white.
h It produces carbon dioxide when added to a carbonate.

8 a Look again at the reactions in question 7. For each say whether the sulphuric acid should be concentrated, dilute, or either.
b To make dilute sulphuric acid, the concentrated acid must be added carefully to water, and *never* the other way round. Explain why.

9 The apparatus above was used to investigate the reaction between hydrogen chloride and iron filings.
a Name a suitable substance for solid A.
b Name a suitable substance for liquid B.
c The reaction between iron and hydrochloric acid is exothermic. What would you see in the combustion tube?
d Gas C burns with a squeaky pop. What is it?
e Suggest a name for the product left in the combustion tube after the reaction.
f Write a word equation for the reaction.
g The experiment could be dangerous. Suggest why.

10 When a piece of damp blue litmus paper is placed in a gas jar of chlorine, the litmus changes colour. First the blue colour turns to red, then the red colour quickly disappears, leaving the paper white.
a Explain why the paper turns red.
b Explain why the red colour then disappears.

11 Chlorine gas is passed through two solutions, as shown below.

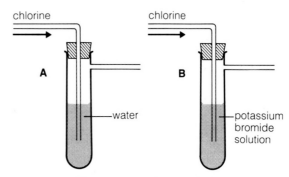

a What would you *see* in each test tube?
b What chemical(s) are formed in each tube?
c Write word and chemical equations for each reaction.
d Which of the reactions is a displacement?
e Write the word equation for a similar displacement reaction, using chlorine.

16.1 Carbon and its compounds

Carbon

A small amount of carbon occurs as the free element in Earth's crust. It can occur as **diamond** and **graphite**. These are **allotropes** of carbon (page 56).

Diamond is a very hard, clear solid that can be cut and polished so that it sparkles in light. Graphite is a dark, greasy solid. Charcoal and soot are forms of graphite. They are made by heating coal, wood, or animal bones in just a little air.

Besides diamond and graphite, carbon also forms allotropes called **fullerenes**. They are nicknamed **bucky balls**, and were first discovered in 1985. Fullerene molecules are hollow clusters of carbon atoms, with as many as 180 atoms in a cluster. They are used to make superconductors.

Diamond is so hard that it can be used to cut stone. This cutting wheel is edged with diamond.

Carbon dioxide

Air contains a small amount of carbon dioxide gas – about 1%. Even though it's only a small percentage, we couldn't live without it, as you'll see in the next Unit.

Properties of carbon dioxide   These are the main ones:

1 It is a colourless gas, with no smell.
2 It is much heavier than air.
3 When it is cooled to $-78\,°C$, it turns straight into a solid. Solid carbon dioxide is called **dry ice**. It sublimes when it is heated.
4 Carbon dioxide does not usually support combustion. That is why it is used in fire extinguishers. (But very reactive metals such as magnesium will burn in carbon dioxide, reducing it to carbon, which appears as soot.)
5 It is slightly soluble in water, forming an acidic solution called carbonic acid. This is a very weak acid. It plays a part in the weathering of rocks, as you will see on page 269.

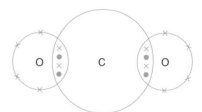

Bonding in carbon-dioxide.

Uses of carbon dioxide   It has several important uses:

1 It is used in fire extinguishers.
2 It is put in drinks like cola and lemonade, to make them fizzy. It is only slightly soluble, so it is bubbled into these drinks under pressure, to make more dissolve. When the bottles are opened it escapes again, and that causes the 'fizz'.
3 Solid carbon dioxide (dry ice) is used to keep food frozen.

Test for carbon dioxide   For this, **lime water** is needed. Lime water is a solution of calcium hydroxide in water. Carbon dioxide makes it go milky, because a fine white precipitate forms. But when more carbon dioxide is bubbled through, the precipitate disappears. The precipitate is calcium carbonate, from this reaction:

$$Ca(OH)_2\,(aq) + CO_2\,(g) \longrightarrow CaCO_3\,(s) + H_2O\,(l)$$

It disappears again because it reacts with more carbon dioxide to form calcium hydrogen carbonate, which is soluble:

$$CaCO_3\,(s) + CO_2\,(g) + H_2O\,(l) \longrightarrow Ca(HCO_3)_2\,(aq)$$

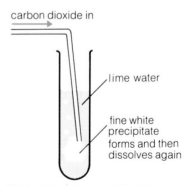

carbon dioxide in

lime water

fine white precipitate forms and then dissolves again

The test for carbon dioxide: it turns lime water milky. See the tests for different gases on page 293.

Carbonates

Carbonates are compounds containing the carbonate ion CO_3^{2-}.
Two important carbonates are:
- calcium carbonate, $CaCO_3$. It occurs naturally in rocks as limestone, chalk, and marble.
- sodium carbonate, Na_2CO_3. It is also known as **washing soda**, from its use in softening hard water.

Limestone and chalk are used for making quicklime, mortar, and cement, as you saw on page 164. Sodium carbonate is mixed with limestone and sand to make glass (page 58).

The French thinker Voltaire, sculpted in marble. This is formed from limestone, as you'll see on page 274.

The properties of carbonates

1 They are insoluble in water, except for sodium, potassium, and ammonium carbonates.
2 They react with acids to form a salt, water, and carbon dioxide.
3 Most of them break down on heating to carbon dioxide and an oxide. For example:

$$CaCO_3(s) \rightleftharpoons CaO(s) + CO_2(g)$$
calcium carbonate   calcium oxide   carbon dioxide
(limestone)   (quicklime)

But sodium and potassium carbonates do not break down on heating. (The more reactive the metal, the more stable its compounds.)

Organic compounds

Organic compounds are carbon compounds found in living things, or derived from living things. There are many thousands of these compounds – far more than all the other compounds put together. They include:

- the proteins, carbohydrates, and fats found in your body
- the hundreds of different compounds found in crude oil and other fossil fuels
- plastics and pharmaceuticals made from compounds found in crude oil.

The study of these carbon compounds is called **organic chemistry**. Most of this section and the next are devoted to organic chemistry.

Built up from organic compounds, and at least 20% carbon!

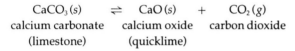

1 What are *allotropes*?
2 Name two allotropes of carbon and give *two* properties of each, to show how different they are.
3 Explain in your own words why carbon dioxide can be used to put out fires.
4 Sodium carbonate is an *inorganic* compound.
  a What do you think that means?
  b Name another inorganic compound.

5 On heating, most carbonates break down to an oxide and carbon dioxide. Name this type of reaction.
6 Write an equation to show what happens when:
  a lead(II) carbonate is heated
  b limestone reacts with dilute hydrochloric acid.
7 What is an *organic* compound?
8 Name an organic compound that contains just carbon and hydrogen, is a gas, and is used as a fuel.

16.2 The carbon cycle

Somewhere in your body there's a carbon atom that may once have been in a sea shell, or a geranium, or the body of Julius Caesar. In fact your carbon atoms have been everywhere and done everything – all courtesy of the **carbon cycle**.

The carbon cycle is all the processes through which carbon circulates between the atmosphere, the ocean, and living things. Carbon dioxide is the key to it:

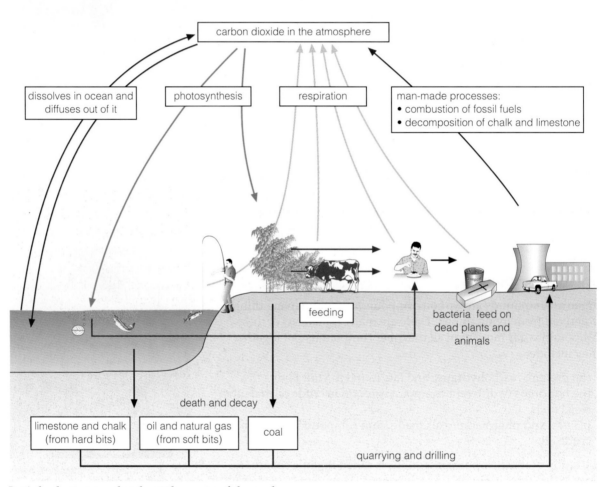

Let's look at more closely at the parts of the cycle.

1 Photosynthesis In this process plants take in carbon dioxide and turn it into a sugar called **glucose**. The reaction is endothermic. The energy for the process comes from sunlight. The green **chlorophyll** in the plants acts as a catalyst:

$$6CO_2\,(g) \ + \ 6H_2O\,(l) \quad \xrightarrow[\text{energy from sunlight}]{\text{chlorophyll}} \quad C_6H_{12}O_6\,(s) \ + \ 6O_2\,(g)$$

carbon dioxide   water   glucose   oxygen

Photosynthesis takes place in *all* plants, from tiny single-celled algae floating on the ocean to massive trees deep in the rain forest. They use the glucose to make starch and other carbon compounds they need. (Algae don't need much!)

2 Feeding   Plants get eaten by animals, which get eaten by other animals. So carbon is carried along the food chain by means of digestion. Bacteria and fungi enjoy the final feast when they feed on dead remains.

3 Respiration   This is the process that gives living things their energy. It goes on in every living cell – inside you, your cat, and the grass in your garden. It releases carbon dioxide:

$$C_6H_{12}O_6\,(s) \ + \ 6O_2\,(g) \ \longrightarrow \ 6CO_2\,(g) \ + \ 6H_2O\,(l) + \text{energy}$$

glucose   oxygen   carbon dioxide   water   energy

Fungi feasting on dead wood

4 Limestone and chalk   Algae take in calcium carbonate to make tiny protective casings. Fish and other sea creatures use it for bones and shells. When they die, their remains fall as sediment to the ocean floor. Then, over millions of years, pressure and heat turn bones and shells into limestone, and the algae casings into chalk.

Some limestone is formed in a simpler way, by precipitation from warm shallow sea water. The tiny amount of calcium carbonate dissolved in the water comes out of solution and crystallizes (page 273).

5 The fossil fuels   Oil and natural gas are formed by the decay of the soft bits of ocean plants and animals, and coal by the decay of vegetation. Again, these processes take millions of years of pressure and heat. You can find out more about them in the next Unit.

6 Man-made processes   It takes millions of years for carbon dioxide to get locked up in limestone, chalk, and the fossil fuels. It takes just minutes for us to release it again:
- When limestone and chalk are heated in lime kilns and cement works, they decompose to form calcium oxide and carbon dioxide. (See page 164.)
- The burning of fossil fuels in plenty of air gives carbon dioxide.

The second process is the one that's causing problems. We burn huge amounts of fossil fuels. The more we burn, the more we disturb the natural balance of carbon dioxide in the atmosphere. It is a **greenhouse gas**, which means it traps heat around Earth. Too much carbon dioxide may lead to **global warming**. See page 178 for more.

Coal-burning power stations unlock the carbon dioxide that got tied up in plants by photosynthesis.

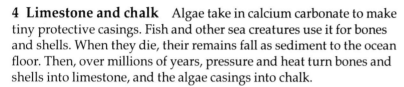

1 What is *the carbon cycle*?
2 Write the equation for:
  i photosynthesis   **ii** respiration.
  Now compare the equations. What do you notice?
3 Does photosynthesis take place at night? Why?
4 Plants take in carbon dioxide for photosynthesis and give it out from respiration. Do they give out the same amount as they take in? Explain.
5 Is respiration *endothermic* or *exothermic*? Explain.
6 Write a balanced equation for the reaction that takes place in kilns when limestone is heated.
7 Natural gas is methane, CH_4. It burns to form carbon dioxide and water vapour. Write a balanced equation for the reaction.
8 Explain how humans are upsetting the natural balance of carbon dioxide in the atmosphere.

16.3 The fossil fuels

Most of the energy used to cook food, drive cars, and keep us warm comes from coal, oil, and gas.

1 Coal, oil, and gas are **fossil fuels**. That means they are the remains of plants and animals that lived millions of years ago.
2 They are made of **organic** compounds, based on carbon.
3 They are used as fuels because they give out plenty of heat energy when they burn.
4 They produce carbon dioxide, water vapour, and energy when they burn. For example, North Sea gas (methane) burns like this:

$$CH_4(g) + 2O_2(g) \longrightarrow CO_2(g) + 2H_2O(l) + \text{energy}$$

Coal

Coal is formed from the remains of lush vegetation that once grew in warm shallow coastal swamps. These are the stages in the process:

- The dead vegetation collects in the bottom of the swamp. It may start to decay. But decay soon stops, because the microbes that cause it need oxygen, and the oxygen dissolved in the still, warm water is quickly depleted.
- The vegetation gets buried under debris.
- Over hundreds of thousands of years, the environment changes. Seas flood the swamps. Heavy layers of sediment pile up on the dead vegetation, squeezing out gas and water and turning it into **peat**.
- As the peat gets buried deeper, the increasing heat and pressure compress it into coal.
- As the process continues, the coal gets harder and more compact. Its carbon content increases, giving different types of coal.

Oil and natural gas

These are formed from the soft remains of sea plants and animals that fall to the ocean floor. These are the stages in the process:

- The remains may start decaying. But decay soon ceases because the dissolved oxygen gets depleted. They get buried under sediment which prevents any further decay.
- The deeper the remains get buried, the higher the temperature and pressure. Chemical changes begin.
- First they turn into a solid waxy substance called **kerogen**. At depths of about 2–4 km, and temperatures of 50–100 °C, this starts to break down into the simpler compounds that make up crude oil and gas.
- These substances are less dense than water. They start to migrate upwards through permeable rock such as limestone.
- Eventually they hit impermeable rock. They can go no further. This environment is an **oil trap**. The rock they collect in is called the **reservoir rock**. They collect in it like water in a sponge.

Is wood a fossil fuel?

- The fossil fuels are the remains of organisms that lived millions of years ago.
- So wood is *not* a fossil fuel.
- But like the fossil fuels it is made of organic compounds.
- Unlike the fossil fuels it is a renewable resource. Wood grows quite fast.

From peat to hard coal

	Name	Carbon content	
	Peat	60%	
pressure	Lignite	70%	
heat	Bituminous coal	80%	hardness
	Anthracite	95%	

- As carbon content increases so does energy given out per tonne.
- But hard coal tends to have a higher sulphur content.

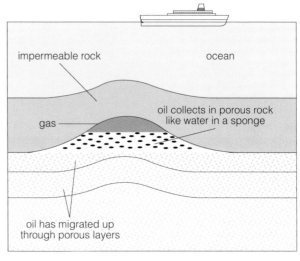

impermeable rock   ocean

gas

oil collects in porous rock like water in a sponge

oil has migrated up through porous layers

Oil and gas in the ocean bed

The problems with burning fossil fuels

We burn huge quantities of fossil fuels. Unfortunately this causes many environmental problems. For example:

- Fossil fuels may contain sulphides. Iron(IV) sulphide (FeS_2) is often found with coal. Sulphides burn in air to form sulphur dioxide, which causes acid rain. Since power stations burn huge quantities of coal, they are major contributors to acid rain.
- The burning of fossil fuels upsets the natural balance of carbon dioxide in the atmosphere. This could lead to **global warming**.

See Unit 11.3 for more on the pollution caused by fossil fuels.

How long will they last?

Oil, gas, and coal are still being formed. But the processes that form them are very slow compared with the speed at which we're using them up. Experts say we're depleting them 100 000 times faster than they are being formed. This means they'll run out eventually. We therefore think of them as **finite** and **non-renewable** resources. It makes sense to use them very carefully.

> **How long do they take to form?**
>
> - Known sources of oil and gas are at least 1–2 million years old. So it takes at least that long to form them.
> - It takes 10 000 years *just to grow enough plant material* to form a 1 metre layer of soft coal.

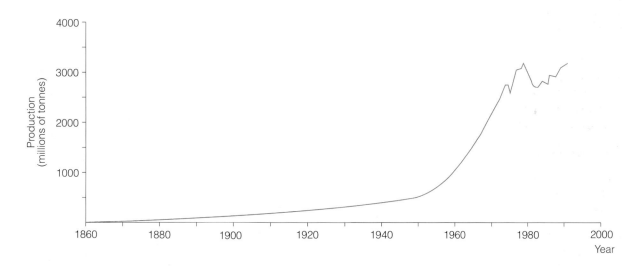

It is difficult to estimate how much of each resource remains and can be recovered. Many areas of the world have not yet been explored. Technology is also improving, so a resource that's not recoverable now may become so later. But many experts believe that 60 years from now, oil will be in short supply. By the end of the 21st century only coal will be abundant.

How oil production has changed over the years. Can you explain the sharp fall around 1979?

---

1 What does the term *fossil fuel* mean?
2 Wood is made from living things. Is it a fossil fuel?
3 As coal gets harder, its carbon content increases. Does that mean more carbon gets added? Explain.
4 Describe in your own words how coal is formed.
5 Explain why gas is found with oil.

6 a What is the starting material for oil?
   b What conditions are needed to form oil?
   c Why does it migrate upwards?
7 Draw a labelled diagram showing the environment in which oil is formed.
8 Fossil fuels are a *non-renewable* resource. Explain.

16.4 Chemicals from oil (I)

Oil is a mixture of hundreds of different carbon compounds. Their molecules are chains or rings of carbon atoms with other atoms bonded on. For example:

This molecule has a chain of 5 carbon atoms.

Here a chain of 6 carbon atoms has formed a ring.

Here 6 carbon atoms form a **branched** chain.

In these examples, the molecules contain only carbon and hydrogen atoms. So the compounds are called **hydrocarbons**.
A hydrocarbon contains only hydrogen and carbon atoms.
Most of the compounds in oil are hydrocarbons.

Using the compounds from oil

The compounds obtained from oil have thousands of different uses. That is why countries with a lot of oil to sell tend to get rich!

- Some are used as fuels or converted into fuels.
- Some are used as the starting point for detergents, dyes, drugs, paints, and cosmetics.
- Some are the starting point for polythene, PVC, and other plastics.

Powered by hydrocarbons from crude oil

Getting oil from Earth

As you saw in the last Unit, oil is found in traps in reservoir rock. These may be 3 or 4 kilometres below ground – a long way to drill!

In places like Kuwait the oil is on land. But in the UK it lies below the North Sea, which adds extra problems. To get at the oil trap, a hole is bored into the sea bed using a giant drill hung from a **drilling platform**. The hole is lined with steel.

The trap must be drilled very carefully because the oil and gas are at very high pressure. If they are not controlled they can shoot to the surface and wreck the platform. But once the pressure drops they need to be pumped out.

The oil and gas are led to the surface of the sea bed, and into a series of pipes that run along it. These carry it to an offshore **tanker terminal**, or all the way to the **refinery** on shore.

An oil rig in the North Sea

At the refinery, the oil goes through a series of processes to turn it into useful materials. This is called **refining** the oil. For example it is **separated** into groups of compounds, and some compounds are **cracked** and some **reformed**, as you'll see on the following pages.

Separating the fractions

Some oil compounds have small molecules with only a few carbon atoms. But some have as many as 50 carbon atoms. To make the best use of it, the oil is separated into groups of compounds that have a similar number of carbon atoms. This is the first step in refining the oil. The groups of compounds are called **fractions**.

Separating is carried out by **fractional distillation** in a tall tower, which is kept very hot at the base, and cooler towards the top. Crude oil is pumped in at the base, and the compounds start to boil off. Those with the smallest molecules have the lowest boiling points. So they boil off first and rise to the top of the tower. Others rise only part of the way, depending on their boiling points, and then condense.

An oil tanker jetty

cool (25 °C)

fractions out

crude oil in

solid

solid

very hot (over 400 °C)

Name of fraction	Number of carbon atoms	What the fraction is used for
Gas	C_1 to C_4	Separated into the fuels methane, ethane, propane, and butane.
Gasoline	C_5 to C_6	Blended with other fractions to make petrol.
Naphtha	C_6 to C_{10}	Starting point for many chemicals and plastics.
Kerosene	C_{10} to C_{16}	Jet fuel. Detergents.
Diesel oil	C_{16} to C_{20}	Fuel for cars and large vehicles.
Lubricating oil	C_{20} to C_{30}	Oil for cars and other machines.
Fuel oil	C_{30} to C_{40}	Fuel for power stations and ships.
Paraffin waxes	C_{40} to C_{50}	Candles, polish, waxed papers, waterproofing, grease.
Bitumen	C_{50} upwards	Pitch for roads and roofs.

boiling points and viscosity increase

As molecules get larger, the fractions get less runny, or more **viscous**: from gas at the top of the tower to solid at the bottom. They also get less **flammable**. Grease burns less easily than petrol, for example. So the lower fractions are not used as fuels.

Once the fractions are separated, they need further treatment before they can be used. You can find out about this on the next page.

1 Copy and complete: Oil is the remains of tiny marine .... and .... It is a .... of compounds. These are mostly .... which means they contain just .... and ....

2 Oil is *refined* before use. What does that mean?

3 Name the fraction where the compounds have the:
a smallest molecules   b lowest boiling points.

4 a The bitumen fraction is the most viscous and has the highest boiling point. Why is this?
b Why wouldn't you use bitumen as a fuel?

16.5 Chemicals from oil (II)

You saw in the last Unit how crude oil is separated into **fractions** by fractional distillation. But that's not the end of the story. The fractions all need further treatment before they can be used.

1 They contain impurities. These are mainly sulphur compounds. If left in fuel, they will burn to form poisonous sulphur dioxide gas.
2 Some fractions are separated into single compounds or smaller groups of compounds. For example, the gas fraction is separated into methane, ethane, propane and butane.
3 Part of a fraction may be **cracked**. Cracking breaks down molecules into smaller ones, as you saw on page 87.

One of the tasks carried out at this refinery is the removal of sulphur-based impurities.

Cracking

Cracking is very important for two reasons:

1 It lets you turn long-chain molecules into more useful shorter ones. For example you may have plenty of naphtha, but not enough gasoline for making petrol. So you could crack some naphtha to get molecules the right size for petrol.
2 It *always* produces short-chain compounds like ethene and propene, which have a double bond between carbon atoms. The double bond makes these compounds **reactive**. So they can be used to make plastics and other substances.

This is the kind of reaction that takes place when you crack naphtha:

decane, $C_{10}H_{22}$
from naphtha fraction

$\downarrow$ ~540 °C, catalyst

pentane, C_5H_{12}
suitable for petrol

propene,
C_3H_6

ethene,
C_2H_4

This complex structure is the cracking plant at Esso's Fawley refinery.

When a catalyst is used, the process is called **catalytic cracking**. Now look at the reaction below. It shows how you can even crack ethane, already a small molecule, to give ethene, by mixing it with steam at high temperatures:

ethane

steam
>800 °C

ethene   hydrogen

In fact you can crack any hydrocarbon molecule that has two or more carbon atoms, with single bonds between them.

From cracking to plastics

The two cracking reactions shown opposite produced ethene.
Look how ethene is used to make plastics:

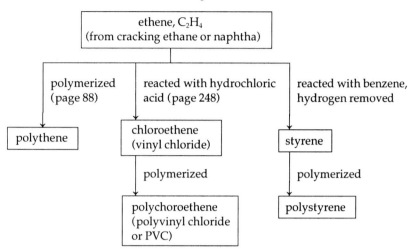

Other compounds from the cracking of naphtha and gas oil are also
polymerized to make plastics. You can find out more about plastics
in the next Unit.

Reforming

Look at these hydrocarbons. Both have the same formula, C_6H_{12}.
Compounds with the same formula but different molecular
structures are called **isomers** (page 253).

Both compounds burn well, and are used in petrol. But the one with
the straight chain will **ignite** or catch fire more easily. In fact it may
ignite a little too soon in the hot cylinder of a car engine, so the
engine will not work as efficiently as it should.

For this reason, high-quality petrol contains mainly hydrocarbons
with branched chains. These are made by taking molecules with
straight chains from the gasoline and naphtha fractions, and making
the chains branch, using heat or a catalyst. This process is called
reforming.

The higher the star rating of petrol, the
more branched chains it contains.

1 What is *cracking*? Why is it so important?
2 **a** A straight-chain hydrocarbon has the formula C_5H_{12}.
Draw a diagram of one of its molecules.
 b Now show what might happen when it is cracked.

3 **a** What does *reforming* mean?
 b Why are compounds reformed to make petrol?
4 Take the molecule you drew for question 2 and show
how it might look when reformed.

16.6 Plastics

Plastics are among the many important materials made from crude oil. As you saw in Unit 6.2 they are made by polymerization.

- The starting point is a small molecule called a **monomer**.
- A monomer for addition polymerization is made by cracking a hydrocarbon obtained from oil, and always has a double bond.
- You may first want to alter the monomer. For example, to make PVC, ethene is first changed into vinyl chloride (chloroethene) by reaction with hydrochloric acid.
- Then addition polymerization is carried out. It usually needs heat, pressure, and a catalyst. The double bonds break and the monomers join up to form molecules with very long carbon chains. The new compounds are called **polymers** or plastics.

Different monomers for different plastics

To make different plastics, just start with different monomers!

Monomer	Polymer	Uses	Monomer	Polymer	Uses
$H_2C=CH_2$ ethene	$(-CH_2-CH_2-)_n$ poly(ethene) or polythene	plastic bags plastic bottles dustbins cling film	$CH_3CH=CH_2$ propene	$(-CH(CH_3)-CH_2-)_n$ polypropene	plastic crates synthetic ropes
$H_2C=CHCl$ chloroethene or vinyl chloride	$(-CH_2-CHCl-)_n$ polyvinyl chloride (PVC)	rain coats wellies credit cards curtain rails	$C_6H_5CH=CH_2$ styrene	$(-CH(C_6H_5)-CH_2-)_n$ polystyrene	plastic cups fast-food cartons packaging foam

Changing the reaction conditions

Look what happens in making polythene:

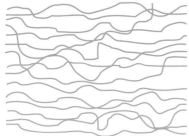

At low temperature, medium pressure, and using a catalyst, you get long chains like these. They are packed close together so the polythene is quite **dense**.

At high temperature and very high pressure, but no catalyst, you can make the chains branch. Now they can't pack so closely, so the polythene is 'lighter'.

So by choosing the right conditions you can change the density of the polythene, and make it 'heavy' or 'light' to suit your needs.

The dense polythene is called **high-density** polythene, and the 'light' one **low-density** polythene. Which type is used for dustbins?

Thermoplastics and thermosets

Some plastics are thermosoftening plastics or **thermoplastics**. Some are thermosetting plastics or **thermosets**. What's the difference?

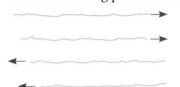

When you heat a **thermoplastic** it gets soft and runny: the heat overcomes the forces between the chains so they can slide over each other. On cooling, it hardens into its new shape.

But when a **thermoset** is first made, strong bonds form between the chains. This is called **cross-linking**. The solid sets into shape and stays hard, even when you heat it.

This shows a white polythene bowl and a red melamine bowl which were put in an oven for just a few seconds. Which is a thermoplastic and which is a thermoset?

Thermosets Because of the cross-linking, thermosets:

- will char or break down at high temperatures, rather than melt.
- are rigid and will break rather than bend. Dropping them may cause the giant irregular structure to break at its weak points.
- are moulded into shape *while they are being made*, because the shape can't be changed later.

Thermoplastics Because there is no cross-linking, thermoplastics:

- are flexible and do not easily break.
- stretch under tension, because the molecules slide over each other.
- melt at quite low temperatures.
- can be moulded into shape *after* they are made.

Moulding thermoplastics These are some of the methods used:

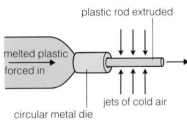

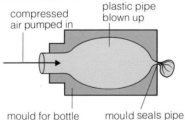

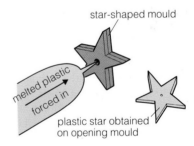

Extrusion. Melted plastic is forced through a shaped nozzle or die, then cooled to harden it. Used for things like pipes and curtain rails.

Blow moulding. Plastic pipe is clamped into a mould, then blown up using compressed air until it fits the mould. Used for plastic bottles.

Injection moulding. Melted plastic is injected into a mould at high speed, then allowed to cool. Used for things like toys, furniture, and bottle caps.

---

1 a What is a *monomer*? b All monomers for addition polymerization have one thing in common. What is it?
2 What type of reaction is used to make a plastic?
3 'Plastic made from the same monomer is always exactly the same.' True? Explain your answer.

4 Name two uses for: a polythene b PVC c polystyrene d polypropene.
5 What is the difference between a thermoplastic and a thermoset? Which type of plastic is polystyrene?
6 Explain why thermosets shatter when hammered.

Questions on Section 16

1 Respiration and photosynthesis are two important processes in the carbon cycle. Which of them:
a produces carbon dioxide?
b releases energy?
c is catalysed by chlorophyll?
d makes food in the leaves of plants?
e is similar to the burning of fuels?
f requires sunlight to work?
g releases oxygen into the atmosphere?
h produces water?
i is endothermic?
j continues at all times?
k takes place in both plants and animals?

2

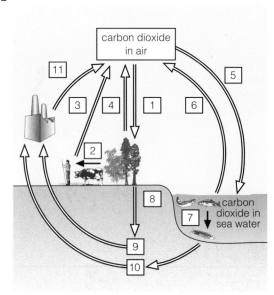

This diagram shows the carbon cycle. The numbers 1–11 represent missing labels.
a Write the numbers 1–11 as a list.
b Beside them write the correct label chosen from this list. (You may use a label more than once.)
respiration   combustion of fossil fuels
comes out of solution   photosynthesis   coal
oil   feeding   dissolves   sediment

3 Fossil fuels are the world's main source of energy.
a What is a *fossil fuel*? Name three.
b Explain how coal was formed.
c Explain how oil was formed. Draw a diagram to show where it is found in rocks.
d Is wood a fossil fuel? Explain.
e 'Fossil fuels are stores of solar energy.' Is this true? Explain.
f Fossil fuels are considered a non-renewable resource. Why is this?
g 'We must become less dependent on fossil fuels in the 21st century.' Give two reasons why, and suggest a list of steps to achieve this aim.

4 Crude oil is a mixture of hydrocarbons, each with a different boiling point.
a What is a *hydrocarbon*?
b Crude oil is an important raw material. Why?
c How is the mixture separated?
d A simple separation can be carried out in the laboratory, using this apparatus:

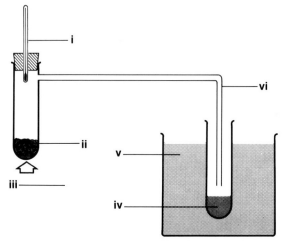

Copy the diagram and label it using these labels:
mineral wool soaked in crude oil   delivery tube
thermometer   heat   water   fraction
e i What is the purpose of the mineral wool?
  ii What is the purpose of the water?
  iii Why is the thermometer placed where it is?

5 In the experiment in question 4, a crude oil sample was separated into four fractions. These were collected in the temperature ranges shown below:

Fraction	Temperature range/°C
A	25–70
B	70–115
C	115–200
D	200–380

Which fraction:
a has the lowest range of boiling points?
b burns most readily?
c has molecules with the longest chains?

6 Crude oil must be refined before it can be used. Refining involves three main processes: **separating**, **cracking**, and **reforming**.
a Explain how fractional distillation is used to separate groups of products from crude oil.
b Explain why cracking improves the range of products obtained from crude oil.
c i What happens to the molecules in the reforming process?
  ii Which particular product makes use of compounds obtained from reforming? Why?

7 The diagram below represents the process used to separate crude oil into different fractions.

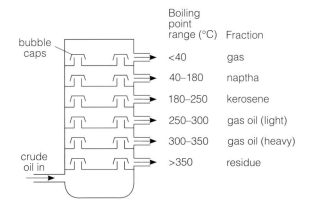

	Boiling point range (°C)	Fraction
	<40	gas
	40–180	naptha
	180–250	kerosene
	250–300	gas oil (light)
	300–350	gas oil (heavy)
	>350	residue

a Name the process.
b This process is based on the fact that different compounds have different _____ _____.
What are the missing words?
c i Using the terms *evaporation* and *condensation*, explain how naphtha is produced.
  ii Is naphtha likely to be a single compound or a group of compounds? Explain your answer.
d How do the bubble caps help in the process?
e Write down one use for each fraction obtained.
f i A hydrocarbon has a boiling point of 200 °C. In what fraction will it be found?
  ii Are the carbon chains in its molecules shorter or longer than those found in naphtha?
  iii Is it more, or less, viscous than naphtha?

8 This shows how hydrocarbons are cracked:

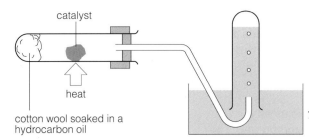

a What is *cracking*?
b What two things are needed to crack hydrocarbons?
c Why should the first tube of gas that is collected be discarded?
d Is the gaseous product soluble or insoluble in water? Explain your answer.
e Ethane, C_2H_6, can be cracked to give ethene, C_2H_4, and hydrogen. Write an equation for this reaction.

9

Name	Formula
ethene	$CH_2 = CH_2$
ethanol	C_2H_5OH
propane	C_3H_8
vinyl chloride	$CH_2 = CHCl$
styrene	$C_6H_5CH = CH_2$
chloropropene	$CH_3CH = CHCl$

a What is a *monomer*?
b Name the type of reaction in which monomers add on to each other to make plastics.
c Which of the compounds above could be used as monomers for making plastics?
d For each compound you chose in **c**:
  i write an equation for the polymerization, using the method shown on page 89.
  ii suggest a name for the polymer formed.
e The polymers obtained from the monomers above are all **thermoplastics**. What special properties do thermoplastics have?
f Some plastics are thermosetting plastics.
  i Name one.
  ii What special property do thermosetting plastics have?
g The use of plastics can cause environmental problems. Name two of these problems.
h Name one other problem caused by our dependency on plastics.

10 'Teflon' is the commercial name given to the polymer obtained from the monomer tetrafluoroethene. The monomer has this structure:

$$\underset{F}{\overset{F}{\diagup}} C = C \underset{F}{\overset{F}{\diagdown}}$$

a What feature of the monomer makes polymerization possible?
b Name the type of polymerization that occurs.
c Show the structure of the repeating unit in the polymer.
d What is the chemical name for this polymer?
e Write an equation for the reaction, like those on page 89.

11 Three plastics A, B, and C have these structures:
  A long-chain molecules with no side chains
  B long-chain molecules with side chains
  C long-chain molecules with cross-linking between the chains
a Draw diagrams to show these structures.
b Which diagram could represent:
  i a thermosetting plastic?
  ii a low-density thermoplastic?
  iii a high-density thermoplastic?
c For each choice in **b**, explain why the structure gives rise to that particular property.

17.1 The alkanes

As you saw on page 239, **organic compounds** are carbon compounds found in living things, or based on compounds from living things. In their molecules, carbon atoms always form the spine.

There are thousands of organic compounds – many more than all the inorganic compounds put together. In order to make sense of them, they are arranged in families of compounds with similar properties. The simplest family is the **alkanes**.

The alkanes are **hydrocarbons**: they contain only carbon and hydrogen. This table shows the first four members of the family:

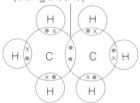

one shared pair of electrons (a single bond)

The bonding in ethane

Name	Methane	Ethane	Propane	Butane
Formula	CH_4	C_2H_6	C_3H_8	C_4H_{10}
Structure of molecule	H │ H—C—H │ H	H H │ │ H—C—C—H │ │ H H	H H H │ │ │ H—C—C—C—H │ │ │ H H H	H H H H │ │ │ │ H—C—C—C—C—H │ │ │ │ H H H H
Number of carbon atoms in chain	1	2	3	4
Boiling point	$-164\,°C$	$-87\,°C$	$-42\,°C$	$-0.5\,°C$

boiling point increases with chain length

Note that:
- the chain grows by one carbon atom each time.
- the family can be described by the general formula C_nH_{2n+2}, where n represents a number. For the first member of the family n is 1, for the second it is 2 and so on.
- a family of compounds like this, which fits a general formula and has similar chemical properties, is called a **homologous series**.

Things to remember about the alkanes

1 They are found in oil and natural gas. Natural gas is mostly methane, with small amounts of ethane, propane, and butane. The mixture of compounds in oil includes alkanes with chains of up to 50 carbon atoms.
2 The longer the chain, the higher the boiling point. So the first four alkanes are gases at room temperature, the next twelve are liquids, and the rest are solids.
3 In an alkane molecule, each carbon atom forms four single covalent bonds.
4 Alkanes burn well in a good supply of oxygen, forming carbon dioxide and water vapour and giving out plenty of heat. So they are used as fuels. For example, methane burns like this:

$$CH_4\,(g)\ +\ 2O_2\,(g)\ \longrightarrow\ CO_2\,(g)\ +\ 2H_2O\,(l)\ +\ heat$$

Alkanes as fuels

- Methane is North Sea gas. We use it for cooking and heating.
- Propane and butane are used as camping gas.
- Calor gas is mainly butane.
- Petrol contains butane, liquid alkanes, and other hydrocarbons, as well as organic additives to make it burn better. (It contains over 300 hydrocarbons altogether.)

5 But if there isn't enough air, they undergo **incomplete combustion** when they burn, producing carbon monoxide and soot, and giving out *less* heat. See page 101 for more.

6 Alkanes also react with chlorine and bromine. For example:

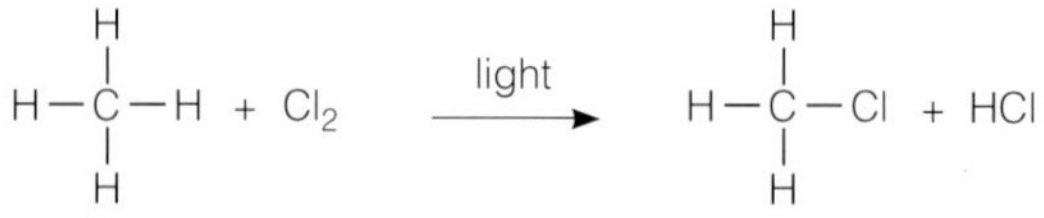

H—C—H + Cl₂ →(light) H—C—Cl + HCl

methane   chloromethane

The reaction needs light. It is called a **substitution** reaction. Can you see why? It can continue until all the hydrogen atoms have been replaced by chlorine, giving tetrachloromethane, CCl_4.

Boiling point and chain length

When a liquid boils to form a gas, its molecules separate. You supply the energy for this by heating.
- The longer the chains in hydrocarbon molecules, the more strongly they are attracted to each other.
- The more strongly they are attracted to each other, the more heat energy is needed to separate them.
- So the longer the chains, the higher the boiling point will be!

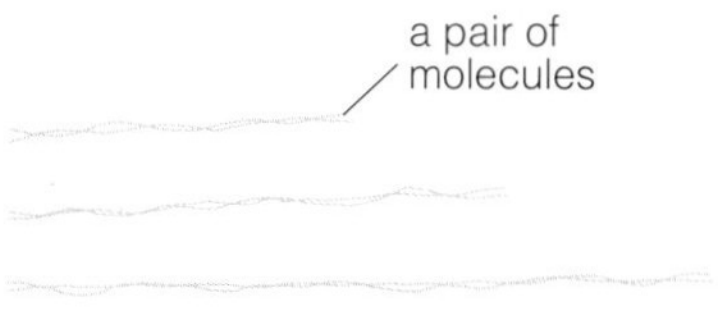

Isomers

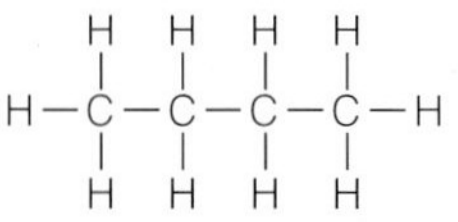

butane

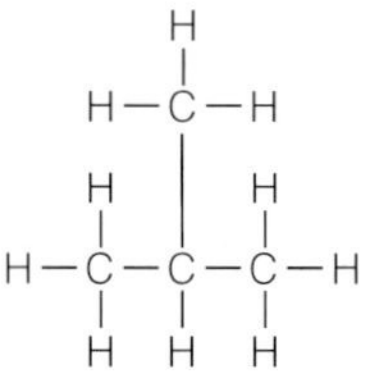

2–methylpropane

Compare these two molecules. Both have the formula C_4H_{10}. But they have different structures, which means the two compounds are different. They are **isomers**. **Isomers are compounds with the same formula but different structures.**

The more carbon atoms in a chain, the more isomers there will be. There are 75 isomers with the formula $C_{10}H_{22}$!

The properties of isomers Since isomers have different structures, they also have slightly different properties. For example, branched-chain alkanes:
- have lower boiling points than straight-chain alkanes. The branches make it more difficult for the molecules to get close and attract each other.
- burn a little less readily in oxygen. As you saw on page 247, this can make them more useful in petrol!

Chlorine and methane

The reaction between them can be explosive in sunlight. Depending on the amount of chlorine present, all the hydrogen atoms can be replaced one by one, giving:

1 chloromethane CH_3Cl
2 dichloromethane CH_2Cl_2
3 trichloromethane $CHCl_3$
   and finally . . .
4 tetrachloromethane CCl_4

These reactions are used to make solvents. Bromine reacts in a similar way, but more slowly. Why?

---

a pair of molecules

as molecules get longer

force of attraction between them increases

Boiling points of some isomers

	Boiling pt (°C)
C_4H_{10}	
Straight chain	0
Branched chain	−10
C_5H_{12}	
Straight chain	36
Branched chain (1)	28
Branched chain (2)	9.5

---

1 The fifth alkane in the alkane family is pentane.
   a What is its formula?   b Draw a pentane molecule.
   c In what state is pentane at room temperature?
2 a Why are alkanes like methane and butane used as fuels? Give *three* reasons.
   b Are all alkanes used as fuels? Explain.

3 Explain why the boiling points of alkanes increase as the chain gets longer.
4 The compound C_5H_{12} has *three* isomers.
   a Draw the structures of these three isomers.
   b The table above shows their boiling points. Match these to your drawings and explain your choice.

17.2 The alkenes

In this Unit we look at another family of hydrocarbons, the **alkenes**. The first three members of the alkene family are:

| ethene | propene | but -1- ene |

Note that the alkenes fit the general formula C_nH_{2n}. Like the alkanes, they form a homologous series.

Things to remember about the alkenes

1 They are made from alkanes by **cracking** (page 87).
2 An alkene molecule contains one **double bond** between carbon atoms. (Alkanes have only single bonds.) The bonding in ethene is shown on the right.
3 Alkenes are much more reactive than alkanes. This is because the double bond can break to form single bonds and add on other atoms. For example, ethene reacts with hydrogen like this:

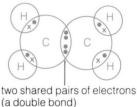

two shared pairs of electrons (a double bond)

The bonding in ethene

$$C=C \ (g) + H_2(g) \xrightarrow[\text{a catalyst}]{\text{heat, pressure}} H-C-C-H \ (g)$$

ethene   ethane

This is called an **addition** reaction. Can you see why?
By addition of hydrogen, an alkene is converted into an alkane.
4 Alkenes undergo a similar addition reaction with water to form compounds called **alcohols**. For example:

$$C=C \ (g) + H_2O(g) \xrightleftharpoons[\text{a catalyst}]{\text{heat, pressure}} H-C-C-OH \ (l)$$

ethene   ethanol

5 Alkene molecules can also add on *to each other* to form long-chain molecules called **polymers**. The reaction is called a **polymerization**. For example, ethene polymerizes to form the plastic **polythene**. The reaction can be summarized like this:

$$n\left(C=C\right) \xrightarrow[\text{a catalyst}]{\text{heat, pressure}} \left(-C-C-\right)_n$$

where n can be up to thousands. There's more about polymerization in Units 6.2 and 16.6.
6 Because the double bond allows them to add on more atoms, alkenes are said to be **unsaturated**. The alkanes don't have a double bond and can't add on more atoms, so they are **saturated**.

Professor Guilio Natta, who, with Fritz Ziegler, developed the Ziegler–Natta catalysts for the polymerization of ethene. These allow reaction to take place at low temperatures (50°C) and medium pressures (10 atmospheres).

An oil well in Kuwait, set alight by Iraqi troops at the end of the Gulf War in 1991. Crude oil contains many unsaturated hydrocarbons, and they don't burn cleanly.

A test for unsaturation

How do you tell whether a hydrocarbon is an alkane or an alkene? See how it reacts with bromine. For example, ethene reacts like this:

$$H_2C=CH_2 \ (g) + Br_2(aq) \ (\text{orange}) \longrightarrow \ H_2BrC-CBrH_2 \ (l)$$

ethene
(colourless)

1,2–dibromoethane
(colourless)

The bromine is usually dissolved in water first, forming an orange solution called **bromine water**. When this is shaken with any unsaturated hydrocarbon an addition reaction takes place, and the orange colour disappears.

Are there isomers in the alkene family?

The answer is yes. Not only can the chains branch in different ways, but the double bond can be in different positions.

but–1–ene

but–2–ene

Look at these compounds. They have the same formula C_4H_8, but the double bond is in a different place in each. So they are isomers.

---

1 What is the general formula of the alkenes?

2 Write a formula for an alkene with 10 carbon atoms.

3 a Name the two simplest alkenes.
  b Now draw their structures.

4 What makes the alkenes behave so differently from the alkanes?

5 Propene reacts with hydrogen to form propane.
  a What is this type of reaction called?
  b Write an equation for the reaction.

6 Propene can *polymerize*. What does that mean?

7 a Propene is *unsaturated*. What does that mean?
  b Write an equation for its reaction with bromine.

17.3 The alcohols

Here we look at a third family of organic compounds: the **alcohols**. This table shows the first five members. Alcohols contain the **–OH** group. This is called a **functional group** because it determines their reactions. It is the reason why they all react in a similar way.
Note the -1- in three of the names. This tells you that the –OH group is attached to the end carbon atom.

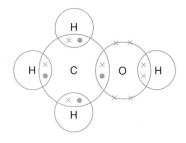

Bonding in methanol.

Name	Methanol	Ethanol	Propan-1-ol	Butan-1-ol	Pentan-1-ol
Formula	CH_3OH	C_2H_5OH	C_3H_7OH	C_4H_9OH	$C_5H_{11}OH$
Structure of molecule	H–C–O–H (with H above and below C)	H–C–C–O–H	H–C–C–C–O–H	H–C–C–C–C–O–H	H–C–C–C–C–C–O–H
Carbon atoms in chain	1	2	3	4	5
Boiling point (°C)	65	78	97	117	137

Ethanol, an important alcohol

Ethanol is the best known of all the alcohols.
- It's the ingredient in alcoholic drinks that makes people drunk!
- It is a good solvent, dissolving many compounds that are insoluble in water.
- It evaporates easily. That makes it a good solvent for use in things like glues, printing inks, deodorants, and aftershave.
- It is the starting point for many chemicals. For example, it is used in making sweet-smelling liquids called **esters** which are used as solvents, in food flavourings, and as fragrances in beauty products.
- It is also growing in importance as a fuel. (See the next Unit.)

Things to remember about ethanol

1 It is a clear, colourless liquid that boils at 78 °C.
2 It is **miscible** with water – it mixes completely with it.
3 It burns well in oxygen, giving out plenty of heat:

$$C_2H_5OH\,(l) + 3O_2\,(g) \longrightarrow 2CO_2\,(g) + 3H_2O\,(l) + heat$$

This reaction is the basis of its use as a fuel.

4 It reacts with sodium like this:

$$2C_2H_5OH\,(l) + 2Na\,(s) \longrightarrow 2C_2H_5ONa + H_2\,(g)$$
$$\text{ethanol} \qquad\qquad\qquad \text{sodium ethoxide}$$

Note that this is just like the reaction between *water* and sodium, but less vigorous. In each case an O–H bond is broken and hydrogen is displaced by sodium.

Humans have been drinking wine for hundreds of years, on social occasions.

Compare these compounds ...

Na^+OH^- sodium hydroxide

$C_2H_5O^-Na^+$ sodium ethoxide

They are both ionic compounds. $C_2H_5O^-$ is called the **ethoxide ion.**

5 Ethanol can be **dehydrated** to form ethene by passing its vapour over heated aluminium oxide. The aluminium oxide acts as a **catalyst** (see page 128):

$$H-\underset{\underset{H}{|}}{\overset{\overset{H}{|}}{C}}-\underset{\underset{H}{|}}{\overset{\overset{H}{|}}{C}}-OH \quad \xrightarrow[\text{Al}_2\text{O}_3, \text{ heat}]{-H_2O} \quad \underset{H}{\overset{H}{>}}C=C\underset{H}{\overset{H}{<}}$$

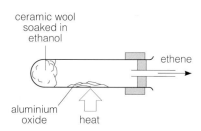

The dehydration of ethanol to give ethene

Compare this with reaction 4 on page 254. What do you notice?

6 If ethanol is left standing in air, it will be **oxidized** with the help of bacteria, forming **ethanoic acid** which has a sour taste. This reaction is the reason why wine left open in air goes sour:

$$H-\underset{\underset{H}{|}}{\overset{\overset{H}{|}}{C}}-\underset{\underset{H}{|}}{\overset{\overset{H}{|}}{C}}-OH \quad \xrightarrow{[O]} \quad H-\underset{\underset{H}{|}}{\overset{\overset{H}{|}}{C}}-C\underset{O-H}{\overset{O}{<}}$$

ethanol   ethanoic acid

Ethanol can be oxidized to ethanoic acid more rapidly by warming it with acidified potassium dichromate solution.

Ethanol in alcoholic drinks

Alcoholic drinks are a weak solution of ethanol. The ethanol in them is made by **fermentation**, as you'll see in the next Unit. It has these effects on the body:

- Even just one drink impairs your co-ordination and judgement. It is a factor in about 40% of car accidents.
- It can make you aggressive. It is a factor in about 40% of arrests.
- Excessive drinking causes depression and other mental and emotional disorders.
- It can lead to ulcers, high blood pressure, brain and liver damage, and cancer of the mouth, throat, and gullet. People who smoke as well are at greater risk from these cancers.

Women are at greater risk than men from the harmful effects of alcohol. One reason is that men's bodies have a higher water content. So alcohol in their body fluids is more diluted.

Drinking during pregnancy means the baby is drinking too. Alcohol can damage the baby's brain and heart, and slow down its growth so that it is born underweight.

How much alcohol is in it?

Alcohol content is measured in units of alcohol. All these contain one unit:

- half a pint of ordinary beer
- a small glass of sherry
- a small whisky, gin, or vodka
- a glass of wine

But note that:
- a half of extra-strong lager contains nearly three units
- a half of low-alcohol beer contains less than half a unit.

It takes about one hour for the body to rid itself of the effects of one unit.

---

1 Why do alcohols all react in a similar way?

2 Write the formula and draw the structure for:
a methanol **b** ethanol **c** propan-1-ol.

3 a In *propan-1-ol*, what do you think the 1 shows?
b There is another isomer with the same formula as propan-1-ol. Draw its structure and suggest a name for it.

4 Look at the boiling points in the table on page 256.
a What do you notice? **b** Suggest a reason.

5 Try to work out a general formula for the alcohols.

6 Give three reasons why ethanol is an important compound.

7 Write an equation for the combustion of: **a** ethanol **b** methane. Compare them. What do you notice?

8 What type of reaction is used in breathalysing?

9 Give: **a** two harmful social effects **b** two harmful physical effects **c** two harmful economic effects of alcohol.

17.4 The manufacture of ethanol

There are two ways of making ethanol in industry.

1 Ethanol by fermentation

Look what happens when you mix glucose (a sugar) and yeast in the absence of air:

$$C_6H_{12}O_6\,(aq) \xrightarrow[\text{in yeast}]{\text{an enzyme}} C_2H_5OH\,(aq) \ + \ 2CO_2\,(g)$$

glucose   ethanol   carbon dioxide

An enzyme in the yeast breaks the glucose down into ethanol and carbon dioxide. The process is called **fermentation**. It is exothermic. The yeast cells use the energy to multiply.

Any substance that contains sugar, starch, or cellulose can be fermented. (Starch and cellulose break down to give glucose.) This means ethanol can be made from sugar cane, corn, potato peelings, trees, waste paper, straw, and even grass.

In several US states, farmers now grow corn to make ethanol, rather than for flour. This is what happens:
- The corn is ground, treated with enzymes to break the starch in it down to glucose, and then fermented with yeast.
- Fermentation is a **batch process**. A batch of material is prepared and left to ferment. Fermentation takes 20–60 hours.
- The fermented liquid contains a mixture of substances. The ethanol must be separated by fractional distillation (page 22).

2 Ethanol from ethene

Ethanol can also be made by the hydration of ethene:

ethene $(g) + H_2O(g)$ $\xrightleftharpoons[\text{a catalyst (phosphoric acid)}]{570\,°C,\ 60–70\ atm}$ ethanol (l)

- The ethene is obtained by cracking long-chain alkanes from oil.
- The hydration is exothermic and reversible. High pressure and a low temperature would give the best yield. In practice it is carried out at 570 °C to give a decent rate of reaction.
- It is a **continuous process**: ethene is fed in at one end of the reaction container, 24 hours a day, and ethanol comes out the other end. Unreacted ethene is recycled.

Continuous process

Batch process

Ethanol by fermentation

- In 1995 1.5 billion gallons of industrial ethanol were made in the US, mainly from corn.
- Brazil has the capacity to produce 4 billion gallons a year from sugar cane.
- In California some is made using waste liquid from the wine industry and from cheese whey.
- The ethanol from these sources is mainly for use as car fuel.

Methylated spirits is ethanol plus a little methanol and some colouring. Methanol is poisonous, and even low doses can cause blindness. Don't drink it!

Comparing the two methods

Ethanol by fermentation	Ethanol from ethene
Advantages	*Advantages*
• It uses renewable resources like corn and sugar cane.	• The reaction is fast.
• It's a good way to use waste organic material.	• The plant runs continuously. No time is wasted shutting down and restarting.
Disadvantages	• It doesn't need such large reaction vessels.
• A large volume of material is needed to make 1 litre of ethanol. So fermentation tanks must be big and you need lots of them. This costs money.	• The process produces pure ethanol.
• Fractional distillation uses energy, which costs money.	*Disadvantages*
• Fermentation is slow, and tanks are shut down between batches which wastes time.	• Ethene is made from oil which is a non-renewable resource.
• When ethanol reaches a certain concentration, the yeast cells become inactive. This limits the amount of ethanol you can obtain per batch.	• Energy is needed to make steam, and to achieve the right temperature and pressure for the reaction.
	• Under the reaction conditions, a high percentage of ethene remains unreacted. You must keep recycling it to obtain a decent yield.

Which method will a manufacturer choose?

It depends mainly on the cost and availability of the raw materials.
In countries where oil is cheap they'll probably go for ethene.

- But fermentation is increasingly important in countries that depend on imported oil, such as Brazil and the United States. The US spends $160 million *a day* on imported oil. The more ethanol is used as a fuel, the less need for imported oil.
- Ethanol causes less pollution than petrol, which is an added bonus. The oxygen in it helps to promote complete combustion. This means lower levels of carbon monoxide and carbon particles in the exhaust. Since ethanol contains no sulphides, sulphur dioxide isn't a problem.
- To encourage its use, the US government has reduced the tax on fuels containing ethanol, and offers loans and subsidies to companies who make it.
- So American manufacturers have begun making **flexible-fuel cars**. They run on petrol, ethanol, or a mixture of the two.

> **What's in it?**
>
> In the US there are two main fuel mixes containing ethanol:
> - gasohol which is 10% ethanol and 90% petrol
> - E85 which is 85% ethanol and 15% petrol.
>
> In Brazil:
> - millions of vehicles run on pure ethanol
> - others run on a mix of 20% ethanol, 80% petrol.

Ethanol in the drinks industry

The ethanol in alcoholic drinks is made by fermentation. Grapes are used for wine, and barley for beer. The mixture of things in the fermented liquid gives the drink its flavour, while ethanol provides the buzz.

---

1 **a** What is *fermentation*? **b** Write an equation for it.
2 Fermentation is a *batch process*. Explain.
3 The hydration of ethene is a *continuous process*. The reaction is *reversible*. **a** Write an equation for the reaction. **b** Now explain the words in italics.

4 Give two advantages and two disadvantages of making ethanol: **a** by fermentation **b** from ethene.
5 What is *gasohol*?
6 Write down two advantages of ethanol as a fuel.
7 In making wine, the ethanol is not separated. Why?

17.5 Carboxylic acids and their compounds

Now we look at a family of organic acids: the **carboxylic acids**. All the members of this family have the functional group **–COOH** which is called the **carboxyl group**. Here are the first five members:

Name of acid	Methanoic	Ethanoic	Propanoic	Butanoic	Pentanoic
Formula	HCOOH	CH_3COOH	C_2H_5COOH	C_3H_7COOH	C_4H_9COOH
Structure of molecule					
Carbon atoms in chain	1	2	3	4	5

The general formula for the family is $C_nH_{2n+1}COOH$. Do you agree?

Things to remember about the carboxylic acids

1 Their solutions affect indicators, just like the inorganic acids do.
2 Like inorganic acids, their solutions contain H^+ ions. When ethanoic acid is mixed with water, it forms ions like this:

$$CH_3COOH\,(aq) \rightleftharpoons H^+\,(aq) + CH_3COO^-\,(aq)$$
the ethanoate ion

Note that this is a reversible reaction. Only some of the acid forms ions. So carboxylic acids are **weak** acids (page 141).
3 Like inorganic acids, they react with metals, alkalis, carbonates, and hydrogen carbonates to form salts. For example:

$$CH_3COOH\,(aq) + NaOH\,(aq) \longrightarrow CH_3COONa\,(aq) + H_2O\,(l)$$
ethanoic acid   sodium   sodium ethanoate
              hydroxide

Note that compounds from ethanoic acid are called **ethanoates**. What will compounds from methanoic acid be called?
4 They react reversibly with alcohols to form sweet-smelling **esters**. Concentrated sulphuric acid is a catalyst for the reaction:

$$CH_3COOH\,(l) + C_2H_5OH\,(l) \overset{conc.H_2SO_4}{\rightleftharpoons} CH_3COOC_2H_5\,(l) + H_2O\,(l)$$
ethanoic acid   ethanol   ethyl ethanoate   water

Esters are manufactured for use as solvents, flavourings, and fragrances for perfumes and beauty products.

Compare the structures of the compounds formed in 3 and 4:

sodium ethanoate, a salt   ethyl ethanoate, an ester

Vinegar is a solution of ethanoic acid in water.

The smell and flavour of green apples comes from natural esters. Artificial esters are used for the shampoo.

Some natural esters: vegetable oils

Vegetable oils like olive oil, palm oil and sunflower oil are in fact mixtures of natural esters. They are formed from **fatty acids** and an alcohol called **glycerol**.

- Fatty acids are carboxylic acids with long chains of carbon atoms. They are called 'fatty' because the long chains repel water, making the acids immiscible with water.
- Glycerol (or propane-1,2,3-triol) has *three* -OH groups. This is how fatty acids and glycerol react:

$$3C_{15}H_{31}COOH \ + \ \begin{matrix} HO-CH_2 \\ | \\ HO-CH \\ | \\ HO-CH_2 \end{matrix} \longrightarrow \begin{matrix} C_{15}H_{31}COO-CH_2 \\ | \\ C_{15}H_{31}COO-CH \\ | \\ C_{15}H_{31}COO-CH_2 \end{matrix} \ + \ 3H_2O$$

a fatty acid   glycerol   a natural ester found
(palmitic acid)   in palm oil

All these oils are mixtures of natural esters. Are they mostly saturated or unsaturated? How can you tell?

Unsaturated oils   In some vegetable oils, the long carbon chains contain double bonds. These oils are **unsaturated**. If the chains contain more than one double bond, the oils are **polyunsaturated**. Unsaturated oils tend to be liquid at room temperature, while saturated oils are generally solid.

Hardening of oils   Unsaturated oils react with hydrogen in an addition reaction. Double bonds break and the hydrogen adds on. The process is called **hardening** because it turns liquid oils into solid ones. It is used in making margarine from vegetable oils.

Soaps from oils   When oils are reacted with a solution of sodium hydroxide, they break down to form glycerol and the sodium salts of their fatty acids. These salts are used as soap. For example:

$$\begin{matrix} RCOO-CH_2 \\ | \\ RCOO-CH \\ | \\ RCOO-CH_2 \end{matrix} (l) \ + \ 3Na^+OH^-(aq) \longrightarrow \begin{matrix} HO-CH_2 \\ | \\ HO-CH \\ | \\ HO-CH_2 \end{matrix} (l) \ + \ 3RCOO^-Na^+(aq)$$

an ester molecule (where R   glycerol   soap – the sodium
stands for the long carbon chain)   salt of the fatty acid

How soaps 'dissolve' grease

- The 'negative' end of the fatty acid chain attracts water molecules.
- The other end attracts grease.
- So the soap is able to surround grease particles like this, and carry them away:

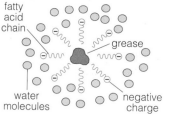

- The attraction between water and the negative end of the chain also explains the 'wetting' effect of soap:

The reaction is called a **hydrolysis** because the ester is first broken down by reaction with water. You can find out more about esters and this hydrolysis reaction on page 303.

The soap you buy is made from a blend of different oils.

1 What do all carboxylic acids have in common?
2 How will ethanoic acid solution affect litmus?
3 **a** Write an equation for the reaction between methanoic acid and sodium hydroxide solution.
  b Name the products that are formed.
4 How are esters made? Give an example.

5 Name three products that contain artificial esters.
6 Which alcohol forms the esters in vegetable oils?
7 What are *fatty acids*?
8 What are *polyunsaturated oils*?
9 What does the *hardening* of oil mean?
10 **a** What is soap?   **b** How is it made?

Questions on Section **17**

1 A homologous series is a family of compounds with a general formula and similar chemical properties. The saturated hydrocarbons form a homologous series with the general formula C_nH_{2n+2}.
   a Explain what the term *saturated* means.
   b Name the series described above.
   c i Write the formula for the hydrocarbon in this series with two carbon atoms, and name it.
      ii Draw a molecule of the compound.
   d i Name a homologous series of unsaturated hydrocarbons, and write a general formula for it.
      ii Write a formula for the member of this series with two carbon atoms, and name it.
      iii Draw a molecule of the compound.

2 Use the information on pages 252 to 255 to answer these questions about the alkanes:
   a Which two elements do alkanes contain?
   b Which alkane is the main compound in natural gas?
   c After butane, the next two alkanes in the series are *pentane* and *hexane*. How many carbon atoms would you expect to find in a molecule of:
      i pentane? **ii** hexane?
   d Write down the formulae for pentane and hexane.
   e Draw a molecule of each substance.
   f Is pentane solid, liquid, or gas at room temperature?
   g Suggest a value for the boiling point of pentane, and explain your answer.
   h Would pentane react with bromine water? Explain.
   i Alkanes burn in a good supply of oxygen. Name the gases formed when they burn.
   j Write a balanced equation for the complete combustion of pentane in oxygen.
   k Name two substances formed during *incomplete* combustion of pentane in air.

3 Hex-1-ene is an unsaturated hydrocarbon. It melts at $-140\,°C$ and boils at $63\,°C$. Its empirical formula is CH_2 and its molecular mass is $84\,g/mol$.
   a i To what family of hydrocarbons does it belong?
      ii What is its molecular formula?
      iii What does the 1 in its name mean?
      iv Draw a molecule of hex-1-ene.
      v Now draw another isomer with this formula.
   b i In what state is hex-1-ene at room temperature?
      ii Make a guess at the boiling point of hept-1-ene, the next member of the series.
   c i Hex-1-ene reacts with bromine water. Write an equation to show this reaction.
      ii What is this type of reaction called?
      iii What would you *see* during the reaction?

4 Propene is one of the many important hydrocarbons made from oil. Like *propane*, it is made up of molecules that contain three carbon atoms. Like *ethene*, it has a double bond.
   a Draw a molecule of propene.
   b How does it differ from a molecule of propane?
   c To which group of hydrocarbons does:
      i propane **ii** propene belong?
   d Write formulae for propane and propene.
   e Which of the two is a *saturated* hydrocarbon?
   f i Explain why propene reacts immediately with bromine water, while propane does not.
      ii What would you *see* in the reaction?
   g Name another reagent that would react immediately with propene but not with propane.
   h Propene is obtained by breaking down longer-chain hydrocarbons. What is this process called?
   i Propene is the monomer for making an important plastic. Suggest a name for the polymer it forms.
   j The reaction which converts propene to a polymer is called an _____ _____.
   What are the missing words?

5 Three hydrocarbons share the molecular formula C_5H_{12}. Their boiling points are $36\,°C$, $28\,°C$, and $10\,°C$.
   a To which family of hydrocarbons do they belong?
   b What name is given to different compounds that have the same molecular formula?
   c i Draw molecules of the three hydrocarbons.
      ii Now assign a boiling point to each structure, and explain your choice.
   d Would you expect the three hydrocarbons to have the same *chemical* properties? Explain.

6 If ethanol vapour is passed over heated aluminium oxide, a dehydration reaction occurs and the gas ethene is produced.
   a Draw a diagram of suitable apparatus for carrying out this reaction in the lab.
   b What is meant by a *dehydration reaction*?
   c Write an equation for this reaction, showing the structure of the molecules.
   d i What will you see if the gas that forms is bubbled through bromine water?
      ii Why would you *not* see this if ethanol vapour was passed through bromine water?
   e In industry, ethanol is made by a reaction which is the reverse of this dehydration reaction: steam and ethene react together in the presence of phosphoric acid at high temperature and pressure.
      i Write an equation for this reaction.
      ii What type of reaction is it?
      iii What is the purpose of the phosphoric acid?
      iv What effect will an increase in pressure have on the reaction? Explain.

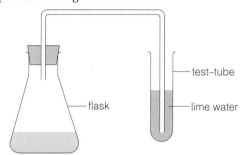

7 Ethanol is a member of the alcohol family.
  a What is the functional group of this family?
  b Write down the formula of ethanol.
  c Ethanol can be made by the fermentation of sugar. This diagram shows apparatus for studying the process. What goes in the flask?

  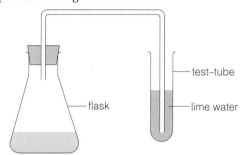

  d Which temperature is best for the reaction?
    i 0 °C ii 10 °C iii 25 °C iv 75 °C
    Explain your answer.
  e i Which gas is released during fermentation?
    ii How could you prove this?
  f Complete the equation for the fermentation:
    $C_6H_{12}O_6 (s) \longrightarrow$
    sugar
  g How long would you expect the reaction to take?
    i 5 minutes ii 5 hours iii 5 days iv 5 months
  h What process would you use to separate reasonably pure ethanol from the mixture?

8 Ethanol is an important compound in industry.
  a Give three uses for it.
  b Describe two methods used to make ethanol in industry, and write equations for the reactions.
  c One of these methods is a *batch process*, and one is a *continuous process*.
    i Explain the terms in italics.
    ii Say which method is which.
  d For each method, give:
    i two advantages ii two disadvantages
      of making ethanol that way.
  e When ethanol is mixed with petrol, it reduces the amount of carbon monoxide and soot from car exhausts. Explain why.

9 Ethanol is a member of the homologous series with the general formula $C_nH_{2n+1}OH$.
  a Name this homologous series.
  b Draw and name the first member of the series.
  c Wine contains ethanol. A glass of wine was found to have a sour taste.
    i What chemical is responsible for this taste?
    ii Explain how it came to be in the wine.
  d Ethanol reacts with sodium.
    i Bubbles of gas form around the sodium. What gas is this?
    ii Write a balanced equation for the reaction.

10 Ethanoic acid is a member of the homologous series with the general formula $C_nH_{2n+1}COOH$.
  a Name this series.
  b What is the functional group of the series?
  c Draw and name the member of the series for which $n = 2$.
  d Ethanoic acid is a *weak* acid. Explain what this means, using an equation to help you.
  e Ethanoic acid reacts with carbonates.
    i What would you *see* during this reaction?
    ii Write a balanced equation for the reaction with sodium carbonate.

11 Ethanoic acid reacts with ethanol in the presence of concentrated sulphuric acid.
  a Name the organic product formed.
  b What type of compound is it?
  c How could you tell that it had formed?
  d What is the function of the sulphuric acid?
  e The reaction is *reversible*. What does this mean?
  f Write an equation for the reaction.
  g i Explain why the reactions between organic acids and alcohols are important in industry.
    ii Give three examples of things you buy that contain products from these reactions.

12 Soaps are salts of fatty acids. They are usually sodium salts.
  a Give one example of a fatty acid.
  b In what way is a fatty acid different from ethanoic acid? In what way is it the same?
  c The fatty acids are obtained from oils of vegetable origin. Name two of these oils.
  d Below is one example of a compound found in vegetable oil, and used to make soap.

$$H_2C - OOC(C_{17}H_{35})$$
$$|$$
$$H\,C - OOC(C_{15}H_{31})$$
$$|$$
$$H_2C - OOC(C_{14}H_{29})$$

    i This oil is in fact an **ester**. Explain why.
    ii To make soap, the oil is reacted with a sodium compound. Which one?
    iii The reaction will produce *four* different compounds. Write down the formulae for them.
    iv Identify which of the four products can be used as soap.
    v One of the four products is an alcohol. Name the alcohol.
    vi How is this alcohol different from ethanol?

13 a Water alone is not much use for removing grease from your hands. Explain why.
   b Draw a diagram to show how soap molecules surround a particle of grease and carry it away.

18.1 A closer look at Earth's structure

As you saw in Unit 10.1, Earth is in layers like an onion.

- The **crust** is made of rock. Compared with the other layers it is thin like the skin of the onion.

- The **mantle** is also made of rock. It extends almost half way to Earth's centre.

- The crust and upper mantle form the **lithosphere**, which is solid.

- Beneath the lithosphere the rock is soft like Plasticine. But it gets harder as you go further into the mantle.

- The **core** has over half Earth's radius. It is made of metal (mainly iron and nickel).

- The **outer core** is liquid and the **inner core** is solid.

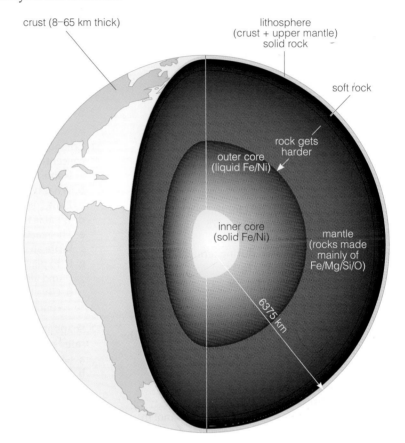

crust (8–65 km thick)

lithosphere (crust + upper mantle) solid rock

soft rock

rock gets harder

outer core (liquid Fe/Ni)

inner core (solid Fe/Ni)

mantle (rocks made mainly of Fe/Mg/Si/O)

6375 km

The conditions inside Earth

Density   Earth started as a molten ball (page 155). It settled into layers of different density as it cooled. The crust is the least dense layer. It is like a scum floating on the denser materials. The density increases as you move towards the core.

Temperature   The deeper you go inside Earth, the hotter it gets: up to 7600 °C at the core. Where does this heat come from?
- Earth began as a hot molten ball and still retains much of that heat.
- The rocks inside it contain radioactive atoms, mainly isotopes of uranium, potassium, and thorium. These eventually **decay** to form stable atoms. This process gives out a huge amount of heat.
- Heat is also generated by friction when huge masses of rock slide past or under each other.

Pressure   The closer you get to the centre of Earth, the greater the pressure, because of the huge mass of rock pressing down. The pressure at the centre is about 3 million times that at the surface.

The higher the pressure, the higher the melting point of a substance. The enormous pressure prevents the inner core from melting or vaporizing. It also explains why the rock in the mantle gets harder rather than softer as you go inwards.

Earth's surface
temperature 25 °C
pressure 1 atm
density 3 g/cm^3
temperature, pressure, and density all increase
Earth's centre
temperature 7600 °C
pressure 3 million atm
density 10.7 g/cm^3

Earth's magnetic field

Earth has a magnetic field around it, as if it were a giant bar magnet. A compass needle always points to the magnetic north pole which is close to the geographic North Pole – at least at the moment!

The source is the liquid metal in the outer core. Because of the spinning of Earth on its axis, and the heating effect of the inner core, the liquid metal is in motion. The moving electrons in it produce a current. The current in turn generates the magnetic field.

The magnetic field is always changing. Just now the magnetic north is drifting west, and the field is weakening. These small changes may be caused by liquid at the surface of the layer swishing and swirling around bumps and hollows on the mantle.

But every so often a much bigger change takes place. The magnetic north pole flips to south, and the field reverses its direction. This has happened around 170 times in the last 75 million years. No one knows when it will happen next, or how long the reversal will last.

The flowing liquid in the outer core is the source of Earth's magnetic field.

The rocks in Earth's crust

We depend on the rocks in Earth's crust for metal ores, building materials, oil and other fuels, and even soil.

You saw on page 160 that rocks are a mixture of different compounds called **minerals**. But no matter what minerals they contain, all rocks can be divided into just three types depending on how they were formed.

Igneous rock is formed when rock melts and cools again. Melting starts around 50–250 km below Earth's surface. On cooling the minerals form crystals.

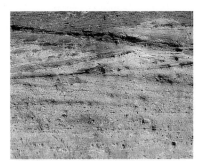

Sedimentary rock is built up from particles compacted and cemented together. This can happen anywhere on Earth's surface, including the sea bed.

Metamorphic rock is formed by the action of high temperatures and pressures on rock, around 10 km or so below Earth's surface.

In the rest of this section we will look at these rocks more closely.

---

1 Draw a labelled diagram to show the different layers that make up Earth.
2 Where does the heat inside Earth come from?
3 What keeps Earth's inner core solid?

4 What causes Earth's magnetic field?
5 'Earth's magnetic field never changes'. Is this true? Explain.
6 Name the three types of rock in Earth's crust.

18.2 Igneous rock

Over 90% of the rock in Earth's crust is **igneous**. But most of it is hidden under a layer of sedimentary rock, or under the oceans. You can see it where sedimentary rock has been eroded away. (The granite on Bodmin Moor in Cornwall is an example.)

How igneous rock is formed

Igneous rock is formed when rock melts, cools, and recrystallizes again. The molten rock is called **magma**. It forms in the hot, soft layer of rock below the lithosphere:

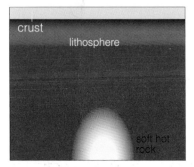

Just a little extra heat from radioactivity, or a small decrease in pressure (due to rock shifting), or a little water coming down (as steam) will make this rock melt.

The magma pushes its way up. If it is very viscous, it moves very slowly. It cools as it travels. If it hardens again below ground, it forms **intrusive** igneous rock.

Granite is the most common example. Slow cooling below ground gives large crystals. The crystals of different minerals interlock randomly.

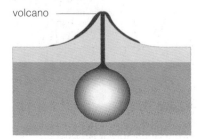

But some magma is very runny. It forces its way to the surface and spews out from a volcano as **lava**. It's often mixed with gases and solid rock.

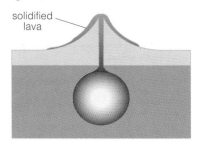

Out in the open the lava cools quickly. It solidifies around the volcano within days or even hours, forming **extrusive** igneous rock.

Basalt is the most common example. Its crystals are tiny because the magma cooled so fast. You'd need a microsope to see them properly.

What makes some magma more runny?

When rock melts, the minerals in it turn to liquid. How viscous (thick and sticky) this is depends on how hot it is and how much water it contains. But it also depends on the minerals.

The more silica the magma contains the more viscous and pale it will be. Granite is over 70% silica. The less silica the runnier it will be, and

---

Igneous rocks provide:
- granite to adorn buildings
- crushed basalt and granite for road building
- pumice as an abrasive
- diamonds, emeralds, topazes
- ores of gold, silver, copper, lead, tin, and zinc (in veins around granite)
- ores of aluminium and iron (left behind after chemical weathering)

Silica

Silica is silicon dioxide, SiO_2. Quartz is mainly silica. So is sand.

therefore the more likely to reach the surface. Basalt is 40–50% silica. It is darker than granite because of the iron and magnesium in it.

The crystal size in igneous rock

When magma cools, the minerals in it start to crystallize one by one. The one with the highest melting point starts first. If cooling is slow, its crystals will grow large before the next crystals start forming. When these start, they grow until they meet the sides of the first crystals and then interlock with them.

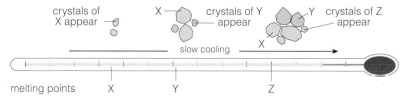

But fast cooling at Earth's surface means crystals of different minerals appear quickly one after the other. None have time to grow large.

Shapes of intrusive igneous rock

All intrusive igneous rock masses are called **plutons**. Geologists also have special names for them depending on their shape and direction:

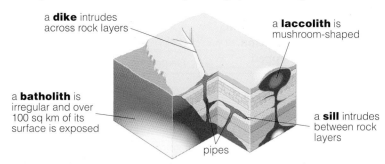

Quartz and copper are found together in this rock.

Igneous rock and metal ores

Ores of valuable metals such as gold, silver, copper, lead, tin, and zinc are often found in and around granite. Why?

When all the minerals have crystallized from magma, a solution of metal ions in water at several hundred °C is left behind. This is called a **hydrothermal solution**. The metal ions in it could not find a place in the crystal structures. The solution runs into cracks in surrounding rock and veins of metal ores are deposited.

1 What is *magma*? What causes it to form?
2 What's the difference between extrusive and intrusive igneous rock? Give an example of each.
3 Some magma is more viscous than others. Why?
4 Basalt is darker in colour than granite. Why?

5 A sample of igneous rock has large crystals. What can you deduce from this?
6 What is: **a** a batholith? **b** a sill?
7 Ores of valuable metals are often found in or near granite. Why?

18.3 Weathering

What is weathering?

Although most of the rock in Earth's crust is igneous, it is mostly hidden from view. Around three-quarters of Earth's land suface is covered by a thin layer of **sedimentary rock**.

The story of sedimentary rock begins with **weathering**, a process in which all rock at Earth's surface gets broken down.

Physical weathering

In **physical weathering** rock gets broken into bits but its chemical composition remains unchanged. This can happen in several different ways.

Freeze–thaw weathering   This occurs in areas where the temperature falls to freezing at night and rises during the day – for example at high altitudes. When the temperature drops to 4 °C, water in the cracks in rock starts to expand. It keeps expanding until it freezes. This expansion exerts great pressure on the rock. When it is repeated night after night, rocks break up over time.

Exfoliation   When rock at Earth's surface gets worn away, a large pluton of igneous rock may be exposed. No longer weighed down from above, this rock is still under pressure from all other directions. So it starts to expand upwards, forming a dome. As it expands it fractures into sheets parallel to the surface. These break up and fall away like layers from an onion. The process is called **exfoliation**. The result of exfoliation is a large curved dome with a stepped appearance.

Biological weathering   When plants grow on rocks, the roots work their way into the cracks and pry bits loose. The cracks grow larger. Eventually the rock breaks up.

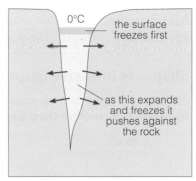

water in a crack in rock

0°C

the surface freezes first

as this expands and freezes it pushes against the rock

Freeze-thaw weathering of rock

The stepped appearance of this granite rock is the result of exfoliation.

Slowly, but surely . . . biological weathering

Chemical weathering

In **chemical weathering** rock is broken down by chemical change. Water always plays a part. For example:

- Carbon dioxide from the air dissolves in rainwater, forming a weak acid called carbonic acid (H_2CO_3). This dissolves limestone rock (calcium carbonate), carrying it away in solution as calcium hydrogen carbonate. That's how limestone caves are formed. (See page 184 for more.)
- Water reacts with many igneous rocks, breaking them down into clays and carrying soluble substances away.
- Water and oxygen react with Fe^{3+} ions in rock to form iron(III) oxide, Fe_2O_3. This reaction is oxidation or **rusting**. The rock turns reddish brown.

Chemical weathering is faster for some types of rock than for others. For example, it is faster for limestone than for sandstone. Like every chemical reaction, it is speeded up by heat. It is much faster in warm, moist climates than cold, dry ones.

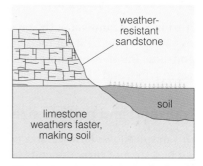

Chemical weathering is faster for limestone than for sandstone.

The economic importance of chemical weathering

Chemical weathering is very important for two reasons:

- It produces clays on which vegetation can grow. A mixture of dead vegetation and clays produces **soil** which contains the minerals plants need. In warm, damp regions soil starts to develop from rock within a few hundred years. Without it we'd have no food crops.
- Soluble ions from rock get carried away, leaving behind a greater concentration of insoluble ions. The result may be a mineable **ore**. For example, the aluminium ore bauxite is formed by the chemical weathering of igneous rocks called **feldspars** in hot, moist climates. (Look at the pictures on page 202.)

The weathering partnership

Physical and chemical weathering work in partnership. Physical weathering breaks rocks into bits so that more surface is exposed to chemical weathering. Chemical weathering breaks it down further:

physical + chemical weathering $\longrightarrow$
    broken rock fragments + clays + soils + solutions

The solutions drain away through the soil to form **groundwater**, or are washed across the surface and into streams and rivers.

Feldspars

- Feldspars are based on a giant structure of silicon and oxygen atoms, like quartz.
- But in feldspars many of the silicon atoms are replaced by aluminium atoms; and sodium, potassium, and calcium ions are trapped in the giant structure.

Products of weathering that get carried away in solution:

calcium ions, Ca^{2+}
potassium ions, K^+
sodium ions, Na^+
magnesium ions, Mg^{2+}
chloride ions, Cl^-
dissolved silica in the form of silicic acid, H_4SiO_4
hydrogen carbonate ions, HCO_3^-

These end up in the soil, rivers, and the sea.

1 Name three different kinds of physical weathering.
2 Which types would you expect to find in the Alps?
3 Explain the process of *exfoliation*.
4 What is the main difference between physical and chemical weathering?
5 Chemical weathering occurs fastest at the sharp edges of rocks. Why? (Hint: see page 124!)
6 Soil depends on chemical weathering. Explain.
7 Many rocks at Earth's surface are red-brown in colour. Suggest a reason.

18.4 From weathering to deposition

In the last Unit you saw that **weathering** is the first step in making sedimentary rock. What happens next? Read on...

Erosion and transport

Erosion is the removal of the products of weathering from where they were formed. Bits of rocks may tumble down a slope in a landslide. But they are more usually carried away or **transported** by water, wind, and glaciers. These are **transport agents**.

Transport and particle size

This graph shows how the size of particles a stream can pick up and carry depends on its speed.

The stream can *pick up* a particle the size of A only if its speed is at least 20 cm/sec. (This is erosion.) It can *carry it* until its speed drops to 1 cm/sec. (This is transport.) At slower speeds A falls to the bottom as **sediment**. (This is deposition.)

Suppose a fast stream slows down slowly and steadily. It will deposit the largest particles first (X). Further on, smaller particles (Y) will be deposited. Even further on, when it has almost stopped moving, the smallest particles (Z) will be deposited. In this way the particles get **sorted** according to size.

What do you think will happen if the stream slows down abruptly?

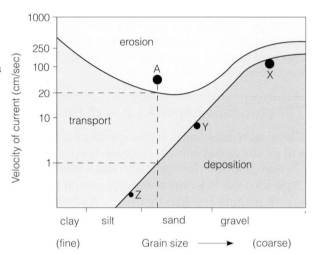

Sedimentary environments

Wherever a glacier melts or a current of wind or water slows down, it drops what it was carrying. So sediments can be deposited anywhere: lakes, rivers, deserts, mountain tops, beaches. This shows where they may be deposited on a river's journey to the sea:

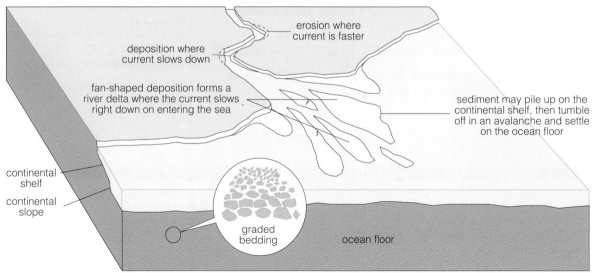

Clues from sediments

By studying a sediment, earth scientists can tell a lot about where it came from and how it was transported.

Particle size   If its particles are close in size, a sediment is **well sorted**. Dust carried by wind is well sorted. If they are all different sizes, the sediment is **poorly sorted**. Poorly sorted sediment on a river bed suggests that the current was strong to begin with, but slowed down abruptly, forcing the river to suddenly drop its load.

Particle shape   A glacier cushions the rock fragments it carries and protects them from collisions, so they remain sharp and jagged. A fast-flowing river flings rocks and stones against each other so they get broken into smaller bits, and smoothed and rounded. The more vigorous collisions a fragment suffers the smoother it will be.

The structure of the sediment   You can also learn a lot from this. For example:

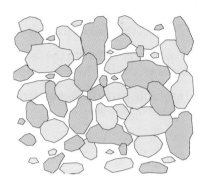

Poorly-sorted sediment

distinct layers of sediment
(e.g. on river bed)

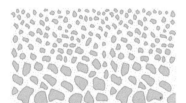

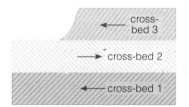

Bedding. Each layer marks a change in the type of sediment or the force with which it was carried. Here the heavy layer on top of a fine one suggests that the river has flooded recently.

Graded bedding is found on ocean floors. River sediment builds up on the continental shelf. If it gets dislodged it tumbles forward into deep still ocean water and settles like this.

Cross-bedding. Layers lie at different angles, reflecting the direction of flow of the wind or water that deposited them. Wind-deposited sand dunes are often like this.

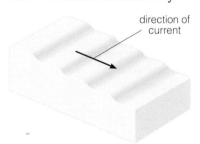

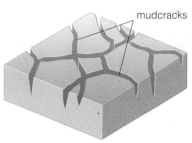

Ripple marks like these suggest that the sediment was shaped by a current of wind or water *after* it had been deposited.

Mud cracks suggest that the watery environment in which this sediment was deposited dried up at some point.

1 What does *erosion* mean?
2 Name the three transport agents.
3 Look at the graph on the opposite page.
   At what speed of current will a particle the size of X be:
   a picked up?   b deposited?

4 Glaciers produce *poorly sorted* sediment.
   What do you think the reason is?
5 Suppose you throw a shovel-full of soil, sand, and gravel into a very deep glass tank full of water. What kind of sedimentary structure will you expect to see?

18.5 From sediment to sedimentary rock

In the last two Units you saw that weathering produces rock fragments, clays and other insoluble substances, and solutions. Insoluble products get carried away by water, wind, and ice, and dumped as sediment. Solutions seep into the earth to form ground-water, or get carried in streams and rivers to the lakes and oceans.

The next stage is to turn the sediments and solutions into sedimentary rock. This can happen in three different ways.

1 Insoluble particles get cemented together. Most sedimentary rock is formed in this way.

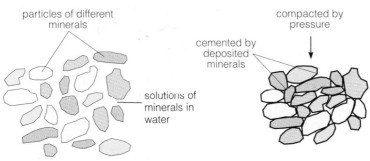

Sediment can be thousands of metres deep. Over millions of years it is slowly compacted by its own weight. The water ...

... gets squeezed out, depositing minerals such as calcium carbonate and silica as it goes. These act as cement.

The result: sedimentary rock. An example is **sandstone** which is sand grains in a silica or carbonate cement.

Sedimentary rock is classified according to particle size. Look at these examples:

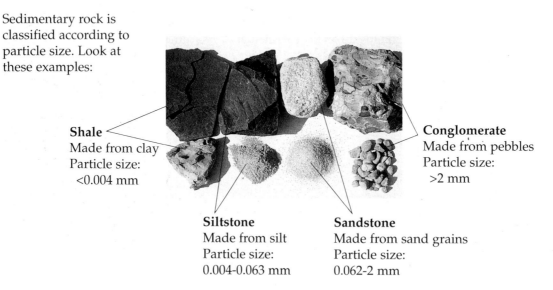

Shale
Made from clay
Particle size:
   <0.004 mm

Siltstone
Made from silt
Particle size:
0.004-0.063 mm

Sandstone
Made from sand grains
Particle size:
0.062-2 mm

Conglomerate
Made from pebbles
Particle size:
   >2 mm

Silt is mostly quartz, so siltstone feels gritty. The tiny flat particles in clay can slide over each other, so shale feels smooth when damp. Geologists check which rock is which by licking them!

2 Solutions first take part in a biological process. Sea animals remove calcium and bicarbonate ions from sea water to build shells and skeletons of calcium carbonate. When they die, these hard parts drop to the sea floor as sediment, which turns into limestone. The soft parts may turn into oil and gas which get trapped in rock.

Tiny single-celled sea plants called **algae** also use calcium carbonate to make protective casings. When they die, the tiny casings fall to the sea floor and the sediment forms **chalk**.

3 Salts crystallize out of solution. Some sedimentary rock is formed by crystallization. Suppose a lake or lagoon starts to dry up. As the water evaporates, the salts in it get more concentrated. They start to form crystals, the least soluble first. Some limestone is formed this way. So are **gypsum** (calcium sulphate), **rock salt** (sodium chloride), and **sylvite** (potassium chloride).

Fossils in sedimentary rocks

Suppose the remains of a plant or animal gets trapped in sediment. When the sediment turns into rock, the remains form a **fossil**.

Some fossils are actual parts of a creature, such as teeth and bones. Some are **trace fossils**, such as footprints and burrows. But most are **casts**. Dead remains decay, leaving behind a hollow that keeps their shape. This acts as a **mould**. It fills up with deposited salts and other substances that harden to form the cast.

If you know when a species lived, its fossil will give you a clue about the age of the rock. Sharks have been around for over 400 million years, so shark fossils are not much help this way. But some plant and animal species were around for only a short time and their fossils can be used to date rock more accurately. They are called **index fossils**.

Making use of sedimentary rocks

They provide us with:
- limestone for cement making
- sandstone for building
- shales for making clays to turn into bricks and ceramics
- oil and natural gas (these get trapped in sedimentary rock)
- coal

Ceramic pots after firing in kiln.

A fossil cast of a tree, at Stanhope in County Durham

A dinosaur footprint in sandstone, in Namibia. Size 6, would you say?

---

1 Name three soluble substances produced by weathering. (Page 269 will help you.)
2 Most sedimentary rock is made from particles compacted and cemented together.
   a Draw a diagram to show how this happens.
   b Where does the cement come from?
3 Explain how this sedimentary rock is formed:
   a limestone **b** chalk.
4 Rock salt is called an *evaporite*. Why?
5 **a** Describe two ways in which fossils are useful.
   b Most fossils are *casts*. Explain what this means.
   c What is a trace fossil? Give an example.

18.6 Metamorphic rock

Igneous rock forms at high temperatures deep inside Earth's crust. Sedimentary rock forms at relatively low temperatures at Earth's surface. Metamorphic rock forms at conditions somewhere in between the two.

How metamorphic rock is formed

Rocks get buried when new sediment forms above them, or during earthquakes and other earth movements. Around 10 km below Earth's surface, temperatures and pressures are high enough to change buried rock without melting it. The change is called **metamorphism.**

> **Making use of metamorphic rocks**
>
> They provide us with:
> - slate for counter tops and floors
> - marble for sculpting statues
> - rubies and sapphires
> - garnet, used as an abrasive and a gemstone
> - talc (used in talcum powder)

increasing pressure from all directions

cemented mineral grains

Here mineral grains in the rock get pressed tightly together from all directions, to produce a more dense and compact structure.

Circulating water speeds up the process. Substances dissolve and recrystallize. Ions migrate from one place to another.

More extreme conditions may cause grains to fuse together to form bigger ones. Higher pressure means larger grains.

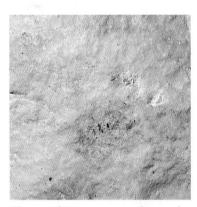

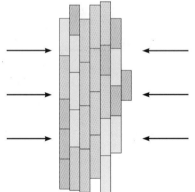

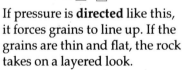

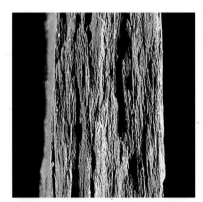

The result is metamorphic rock. **Marble** is an example. It is formed from limestone. But it is harder than limestone. Why?

If pressure is **directed** like this, it forces grains to line up. If the grains are thin and flat, the rock takes on a layered look.

Slate is an example. It is formed from shale (page 272) which is made of flat particles of mica and other minerals.

The alignment of crystals in rocks due to directed pressure is called **foliation**. Slate is foliated. Marble is non-foliated because its grains do not line up easily.

The minerals in the parent rock determine the minerals in the metamorphic rock. Marble, like limestone, is calcium carbonate. But where a parent rock has several minerals, new minerals may form as ions move around.

Which rocks can be metamorphosed – and where?

Any rock can be metamorphosed. Sedimentary rocks are the easiest
to change. (Why?) Even metamorphic rock can go through further
changes with increasing temperatures and pressures:

slate
(a metamorphic rock)

schist
grains have grown larger
(up to 1 cm across)

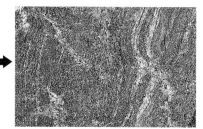

gneiss (pronounced *nice*)
minerals have migrated into
single-mineral layers

If the temperature had risen any further when gneiss was formed,
it would have melted to form igneous rock.

Metamorphic rock is usually found within mountain ranges. It is a
sign of the high pressures and temperatures associated with
mountain building.

In particular gneiss is found within ancient mountain ranges, or
where mountains once stood. The gneiss that forms the Scottish
islands of Lewis and Harris is the oldest rock in Europe.

Contact and regional metamorphism

Rocks *can* metamorphose through heat alone. This happens when
molten magma pushes its way into Earth's crust. The rocks in
contact with it undergo **contact metamorphism**. The change can
extend from a few centimetres to a few kilometres into the rock,
depending on how big and hot the body of magma is.

But where rocks get buried under sediment, or trapped by large
movements in Earth's crust (as in mountain building),
metamorphism may extend over thousands of square kilometres.
This is called **regional metamorphism**.

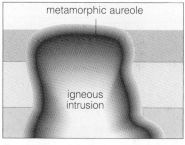

Contact metamorphism. The **aureole** is
the metamorphosed zone around the
intrusion.

1 How is metamorphic rock formed?
2 Name a metamorphic rock made from:
  a limestone **b** shale **c** slate.
3 **a** What is *foliation* in metamorphic rock?
  b What causes it?

4 **a** What is the difference between:
    i slate and schist? **ii** schist and gneiss?
  b Do you think they contain different minerals?
5 Where would you expect to find examples of contact
  metamorphism?

18.7 The rock cycle

At this moment, on the ocean floor, sediment is slowly turning into limestone. Come back a few million years from now, and you may find the limestone buried below ground, turning into marble. Check again a few million years later, and you may find its calcium ions in the lava erupting from a volcano.

All rock in Earth's crust changes continually from one form to another over millions of years. It is recycled. This is summed up in the **rock cycle**.

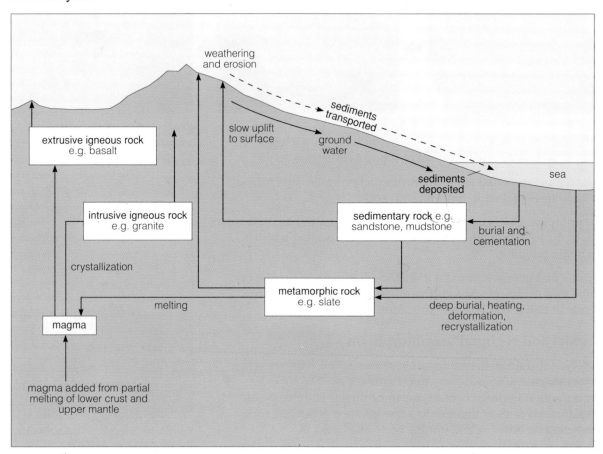

What drives the rock cycle?

The rock cycle is driven by **energy**, usually in the form of heat.
- Energy from the sun is responsible for wind, rain, vegetation, oxygen, carbon dioxide, and other factors that play a part in weathering, erosion, and transport – leading to sedimentary rock.
- The heat inside Earth helps rocks to metamorphose, or melts them into magma.
- It also causes rocks to get buried. Earth's crust and upper mantle form giant slabs called **plates**. These move around slowly, dragged by heat currents in the soft rock below them. When they collide, rock at the edges gets forced upwards to form mountains, or down into the crust to become metamorphic rock or magma. (You can find out more about this in Section 19.)

Note that if rock stays where it was formed, it does not change. For example, igneous rock is stable as long as it remains deep underground. But at Earth's surface it is in a hostile environment: the temperature is low, and it is surrounded by oxygen, water, and weak acids. Under conditions like these, igneous rock breaks down.

Keeping track of changes

The continual recycling of rock means geologists have to be detectives. They look for clues that will tell them where rocks have come from, and how they were formed. Below are examples of the clues they gather. You have met some of these already:

- Big crystals show that a rock cooled slowly from magma deep underground. Small crystals show that it cooled quickly on the surface.
- Cemented particles show it started as sediment.
- Sharp edges show that rock fragments did not suffer much chemical weathering, and were protected from collisions.
- If a rock contains shelly remains of sea animals it was created in a marine environment (even if you find it on a mountain top).
- Sediment is usually deposited in horizontal layers. If layers are slanted, it shows the rock has been folded or tilted at some stage by powerful forces.
- Usually, younger rock lies on top of older rock. If you find it *below* older rock, it shows that powerful forces have turned the rock upside down.

A sample of sedimentary rock called breccia

Some powerful forces were at work here!

1 What is the *rock cycle*?
2 Explain how calcium in the shells of sea animals might end up in magma.
3 Rock changes only if it gets moved to a different environment. Explain this.

4 Look at the first photo above. It shows a sedimentary rock called breccia. What conclusions can you draw from its appearance?
5 Study the second photo above, and then list your conclusions about the history of this rock.

Questions on Section 18

1 Imagine slicing Earth in half and taking a journey from the crust to the centre.
a Draw a diagram showing the cross-section through the centre, and label the layers.
b What happens to:
 i temperature ii density iii pressure
as you move from the surface to the centre?
c Why is the inner core solid and not liquid?
d Why is the overall density of Earth much greater than the average density of the rocks that form the crust?

2 This table shows a selection of common minerals found in rocks.

Mineral	Chemical formula	Hardness (Mohs scale)	Density g/cm³
quartz	SiO_2	7	2.65
halite	$NaCl$	2.5	2.16
calcite	$CaCO_3$	3	2.71
gypsum	$CaSO_4.2H_2O$	2	2.3
pyrite	FeS_2	6	5
diamond	C	10	3.51
a feldspar	$KAlSi_3O_8$	6	2.55

a What are *minerals*?
b What is the difference between a rock and a mineral?
c Which mineral in the table is:
 i a chemical element?
 ii an iron ore?
 iii a hydrated salt?
 iv a sulphide?
 v a form of silica?
d Which mineral is the main component of chalk?
e Which mineral is a silicate?
f Write down the chemical name for each mineral.
g The **Mohs scale** for measuring the hardness of minerals is named after the German mineralogist Friedrich Mohs. The larger the number the harder the mineral. List the minerals above in order of hardness.
h Which mineral cannot be scratched by any others in the list?
i A fingernail has a hardness of 2.6 on the Mohs scale. Which minerals in your list can be scratched by a finger nail?
j Name the two least dense minerals from the table.
k Do you think there is a connection between density and hardness? Use evidence from the table to support your answer.
l Quartz is used to make cutting and sanding discs for power tools. Why?
m Which minerals occur as evaporites from sea water?
n To which group of rocks do evaporites belong?
o Which mineral will dissolve in acidic rain water?

3 A laboratory demonstration was carried out to show how igneous rock forms. The following results were obtained by recrystallizing equal amounts of the chemical **salol** at 0 °C and at 50 °C.

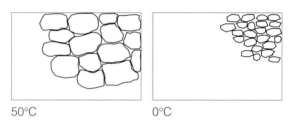

50°C   0°C

a What is meant by *recrystallizing* and how is it carried out in the laboratory?
b What effect does the temperature have on:
 i the speed of crystallization?
 ii the size of the crystals?
c Explain how these results relate to the two main types of igneous rock, extrusive and intrusive.

4 Igneous rock can be classified according to the percentage of silica in it. As the following graph shows, the amount of silica present affects both the crystallizing point and the viscosity of the magma.

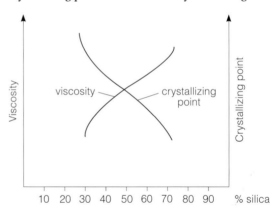

a How are igneous rocks formed?
b Give the chemical name and formula for silica.
c i What happens to the crystallizing temperature as the percentage of silica increases?
 ii Which type of rock would remain molten for longer, one with a high or a low silica content?
d i What effect does the addition of other elements have on the viscosity of the magma?
 ii Explain this effect using the idea of forces between atoms in giant structures.

5 Gabbro is an *intrusive* igneous rock. Andesite is an *extrusive* igneous rock.
a Explain the terms in italics.
b In what way(s) would you expect these rocks to look: i similar? ii different?
c Which of them appears at Earth's surface only after erosion of other rocks?

6

This is a sample of the igneous rock **obsidian**. It is a volcanic *glass*, which means it is not crystalline.
a What is the source of the rock obsidian?
b What type of structure (giant or molecular) does it have?
c How would you describe the arrangement of the atoms in obsidian: *ordered* or *disordered*?
d What kind of cooling produced obsidian: rapid or slow? Explain your answer.
e How do you think obsidian would respond to being struck with a hammer? Explain.
f Suggest how obsidian may have been used by early civilizations.

7 Erosion and transport are part of the rock cycle.
a What is *erosion*?
b What name is given to the movement of rock particles that causes erosion of other rocks?
c List the three main transport agents.
d Explain what would cause each transport agent to drop the load of particles it carries.
e A sediment deposited by wind or water may be *well sorted*. What does that mean?
f Explain how a stream sorts particles.

8 Sedimentary rocks often contain rock fragments that have been transported by rivers and deposited on river and sea beds. The rock fragments can be classified as sand, mud, or pebbles according to their size. All three are affected to different extents by moving water.
a Which of the three types of particle requires most energy to transport it? Why?
b Place the following observations in order of increasing water speed.
  A *the small pebbles move further downstream*
  B *sand is lifted from the sea bed*
  C *mud settles on the river bed*
  D *large pebbles are moved from one place to another*
  E *only pebbles are found*
  F *sand is dropped on the sea bed*
c Explain why the opposite banks of a river do not always look the same. (The drawing on page 270 will help you.)

9 The following diagram represents a geological feature found in many areas.

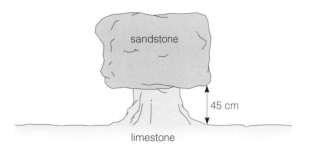

a Which type of rock is:
  i sandstone? **ii** limestone?
b Explain the sequence of events that led to sandstone being deposited on the limestone.
c Why is the limestone under the sandstone a different height from the rest?
d Assume that the rate of weathering of limestone is about 0.025 mm per year.
  i How long has it been since the sandstone was level with the surrounding limestone?
  ii What assumption have you made in this calculation?

10 **Dolerite** is an igneous rock found in Cumbria. In one area it is found in contact with the sedimentary rock sandstone.

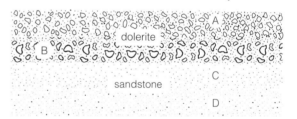

a The crystals of dolerite at A are smaller than those at B. Suggest a reason.
b The sandstone at C has become 'baked' and is much tougher than the sandstone at D. Why?
c From this evidence, which rock is older? Explain.
d Sometimes dolerite forms a **sill**, where lava pushes its way between the planes of sedimentary rock. Make a sketch of a sill.
e Is a sill younger or older than the sedimentary rock above it?
f What characteristic(s) might you expect the sedimentary rock on top of the sill to have?
g Other igneous rocks form as a result of lava flows. They cool and harden. Then sedimentary layers form on top.
  i How would you check whether a piece of igneous rock was a sill or a lava flow?
  ii Why is it likely that the surface of a lava flow will be uneven and full of small holes?

19.1 Another look at Earth's crust

Earth's crust is its surface layer. It varies between 8 and 65 km in thickness. It is made up of **continents** and **ocean basins**.

You may think the continents are big, but the ocean basins are much bigger. They cover 70% of Earth's surface. Like the continents they have mountains and chasms, and these are on a larger scale too!

The continents	metres	The ocean basins	metres
Average height above sea level	800	Average depth	3740
Highest mountain (Mount Everest, Tibet/Nepal)	8843	Highest mountain (Mauna Kea, Hawaii. It's so high it pokes up as an island.)	10 023
Deepest canyon (Colca Canyon, Peru)	3223	Deepest trench (Mariana Trench, Pacific Ocean)	11 033

The continents

All the continents have the same general features.
- They rest on a thick layer of igneous rock, which is mainly granite. This is called the **continental crust** or **granitic crust**.
- Above the granitic crust is a thinner layer of **metamorphic rock**.
- On top of this lies a thin and younger layer of **sedimentary rock**. It covers around three-quarters of the continents.
- Every continent also has chains of **mountains**, stretching for thousands of kilometres. These are made of folded layers of sedimentary and metamorphic rock, often with igneous intrusions (plutons) of granite.

Part of a mountain chain in the Himalayas in Nepal

Although metamorphic and igneous rock are mostly hidden from view, they've been pushed to the surface in some places by powerful forces, or exposed by erosion of sedimentary rock.

The ocean basins

The ocean basins have these features:
- The floor is a layer of basalt, around 4 to 5 km thick. This is called the **oceanic crust** or the **basaltic crust**.
- The oceanic crust is a little more dense than the continental crust: $3.0\,g/cm^3$ compared with $2.7\,g/cm^3$. That is why the continental crust sticks up above sea level!

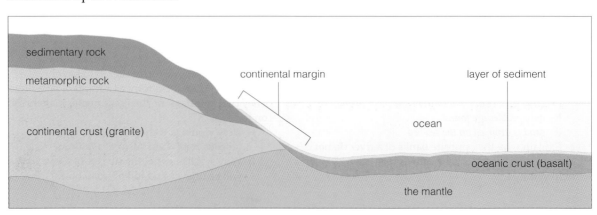

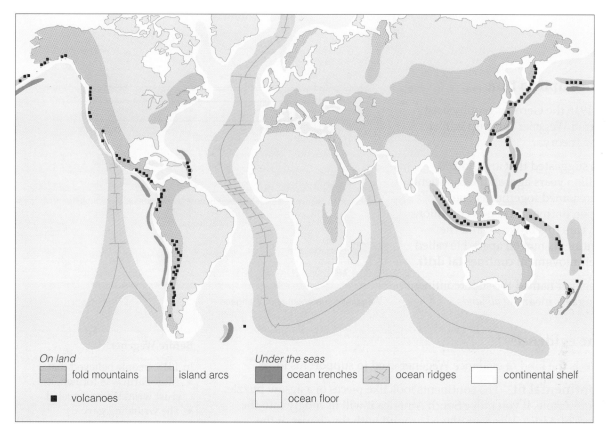

On land

▨ fold mountains ▨ island arcs

■ volcanoes

Under the seas

▨ ocean trenches ⊠ ocean ridges ☐ continental shelf

☐ ocean floor

- The area of transition between the continent and the deep ocean is called the **continental margin**. In some places it is very wide, like a shelf. In other places it is narrow and steep.
- Huge areas of the deep ocean basin floor are flat, and covered by a very thick layer of sediment: bits of bone and shell, dead algae and other microorganisms, and a mud of basalt particles.
- In places the ocean basins are split by deep V-shaped **ocean trenches**. Inside these the pressure is very high and the water very cold, so they are dangerous to explore.
- Chains of volcanoes often run alongside these trenches, on islands or the neighbouring continent. This suggests some connection between volcanoes and trenches.
- Chains of underwater mountains called **oceanic ridges** run round Earth from ocean to ocean. They are made of basalt. Most have rift valleys down the middle. Here basalt erupts from underwater volcanoes, creating new ocean floor.
- Numerous **earthquakes** also occur along oceanic ridges and ocean trenches.

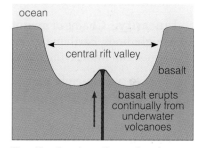

The rift valley down the centre of an oceanic ridge

1 Is it on a continent or in an ocean basin?
  a Earth's highest mountain
  b Earth's deepest chasm.
2 a What is: **i** continental crust? **ii** oceanic crust?
  b Write down another name for each.
  c Identify them as igneous or sedimentary.

3 Draw a diagram showing both kinds of crust.
4 Why is the continental crust raised higher?
5 Where would you find a *continental margin*?
6 What would you expect to find covering the flat ocean floor?
7 What is an *oceanic ridge*? What happens there?

19.2 Are the continents moving?

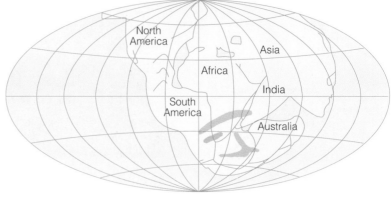

Once upon a time...

In 1915 the German geophysicist Alfred Wegener came up with a new theory.

He suggested that around 200 million years ago, all the continents were joined together as one supercontinent. Then over millions of years this broke up and the continents moved apart. He called their movement **continental drift**.

Wegener named the supercontinent **Pangaea**, meaning *all lands*.

 Glossopteris

Mesosaurus

Wegener's supercontinent, Pangaea

The evidence

This is the kind of evidence Wegener used to support his ideas.

Continental fit   The continents look like pieces of a jigsaw puzzle. For example, if you move South America it will fit snugly into the curve of Africa. Wegener played around with the shapes of the continents until he got Pangaea.

Linearity   Chains of mountains and volcanoes lie roughly along straight lines. Earthquakes also tend to occur along lines. This suggests an underlying pattern like you might find on a jigsaw.

(In 1958, twenty-eight years after Wegener's death, oceanic ridges were discovered. They too lie along lines, as you will see from the map on page 281.)

Fossils   Animals and plants survive only in climates and habitats that suit them. Yet fossils of the same animals and plants have been found on different continents, often in completely different climates.

For example, the small reptile *Mesosaurus* lived 240 million years ago. From its fossil, scientists can tell it swam only in shallow waters. Yet it has been found in Brazil and South Africa, separated by 5000 km of ocean. This suggests that South America and Africa were once joined.

The *Glossopteris* flora are plants that grew in humid temperate areas over 200 million years ago. Today, their fossils are found in rocks in every climate. And the Arctic island of Spitsbergen has fossils of palm trees as well as relatives of the oak and beech. This suggests that Spitsbergen drifted from tropical to temperate to arctic climates.

Rocks   Rocks of the same age, composition, and structure have been found thousands of miles apart. Similar rocks, 390 million years old, have been found in mountains in eastern North America, western Europe, and West Africa. This suggests that North America, Africa, and Europe were once joined, with a long line of mountains running through them.

Before Wegener

In 1915, people believed that:
- After Earth was formed, its crust wrinkled as it cooled.
- The wrinkling gave us mountains and other features.
- Once it had cooled and hardened, the crust changed very little.

Alfred Wegener (1880–1930), who was proved right in the end

There has been a huge increase in marine exploration in the last thirty years. This is the British Antarctic research vessel *James Clark Ross* at work in Antarctica.

The proof

Wegener's ideas were laughed at. No one could see how continents could move through the dense, ancient, and unchanging rock of the ocean basins. Then in 1968, thirty eight years after Wegener's death, the world's oceanic institutes launched a deep sea drilling project, using their ship *The Glomar Challenger*. It made two amazing discoveries:

- The rock in the ocean floor is much younger than the rock in the continents. The oldest continental rock found so far is 3.96 billion years old. The oldest ocean rock is only 0.18 billion years old.

- The further it is from the oceanic ridges, the older the ocean rock. The rock at the top of the ridges is almost new.

These discoveries shattered earlier beliefs. Ocean rock isn't ancient and unchanging after all, but young and mobile! Molten basalt is continually welling up at the oceanic ridges, forming new rock. At the same time the older ocean crust is moving aside, carrying the continents with it. The process is known as **sea-floor spreading**.

And that meant Wegener was right about continents moving.

Over the last thirty years, scientists have developed a theory, now generally accepted, about how and why continents move. They are carried on thick slabs or **plates**. These are dragged along by convection currents in the soft rock below them, whilst basalt wells up in their wake. This is the theory of **plate tectonics**.

You can find out more about it in the next Unit.

After Wegener

Thanks partly to Wegener, we now know that:
- Earth's crust is broken into slabs or plates, which are in continual motion.
- Collisions between plates produce features like mountain ranges and volcanoes.

---

1 What did Wegener call his supercontinent?
2 a What is *continental drift*?
  b Explain how evidence for it is provided by:
    i fossils ii rocks iii the shapes of continents.

3 a What is *sea floor spreading*?
  b What is the evidence for it?
4 Explain how sea floor spreading causes the continents to move.

19.3 Earth's plates

The theory of plate tectonics

The theory of plate tectonics was developed in the late 1960s. This is what it says.

- Earth's **lithosphere** (the crust + the upper part of the mantle) is cracked into a number of large rigid slabs called **plates**.
- Some plates carry just ocean basin. Most have both continent and ocean basin.
- These plates move, floating on the soft rock below the lithosphere. Everything on them moves too.
- They can drift apart (through sea floor spreading), or push into each other, or slide past each other.
- Most earthquakes, volcanic eruptions, and mountain forming take place at the edges or **boundaries** of the plates. That explains the 'linearity' of these events.

As you can see on the map below, there are seven large plates and a number of smaller ones. They are like a crude jigsaw. Britain is on the **Eurasian plate**.

How fast do plates move?

Their speed is measured in different ways. One method uses laser beams bounced off a satellite. The fastest plates (for example the Pacific plate) move at over 10 cm a year. The North American and Eurasian plates are among the slowest. The North American plate is moving west. The Eurasian is moving east. Each year, the distance between New York and London increases by 5–10 centimetres!

A volcano erupting on Big Island in Hawaii. Where there's a volcano, there's plate movement . . .

Earth's plates

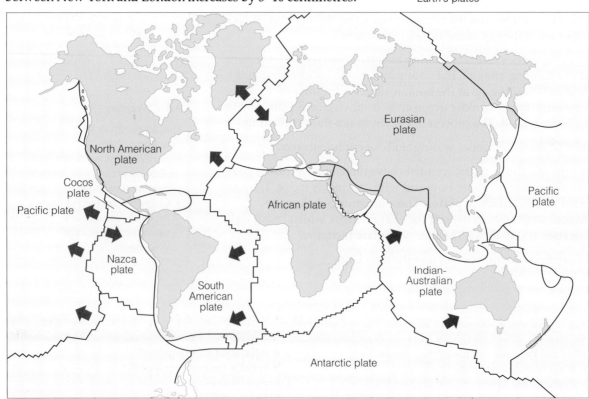

Tracking the plates

Remember what happens when you place a bar magnet on a sheet of paper and sprinkle iron filings around it? When you tap the paper, the filings form curves that follow the magnetic field.

In Unit 18.1 you saw that Earth acts like a giant bar magnet. Basalt contains iron compounds, including magnetite (Fe_3O_4), that act like iron filings. As the molten basalt cools, crystals of these compounds align themselves in a curved path along Earth's magnetic field. Once the basalt hardens, they keep this alignment.

You also saw in Unit 18.1 that the north magnetic pole flips south from time to time, which means Earth's magnetic field switches direction. The basalt in the ocean basins records these reversals:

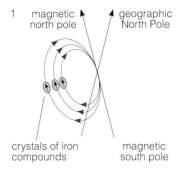

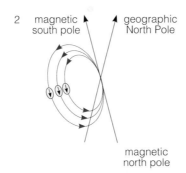

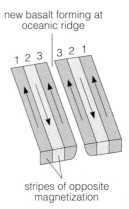

new basalt forming at oceanic ridge

stripes of opposite magnetization

The magnetic North is near the geographic North. In the cooling basalt, the crystals of iron compounds align themselves north like compass needles.

Many thousands of years later, Earth's magnetic field has reversed. Crystals that form now will align themselves in the opposite direction.

Another reversal takes place. Once again it is recorded by the crystals. So the basalt now shows stripes of opposite magnetization.

The stripes exist only because new basalt forms and pushes older basalt outwards. The discovery of stripes confirmed that sea floor spreading really is taking place.

Scientists can check the direction of magnetization in the basalt using a **magnetometer**. By seeing how the direction of magnetization changes, they can tell how a plate has moved.

They have been able to trace the paths of plates all the way back to Wegener's Pangaea, which appears to have broken up over 200 million years ago, in the age of the dinosaurs.

1 What are: **a** Earth's plates? **b** plate boundaries?
2 What are the plates made from?
3 Name a plate that has only ocean on it.
4 What plate are you sitting on?
5 How fast do plates move?

6 Basalt records the direction of Earth's magnetic field, just like a notebook. Explain how.
7 The ocean basins show stripes of opposite magnetization. How does this come about?
8 What is a *magnetometer*?

19.4 A closer look at plate movements

Why do plates move?

Why exactly do the plates move? To answer that, we have to return to Earth's interior. There, deep in the mantle, heat from natural radioactivity makes rock soften.

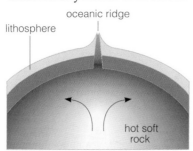

The hotter and softer the rock, the less dense it is. So it rises slowly, forming **convection currents** like you get when you heat very thick porridge.

It reaches the solid plates. A small amount melts and wells up between them. But most is forced sideways below the plates and drags them along.

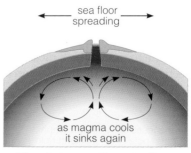

The plates continue to diverge over millions of years. The molten material that welled up between them hardens. The result is **sea floor spreading**.

The different plate movements

If plates drift apart, does that mean Earth's surface is getting larger? No. A drifting plate squashes up against other plates and its boundaries are forced upwards or downwards. In fact plates can move relative to each other in three different ways.

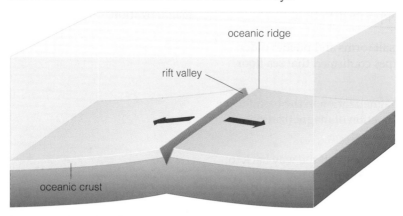

Two ocean plates diverging

1 They can diverge They drift apart, as you saw above, with new ocean crust forming at oceanic ridges. For example, the Eurasian and North American plates are diverging along the mid-Atlantic ridge. There are numerous earthquakes along the ridge as a result of the plate movements.

Since new crust is being formed, the boundary between diverging plates is called a **constructive boundary**.

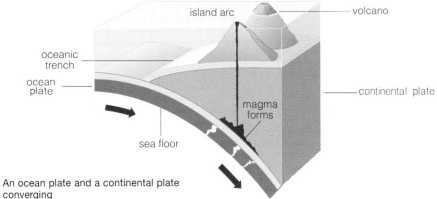

An ocean plate and a continental plate converging

2 They can converge They push up against each other. When an ocean plate and a continental plate converge, the thinner denser ocean plate is driven down or **subducted** below the thicker continental plate. When two *ocean* plates converge, the denser one is subducted.

Since crust is destroyed by subduction, the boundary between the plates is called a **destructive boundary**.

Subduction takes place at **ocean trenches**, and causes earthquakes and volcanoes. As you saw on page 281, chains of volcanoes often lie alongside ocean trenches. They form when subducted material melts and the magma pushes its way up through Earth's crust. Subduction along the Peru–Chile trench off the west coast of South America gives rise to volcanoes in the Andes.

Part of the Turkish town of Adarpazari after the 1999 earthquake.

But what if two *continental* plates converge? Neither is dense enough to subduct. So rock gets forced upwards, forming mountain ranges. That's how the Alps and Himalayas were formed.

Two plates in transform motion

3 They can slide past each other This is happening along the coast of California. The movement is called **transform motion**. No crust is created or destroyed, so the boundary between the plates is a **conservative boundary**. Since no material is subducted, no volcanoes are formed. But the friction between the plates causes earthquakes that can be devastating.

1 Explain in your own words why plates move.
2 What is happening at the mid-Atlantic ridge?
3 Draw a diagram to show what happens when:
  a an ocean plate and a continental plate converge
  b two continental plates converge.
4 For each part of question 3, give two examples of where this is happening. (Use the map on p. 284.)
5 What is *transform motion*?
6 Why are there earthquakes but no volcanoes along the coast of California?

19.5 Shaping the continents

What happens at plate boundaries?

When plates move, the rocks at the boundaries get pushed and pulled by powerful forces. What effect does this have on them? You will find out in this Unit.

Faults

When a rock is put under enough stress it will crack and shift. A **fault** is a crack in rock along which movement has taken place. The direction of movement depends on the direction of the force, as these diagrams show:

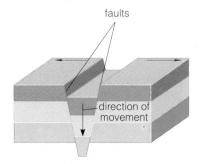

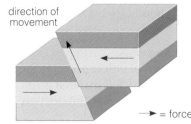

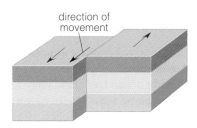

Here a block of rock has dropped down between two faults. This happens when rock is being pulled in opposite directions, for example where two plates are diverging.

Here one block is pushed up over the other at the fault. This happens when rock is being squashed with great force, for example where two plates are converging.

Here one block gets pushed past the other. This happens when rock is subjected to a shearing stress, for example when plates slide past each other.

When a rock is under stress, it stores up **strain energy**. When it gives way, this energy is released. It passes through the surrounding rocks as high-speed waves and causes the vibrations called **earthquakes**. Faulting always gives rise to earthquakes.

When the rock comes to rest, new forces may build up around it and cause it to move again. This means more earthquakes.

Folds

Folds are formed when compressed rocks deform *without* cracking, as shown on the right.

It's like when you press your hands on a table cloth and slide it towards the centre of the table.

Folding mainly takes place at the boundary between two converging plates. The diagram shows how sedimentary layers fold.

Extreme pressure will cause folds to **overturn**, as shown.

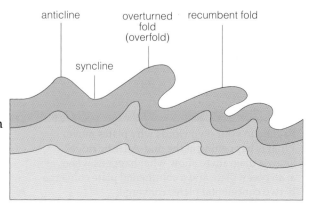

Fold mountains

Many of the world's highest mountain ranges are the result of plate movement. They started as sedimentary rock on the sea floor. As this was squashed upwards by converging continental plates, it got folded, faulted, metamorphosed, and intruded by igneous plutons.

For example, when the Indo-Australian plate converged on the Eurasian plate, an ancient sea called the **Tethys** got squashed upwards. The result is Mount Everest, K2, and the other mountains of the Himalayas. The plates are still converging so the Himalayas are still growing, and causing major earthquakes in China.

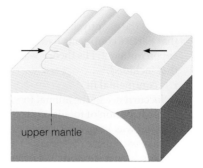
The making of a fold mountain

Volcanoes

As you saw earlier, volcanoes are linked with plate movements.

- Underwater volcanoes erupt non-stop at diverging boundaries, making new ocean floor.
- Subduction at ocean trenches also leads to volcanoes. Subducting rock heats and softens as it slides below the other plate. The water it carries with it superheats, and triggers melting. Magma forces its way to the surface, forming a line of volcanoes parallel to the trench. They may appear in the ocean as **island arcs**, or on a continent, like the volcanoes of the Andes.

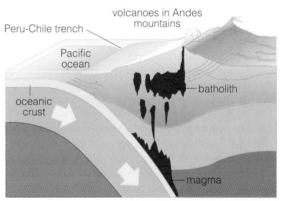
Subduction at ocean trenches leads to volcanoes

More about earthquakes

As you saw earlier, earthquakes are caused by rock movement at faults. About every two minutes there's an earthquake somewhere on Earth. Most are very slight. But around 20 a year cause severe damage.

Earthquakes can cause rockslides and landslides which sweep roads and homes away. In the ocean they can cause giant destructive waves called **tsunami**. But much of the damage they do is indirect.

For example, torn electric cables and fractured gas pipes lead to fires. Burst water mains mean the fire brigade can't operate properly. Blocked roads mean help can't get through. Communications are cut off. People are left homeless, isolated, scared, and hungry.

You can try some exercises about earthquake damage on page 291.

1 What is a *fault*? What causes it?
2 A depression between two faults is called a *graben*. Draw a diagram to show how a graben develops.
3 What is a *fold*? What causes it?
4 Name one type of rock that's likely to fold.

5 Describe in your own words how plate movements can result in mountains.
6 a What are island arcs?   b How do they form?
   c Why are they linear?
7 List five hazards caused by earthquakes.

Questions on Section 19

1 In 1915 the geophysicist Alfred Wegener suggested
that Earth's continents had once been joined as a
supercontinent. His theory was based on evidence
that included:
i 'continental fit'
ii occurrence of the animal fossil *Mesosaurus*
iii distribution of fossils of the flora *Glossopteris*
a What was Wegener's supercontinent called?
b Take each piece of evidence i–iii in turn, and
explain how it supports Wegener's theory.

2 Early in the 20th century two opposing theories were
put forward to explain the features of Earth's crust.
Theory A: Its features are the result of the shrinking
and wrinkling of the crust as it cooled.
Theory B: Earth was once a giant continent which
split apart to form separate continents.
a Which of these theories:
i fits best with the idea that Earth is unchanging?
ii suggests that changes are still going on?
iii fits best with the idea that Earth's crust is a
fixed solid mass?
b Suggest why Theory B, proposed by Alfred
Wegener in 1915, was not accepted by scientists at
that point.
c Describe two key pieces of evidence that were later
discovered, which helped to establish his theory.

3 Plate tectonics is the driving force behind the
recycling of rocks. The key to plate tectonics is the
heat generated in Earth's interior.
a What is meant by the *recycling* of rocks?
b i Name an important source of heat energy in
Earth's interior.
ii Explain how this heat energy leads to the
movement of plates.
c Sedimentary rock can be recycled as metamorphic
rock.
i How does sedimentary rock get pushed down
through Earth's crust?
ii Explain how it becomes metamorphic rock.
d Sedimentary rock can also be recycled as igneous
rock. What further conditions would be needed to
turn it into igneous rather than metamorphic rock?

4 The convection currents in Earth's mantle provide the
energy for plate movement. Draw a simple diagram
to represent the lithosphere and show how plate
movement occurs. Use arrows to represent the
convection current and add these descriptions:
i Heat energy arises from the decay of radioactive
elements in the centre of Earth.
ii Magma rises as it becomes less dense.
iii Plates are carried along on the currents of magma.
iv Magma cools and sinks as it becomes more dense.

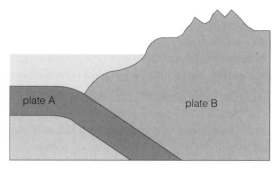

5 a This diagram shows a cross-section through two
plates, A and B. Copy it and add these labels:
*ocean   oceanic crust   continental crust   mountains
oceanic trench   sediments   continental margin*
b i Draw an arrow to show the movement of A.
ii Guess how far A would move in 12 months.
iii What other name is given to the crust in A?
iv Where is the oldest part of plate A? Explain.
c i Why does plate A sink beneath plate B?
ii What scientific name is given to this process?
d Name a mineral you'd find in the sediment.
e Mark on your diagram an area where metamorphic
rock could form.
f Mark a point on your diagram where you think a
volcano might erupt. Explain your reasoning.

6 'Stripes' of magnetic reversal appear in the
oceanic crust. They provide evidence about the past.

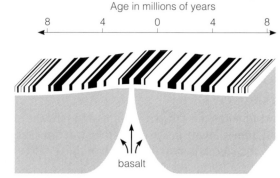

a i Where on this diagram is the oceanic ridge?
ii What process happens here?
iii Where is the oldest rock?
iv Why is the pattern symmetrical?
b i What do the black and white stripes represent?
ii What substance present in basalt is affected by
Earth's magnetic field?
iii This rock provides a permanent record of
changes occurring over millions of years. How?
c From the diagram, estimate the average lifetime of
a magnetic reversal.
d Would you expect this pattern of magnetic
reversals to continue in the future? Explain.

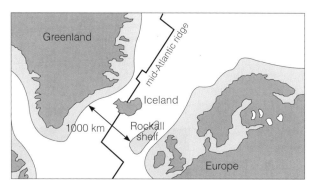

7 Above is a map of the North Atlantic ocean. Greenland and Europe were once joined. Now the North Atlantic ocean is widening by about 2 cm per year.

a What is the cause of this widening?

b Is the plate boundary constructive or destructive?

c What do you find along the mid-Atlantic ridge?

d How was Iceland formed? From what type of rock?

e **i** Estimate how long ago the two continents were joined.

ii Why are distances from the edges of the continental shelves used in the calculation, rather than distances from the coasts?

iii What other assumption have you made in doing the calculation?

f What other tectonic process must be taking place to balance out the fact that the crust is getting larger between Europe and Greenland?

8 Strips of coloured Plasticine can be used to show the movement of rock layers to form fold mountains.

strips of Plasticine

a Draw a diagram to show what you predict will happen if equal pushing forces are applied to both ends at the same time.

b Draw a second diagram to show what you predict will happen if one end is held firmly and a pushing force is applied to the other end.

c Which of the two processes do you think will produce an *overturned fold*?

d **i** What do you expect the time scale for this process to be?

ii What other process will be taking place over this period of time, which will lead to the loss of rocks from the top of the mountain?

iii Using the Plasticine model, draw a diagram of what the top of the fold mountain may look like at the end of the time period.

9 When continents collide, rocks sometimes become folded and sometimes broken.

a Which of these two processes results in:

i land being uplifted? **ii** the formation of a fault?

b Draw diagrams to illustrate the difference between the two processes.

c What is: **i** an *anticline*? **ii** a *syncline*?

d Identify an anticline and a syncline on the appropriate diagram from **b**.

10 Earthquakes can cause great loss of life and millions of pounds worth of damage. Much of the damage is caused by fire.

a What is an earthquake?

b Give three examples of earth movements that result in earthquakes.

c Why don't major earthquakes occur in the UK?

d Imagine a major earthquake hitting your nearest town. List as many ways you can in which it could lead to fires, including fires in homes.

e What other kinds of earthquake damage make fires more difficult to fight?

11 In major earthquakes many buildings are destroyed or damaged beyond repair. Buildings are usually designed to withstand vertical stress (the weight of several storeys) but not the 'whiplash' effect they experience during earthquakes.

In the Great Kwanto earthquake of 1923 in Japan, 120 000 people were killed and 80% of the buildings in Tokyo destroyed. Most were wooden buildings which were consumed by fire. Following this, Tokyo was rebuilt with mainly concrete buildings. Reinforced concrete was used in certain buildings. The height limit for buildings was set at 160 metres.

a What is *reinforced concrete*?

b Explain why each of the above measures should help reduce earthquake damage in Tokyo.

c A building technique called *base isolation* is used to a limited extent in some earthquake zones. The building is mounted on blocks consisting of alternate layers of rubber and steel. Explain why this technique should help to reduce earthquake damage.

12 The main causes of injury during earthquakes are:
- building collapse or damage
- flying glass from broken windows
- falling pieces of furniture such as book cases
- fires from broken gas lines, electrical short circuits, overturned stoves, etc.
- fallen power lines.

Imagine you live in an earthquake zone. Write down:

a five precautions you would take in your home on receiving an earthquake warning.

b three precautions you would take if you were out of doors when an earthquake occurred.

20.1 The development of the periodic table

Imagine your work seems like a confusing mess – and then suddenly it all makes sense! That's how many chemists felt, when the periodic table was first published over 130 years ago.

The periodic table is the chemist's map. It helps you understand the patterns in chemistry. Today we take it for granted. But it took hundreds of years, and the work of hundreds of chemists, to develop.

First, find the elements!

The first job was to find out which substances were elements. Not easy. As you saw on page 6, Plato and Aristotle thought earth, air, fire and water were the only elements. The alchemists opted for mercury, sulphur and salt.

In fact many elements were already in use – for example gold, silver, lead and copper – but they were not classed as elements. Then in 1661, Robert Boyle came up with the definition we still use: *an element is any substance that cannot be broken down into simpler substances*. The show was on the road.

Over the next two hundred years, over 50 elements were discovered. Scientists began to notice patterns in their behaviour, and try to explain them.

The Law of Triads

In 1817 the German scientist Johann Dobereiner noticed that calcium, strontium and barium had similar properties, and that the atomic weight of strontium was halfway between the other two. He found the same pattern with chlorine, bromine and iodine, and lithium, sodium and potassium. So he put forward the **Law of Triads**: *Elements exist in threesomes or triads with similar properties; the atomic weight of the middle one is the average of the other two.*

Soon, scientists found that they could expand the groups. For example, fluorine seemed to fit the halogen group.

The Law of Octaves

By 1863, 56 elements were known. John Newlands, an English chemist, noted that there were many pairs of similar elements. In each pair, the atomic weights differed by a multiple of 8. So he produced a table with the elements in order of increasing atomic weight, and put forward the **Law of Octaves**: *an element behaves like the eighth one following it in the table.*

This was the first table to show a **periodic** or repeating pattern of properties. But it was not widely accepted because there too many inconsistencies. For example he put copper and sodium in the same group, even though they have very different properties.

The periodic table

Dmitri Ivanovich Mendeleev was born in Siberia in 1834, the youngest of 17 children. By the time he was 32 he was a Professor of Chemistry.

A few elements, such as copper, have been known for thousands of years. But most were discovered in the last 400 years. This painting shows Henry Brand, the German alchemist, on his discovery of phosphorous in 1659. He extracted it from urine, by accident, during his search for the elixir of life. To his amazement it glowed in the dark.

John Newlands

Mendeleev collected a huge amount of data about the elements, some by research, and some by writing to scientists around the world. He made a card for each of the known elements (by then 63), with all the data written on. He arranged the cards on a table, first in order of increasing atomic weight, then into groups with similar behaviour, resulting in the periodic table, published in 1869.

Mendeleev's table was a big improvement on Newland's. He ironed out the inconsistencies by leaving gaps for elements not yet discovered. He named three of them **eka-aluminium**, **eka-boron** and **eka-silicon**, and predicted their properties. Soon **gallium**, **scandium** and **germanium** were discovered. They fitted his predictions – and at last scientists accepted the idea of periodicity. His table became the blueprint for our modern one.

Atomic structure and the periodic table

Mendeleev thought the pattern in his table was related to atomic weight. He was puzzled. He'd been forced to swop some elements (like tellurium and iodine) to get them into the right families.

Then in 1911 Ernest Rutherford, a New Zealander working in Cambridge, showed that atoms have a charged nucleus. The charge is due to protons. It became clear that **proton number** or **atomic number** (and not atomic weight) was the key factor in deciding an element's position in the table.

In 1914 the Danish scientist Neils Bohr put forward the first atomic model for hydrogen. It showed an electron circling the nucleus. Over the next 20 years we learned that the number of protons in an atom dictates the number of electrons. The electrons are arranged in shells. The outer electrons dictate how the element behaves. The periodic table made sense!

Today

The periodic table has grown a lot since Mendeleev. Today 115 elements are known. But not all are found in nature. More than 20 were made artificially in nuclear reactions and existed for only a few seconds before decay.

Dmitri Mendeleev in the chemical laboratory of the University of St. Petersburg.

Atomic weight

- In the past scientists worked out the **atomic weight** of an element by comparing it to hydrogen (atomic weight 1).
- Then in 1961 the carbon-12 isotope became the standard: $^{12}_{6}C$.
- Since then we have used the term **relative atomic mass** instead of atomic weight.

See page 62 for more!

The periodic table on page 330 will help for these questions.

1 This is some modern data about elements.

Element	Atomic number	Mass number of main isotope
Li	3	7
Na	11	23
K	19	39
Rb	37	85

  a In what way does it fit:
    i the Law of Triads? **ii** the Law of Octaves?
  b Rubidium is the odd one out. Suggest a reason.

2 **a** Suggest reasons why Mendeleev thought that the behaviour of the elements depended on their atomic weight.
  b Today, what do we know to be the key factor in deciding which group an element belongs to?
  c Which is the main factor in deciding how an element behaves?

3 Give the name and atomic number of the element named after:
  a Mendeleev   **b** Rutherford   **c** Neils Bohr
  d Europe   **e** America

20.2 Radioisotopes and radioactivity

You saw on page 31 that many isotopes are unstable. Let's take carbon as an example. It has three isotopes, ^{12}C, ^{13}C and ^{14}C. They are called carbon-12, carbon-13 and carbon-14.

$^{12}_{6}$C
6 protons
6 electrons
6 neutrons

An atom of carbon-12. Almost 99% of carbon atoms are like this.

$^{13}_{6}$C
6 protons
6 electrons
7 neutrons

An atom of carbon-13. Just over 1% of carbon atoms are like this.

$^{14}_{6}$C
6 protons
6 electrons
8 neutrons

An atom of carbon-14. A tiny % of carbon atoms are like this.

In carbon-14, the extra neutrons make the nucleus unstable. Sooner or later, every carbon-14 atom throws a particle out of its nucleus, and turns into a *nitrogen* atom. This incredible process is called **decay**.

Carbon-14 is said to be **radioactive**. It is called a **radioisotope. Radiation** is the term used for the particles it gives out on decay. Sooner or later, all radioisotopes become stable atoms by giving out radiation.

Radiation

Radioisotopes may give out these forms of radiation when they break down:

Alpha particle	Beta particle	Gamma rays
alpha particle / radioactive substance	beta particle	gamma ray
2 protons + 2 neutrons Mass: 4 units Shoots out at high speed but slows down quickly in air.	1 fast electron Mass: almost none Forms when a neutron turns into a proton and an electron: $n \rightarrow p + e$ The electron is thrown out. But the mass of the isotope does not change!	Usually given out at the same time as alpha and beta particles. High energy waves that travel at the speed of light.

Some examples of decay

1   When carbon-14 decays to nitrogen, it does so by emitting a beta particle. Like this:

$^{14}_{6}$C $\rightarrow$ $^{14}_{7}$N + a beta particle

Since the atom gains a proton, it is charged at first. But it rapidly picks up an electron from its surroundings and the charge is cancelled.

2   The radioisotope americium-241 is used in smoke alarms. It has 95 protons and decays by giving out an alpha particle. That leaves 93 protons, which means it has turned into neptunium:

$^{241}_{95}$Am $\rightarrow$ $^{237}_{93}$Np + an alpha particle

Can you see why the mass number has dropped to 237?

How far can radiation travel?

Alpha particles can't get through your skin. But other radiations can! Look at this diagram.

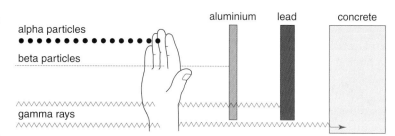

A large burst of radiation will kill millions of cells in your body. It causes **radiation sickness**: vomiting, tiredness, loss of appetite, hair loss, bleeding gums, and usually death within weeks. Small doses of radiation over a long period of time may lead to cancer.

Making use of radioisotopes

Although radioisotopes are dangerous, they are also useful. For example:

As tracers Engineers check oil and gas pipes for leaks by adding radioisotopes to the oil or gas. At a leak, radiation is detected using a Geiger counter. Radioisotopes used in this way are called **tracers**.

In cancer treatment Gamma rays can cause cancer. But they are also used in **radiotherapy** to *cure* cancer – because they kill cancer cells very easily. Cobalt-60 is the usual source of gamma rays for this. The beam of rays is controlled so that it hits its target exactly.

As fuels In nuclear power stations, large radioisotopes such as uranium-235 are used as fuel for making electricity. This time the radioisotopes are bombarded with neutrons, which split them into smaller atoms:

Checking for oil/gas pipe leaks using a Geiger counter

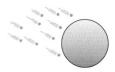

1 The radioisotopes are bombarded with neutrons …

2 … which split them into smaller atoms. This is called **nuclear fission**.

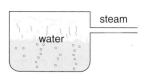

3 The huge amount of heat it gives out is used to turn water into steam …

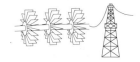

4 … which is used to drive turbines for making electricity.

The splitting is called **nuclear fission**. It gives out a huge amount of heat, which is used to turn water into steam. The steam is used in turn to drive turbines for making electricity.

The periodic table on page 330 will help you answer some of these questions.

1 What is:
  a a radioisotope? b radiation?
  c radioactive decay?
2 Name the 3 types of radiation and say what they are.
3 These radioisotopes decay by giving off an alpha particle. Say which elements they become.
  a polonium-212 b uranium-238
  c plutonium-239
4 These radioisotopes decay by giving off a beta particle. Say which elements they become.
  a hydrogen-3 (tritium) b phosphorus-32
  c iodine-131

5 The radioisotope sulphur-35 decays into an element with a proton number of 17 and a mass number of 19.
  a What is the proton number of sulphur?
  b Name the new element and give its symbol.
  c Name one type of radiation given out during the decay.
6 Give two examples of how radiation can help us.
7 Suggest a reason why:
  a the elements at the bottom of the periodic table often have several different radioisotopes;
  b alpha particles can't pass through your skin, while beta particles can;
  c waste material from nuclear power stations is usually buried deep below ground.

20.3 More about cells and batteries

The simple cell

In Unit 12.4 you learned something amazing ...

1 If you connect two strips of different metals to a bulb – like this ...

2 ...and stand them in an **electrolyte** (a solution that can conduct electricity) ...

3 ... the bulb lights up!

4 That's because one metal – the more reactive one – has a stronger tendency or **potential** to give up electrons than the other.

    We say there is a **potential difference** between them.

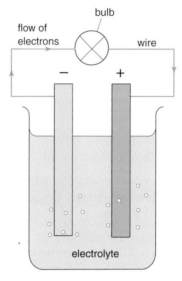

Zn/Cu strips and NaCl solution

5 So reactions take place at the metal strips ...

6 ... and the more reactive metal 'pushes' electrons to the other metal. An electric current flows along the wire, making the bulb light.

This set-up is called a **simple cell**. The metal strips are called **electrodes**. The **voltage** of the cell is the potential difference between them. It is measured in **volts.**

Things to remember about cells

- The more reactive metal becomes the negative electrode because it 'pushes' the electrons out.
- You can change the voltage of the cell by choosing different metals.
  The bigger the difference in reactivity of the metals, the bigger the voltage of the cell.
  For example with zinc and copper electrodes you get 1.1 volts but magnesium and copper give 2.7 volts.

Batteries

Batteries are just portable cells. This diagram shows a torch battery.

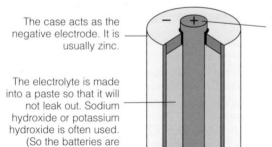

The case acts as the negative electrode. It is usually zinc.

The electrolyte is made into a paste so that it will not leak out. Sodium hydroxide or potassium hydroxide is often used. (So the batteries are called **alkaline** batteries.)

The positive electrode is down the middle.

Manganese(IV) oxide is often used, packed around a rod of carbon. It is suitable because the Mn^{4+} ion can accept electrons to become Mn^{3+}.

The battery 'dies' when reactions at the electrodes stop.

A torch battery

Battery companies are always trying out different electrodes and electrolytes to produce better longer-lasting batteries.

The ideal battery

The ideal battery would be:
- small and light so that it is easy to carry around
- cheap and long-lasting
- safe (won't harm the user or pollute the environment)
- sturdy (won't break when you drop it).

Some of these properties are more important than others.
It depends what the battery will be used for.

Rechargeable batteries

Some batteries are **rechargeable.** When they run down, you
connect them to a source of electricity. It forces the reverse reaction
to take place at each electrode. So the electrodes are almost as good
as new. The battery is ready to work again.

One good example is the car battery. It produces a current to start
the car engine. But once it is running, the engine acts as a source of
electricity to **recharge** the battery. So car batteries can last for years.

Fuel cells

Fuel cells are a special type of battery. They make use of the
potential difference between two **gases**. The electrodes in the fuel
cell do not themselves react – they are **inert**. They just allow current
to pass through them. This is a simple diagram of a fuel cell:

A collection of batteries

A lead-acid car battery

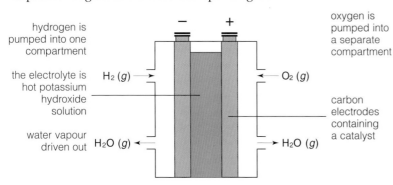

hydrogen is pumped into one compartment

the electrolyte is hot potassium hydroxide solution — H_2 (g) →

water vapour driven out — H_2O (g) ←

oxygen is pumped into a separate compartment

← O_2 (g)

carbon electrodes containing a catalyst

→ H_2O (g)

The fuel cell reactions

At one electrode:
$2H_2(g) + 4OH^-(aq) \rightarrow 4H_2O(l) + 4e^-$

At the other:
$O_2(g) + 2H_2O(l) + 4e^- \rightarrow 4OH^-(aq)$

Overall:
$2H_2(g) + O_2(g) \rightarrow 2H_2O(l)$
So hydrogen and oxygen combine to give water.

Fuel cells are used to power space craft and are being tried out as
car 'engines' and to provide electricity in homes. A big advantage
is that the product – water – does not pollute the environment.

1 Look at these three pairs of metals:
   copper, zinc;   zinc, iron;   silver, zinc.
   a Which pair will give the highest voltage?
   b Which will give the lowest voltage?
   (Page 196 will help.)
2 Suggest a reason why:
   a sodium metal is not used in batteries;
   b gold is not used in torch batteries;
   c iron is not used in torch batteries even though it is
      quite cheap.

3 Suggest a reason why:
   a silver is used in camera batteries but not in
      batteries for children's toys;
   b rechargeable batteries are used in cars;
   c dilute sulphuric acid is used in car batteries – but
      not in torch batteries;
   d torch batteries often carry the label 'No mercury'.
4 a Draw and label a diagram to show a fuel cell.
   b Why are fuel cells used in space craft, instead of
      ordinary batteries? Give at least three reasons.

20.4 Colloids

Think about a glass of milk.
- You can't see through the milk – so it is not a solution.
- You can't see any particles in it – so it is not a suspension.
- It is a special kind of mixture called a **colloid**.

What are colloids?

A colloid is a mixture in which tiny bits of one substance are evenly spread or **dispersed** through another substance.

The dispersed bits are so small that you need a microscope to see them. But they are large enough to scatter light – and that's why you can't see through a colloid clearly.

Make-up remover is a colloid ...

The dispersed material is called the **disperse phase**.

It can be:
- tiny particles of solid
- tiny droplets of liquid
- or tiny bubbles of gas

The main material is called the **continuous phase**.

It can also be liquid, solid or gas.

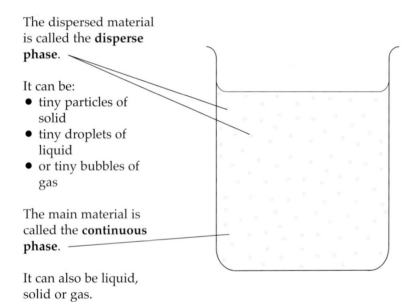

... and so is polystyrene foam.

Different types of colloid

Since each part of a colloid can be liquid, solid or gas, there are many different kinds of colloid. Here are some examples:

Disperse phase	Continuous phase	Colloid is called ...	Example
liquid	liquid	an **emulsion**	milk (fat droplets on water) make-up remover (mainly oil in water)
gas	liquid	a **foam**	hair mousse (air in a liquid) whipped cream (air in cream)
solid	liquid	a **sol**	paint (tiny particles of solid in water or oil)
solid	solid	a **solid sol**	coloured glass
liquid or solid	gas	an **aerosol**	hair spray (liquid in gas) mist (tiny water droplets in air) smoke (tiny ash particles in air)
gas	solid	a **solid foam**	polystyrene foam (air in polystyrene)

Gels

Some solids with large molecules form a special type of colloid called a **gel** when you mix them with liquid.

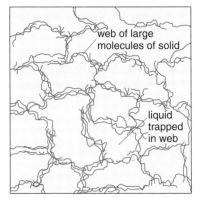

The large molecules attract each other, forming a web. The liquid gets trapped in the web so the mixture thickens.

Wallpaper paste is a gel. It is runny when you first mix it but gets quite thick within 5 or 10 minutes.

Jelly is a gel of gelatine and water with colour and flavour added. It 'sets' as the gelatine web forms.

Colloids are everywhere

Colloids are everywhere. Most sauces and household cleaners are colloids. So are most cosmetics.

Colloids combine the useful properties of their constituent materials. Hair mousse contains liquid, so you can spread it easily through your hair. The air bubbles make it easy to spread just a *small* amount. In polystyrene foam, the combination of air and plastic makes it light but rigid, ideal for packaging and as an insulator.

Most make-up is oil-based. So it is hard to remove with water alone. That's why make-up remover contains oil.

Emulsifiers

To make salad dressing you shake oil and vinegar together. The result is an emulsion. After a time it will separate into two layers again. So the salad dressings you buy in shops have an **emulsifying agent** or **emulsifier** added to stop this happening.

Emulsifiers are made up of long molecules. One end of these is attracted to oil and the other to water. The molecules surround the oil droplets and keep them dispersed.

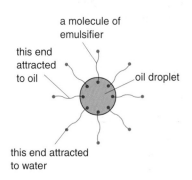

1 Draw a diagram to show what a colloid is. Label the disperse and continuous phases.
2 Which type of colloid do you think this is?
   a foam rubber   b hand cream
   c meringue   d hair gel
   e lipstick   f insect spray
3 Why is it not possible to make a colloid from:
   a ethanol and water?   b oxygen and nitrogen?
4 What does an emulsifier do?

5 Washing-up liquid acts as an emulsifier in greasy water. Draw a diagram to show how it works.
6 Explain how the properties of air and rubber make foam rubber suitable for cushions.
7 When you make wallpaper paste, it starts thin and runny but then thickens as it stands.
   a Draw a diagram to show why it thickens.
   b Is this property useful to us? Explain.

20.5 Products from plants

Plants are like factories.

1 They take carbon dioxide from the air and water from the soil as their raw materials.

2 Using energy from sunlight and chlorophyll as catalyst, they turn them into **glucose**, a sugar.

3 But that's just the start! They polymerise the glucose to make **carbohydrates** such as starch and cellulose.

4 Using glucose and minerals from the soil, they also produce **proteins**, **fats** and thousands of other chemicals – mostly with very large molecules.

5 These are used to make roots, leaves, seeds, petals and so on. You may end up eating them!

(You can find out about the carbohydrates, proteins and fats in food in the next unit.)

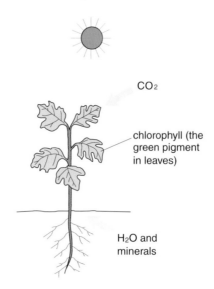

CO_2

chlorophyll (the green pigment in leaves)

H_2O and minerals

What do we take from plants?

Plants do the work – and we reap the benefit! We obtain thousands of products from plants. These are the main groups:

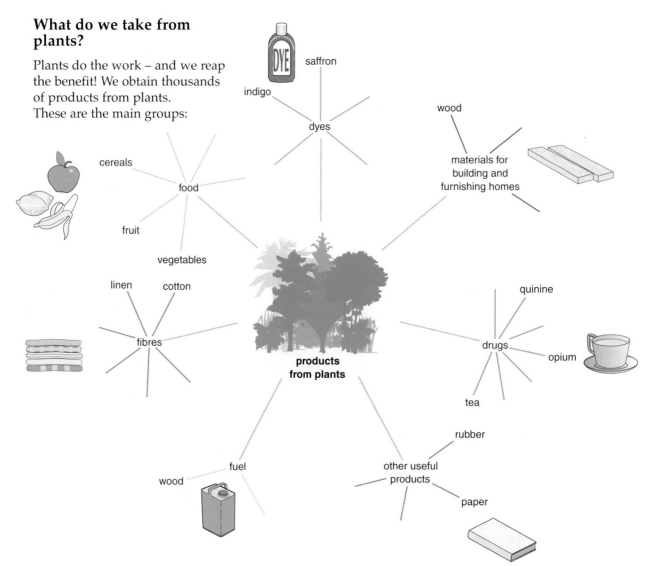

indigo

saffron

dyes

wood

materials for building and furnishing homes

cereals

food

fruit

vegetables

linen

cotton

fibres

quinine

drugs

opium

tea

rubber

products from plants

fuel

wood

other useful products

paper

Some useful natural polymers

Cellulose. It is the carbohydrate that forms the walls of plant cells. We eat it in our food – but we can't digest it. (We call it **fibre**.)

Wood is mainly cellulose. We turn it into paper – and even clothing. The first step is to soak wood chips and add chemicals to break them down into pulp. To make **paper** the pulp is pressed between heavy rollers. To make **rayon** or **viscose** it is reacted with chemicals to give a sticky solution. This is then forced through nozzles to make fibres which are woven into cloth.

Natural rubber. It is made from **latex**, a liquid taken from rubber trees. It is very useful because it is stretchy *and* tough. Today, though, over half the rubber we use is synthetic, made (like plastics) by polymerisation of chemicals obtained from oil.

Woman wearing a rayon blouse

Drugs

A drug is any chemical that changes the way your body works. So alcohol, nicotine (cigarettes) and caffeine (tea, coffee) are drugs!

Around 7000 of the medical drugs in use today come from plants. For example extracts from the rosy periwinkle plant are used to treat cancer. Quinine from the bark of the cinchona tree is used for malaria. (Both these plants grow in the rainforest.)

Other drugs have been developed by studying plant chemicals. For example, around 1760 a British clergyman called Eric Stone tried using an extract from the bark of the willow tree to cure malaria. It didn't cure malaria – but it did reduce fever.

Its active chemical was found to be **salicylic acid**, which could be made in the lab. But it irritated the mouth and stomach. This problem was solved by turning it into **acetylsalicyclic acid** – which we call **aspirin**. It is an **analgesic** which means it kills pain.

The boy is suffering hair loss: a side effect of chemotherapy for the treatment of cancer.

Dyes

Humans have used plant dyes from earliest times – for example **indigo** (a blue dye from the indigo plant) and **alizarin** (a red dye from the madder plant). Today, most dyes are synthetic.

Dyed yarns in a market in Morocco

---

1 a Name the raw materials plants use.
  b Name three types of compounds they produce.
2 a Which type of compound is cellulose?
  b Name one food you eat that is likely to contain cellulose.
  c Cellulose acts as fibre in our diet. Is this good for us? Explain.
  d Many animals are able to digest cellulose and turn it into a liquid humans can drink.
    i Name one of these animals.
    ii From which plant does it obtain most of its cellulose?

3 a Explain what these terms mean:
    i drug **ii** analgesic **iii** chemotherapy
  b Name:
    i two socially acceptable drugs;
    ii two illegal drugs obtained from plants.
4 Many substances once obtained from plants are now made in factories. Dyes are an example.
  a Suggest some reasons for this change.
  b Is it a good thing or a bad thing? Write a list of pros and cons.

20.6 The macromolecules in food

Macromolecules are large molecules built from small building blocks called **monomers**.
- You have already met some man-made macromolecules – the large molecules in polymers like PVC and nylon (Unit 6.2).
- But there are also many natural macromolecules such as the large molecules of proteins, carbohydrates and fats in your food.

Pupils being served a school meal

Carbohydrates

Carbohydrates are made of carbon, hydrogen and oxygen. The building block is **glucose** which plants make in their leaves, using sunlight for energy and chlorophyll as a catalyst:

$$6CO_2 \ (g) \quad + \quad 6H_2O \ (l) \xrightarrow[\text{chlorophyll}]{\text{sunlight}} C_6H_{12}O_6 \ (s) \ + \ 6O_2 \ (g)$$

carbon dioxide   water   glucose   oxygen

Glucose is called a **simple carbohydrate**. Then glucose molecules join by **condensation polymerisation** (page 89) to form **starch**. We show it in a simple way like this:

HO —☐— OH   —O—☐—O—☐—O—☐—O—☐—

a glucose   part of a molecule of starch
molecule

Starch is a **complex carbohydrate**. You eat a lot of it in foods like potatoes, cereals, pasta, rice and bread.

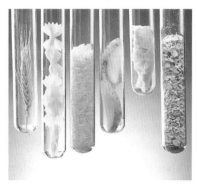

The foods here are rich in starch. The modern diet is now less reliant on natural sources of starch.

Proteins

You eat **proteins** in foods like meat, eggs and cheese. And you have over 100 000 different proteins in your body!

This time the building blocks are **amino acids**, which are made of carbon, hydrogen, oxygen and nitrogen. (Some also contain sulphur.) There are 20 common amino acids – and many less common ones. Their general structure is shown as:

$$R-\underset{\underset{H}{|}}{\overset{\overset{NH_2}{|}}{C}}-C\underset{OH}{\overset{O}{\lessgtr}}$$

where R represents a chain of carbon atoms with hydrogen atoms bonded to it. The amino acids join up by condensation polymerisation to form proteins, like this:

```
      H  O      H  O      H  O      H
      |  ||     |  ||     |  ||     |
   —N—C—C—N—C—C—N—C—C—N—C—
      |  |      |  |      |  |      |
      H  R      H  R      H  R      H  R
```

the linking group in proteins is called the **peptide link** – so proteins are **polypeptides**

At least 100 units join up. They can be different amino acids in any order. That explains why there are so many different proteins!

Fats

The **fats** in food are all **esters** – compounds of carbon, hydrogen and oxygen formed from **fatty acids** and **glycerol** (page 261). These react to give very large molecules. Here R represents the long carbon chain. You can see that the reaction is a condensation:

this group is called an **ester** linkage

Each OH group in a glycerol molecule can react with a *different* fatty acid – which means you can get many different esters. (Butter is a mixture of many esters.)

Digestion

When you digest foods, the macromolecules get broken down again into their building blocks, by reaction with water. (Reaction with water is called **hydrolysis**.)

- Carbohydrates break down to glucose which you use for energy.
- Proteins get broken down to amino acids which you use to build up your body tissues and hair.
- Fats get broken down into glycerol and fatty acids. You turn these into new fats which you store until you need them for energy. (Your body can store limitless amounts of fats!)

These reactions are catalysed by the enzymes in your digestive system, in the presence of acid. (Your stomach contains dilute hydrochloric acid.) They are just the *reverse* of condensation. For example, this shows the hydrolysis of a macromolecule of fat:

Athletes in action

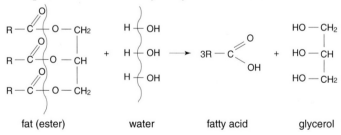

fat (ester)   water   fatty acid   glycerol

---

1 **a** What is a macromolecule?
  b Give two natural and two man-made examples.
2 State the three main constituents of food (apart from water!).
3 **a** Draw a diagram (using blocks) to show the structure of: **i** carbohydrates **ii** proteins
  b Now name the monomer for each.
4 Compare the linking groups in proteins and nylon (page 89). What do you notice?

5 **a** In what way is a fatty acid different from an acid such as ethanoic acid (page 260)?
  b Explain how fats are formed.
6 Compare the structure of fats with the structure of Terylene (page 89) What linking group do they have in common?
7 During digestion, a hydrolysis reaction takes place.
  a Why is it called hydrolysis?
  b What happens in this reaction?

20.7 % yield and % purity

The **yield** is the amount of product you obtain from a reaction. Companies like the maximum possible yield for the minimum cost! For many companies – such as drug and food companies – the **purity** of the product is even more important. Impurities can kill.

In this unit you'll learn how to calculate the % yield from a reaction, and the % purity of the product obtained.

Technicians checking machinery for the recycling of chemicals

% yield

You can work out % yield like this:

$$\% \text{ yield} = \frac{\textbf{actual mass obtained}}{\textbf{calculated mass}} \times 100\%$$

Example The drug aspirin is an ester made from salicylic acid.
1 mole of salicylic acid produces 1 mol of aspirin:

$$\underset{\text{salicylic acid}}{C_7 H_6 O_3} \xrightarrow{\text{chemicals}} \underset{\text{aspirin}}{C_9 H_8 O_4}$$

In a trial, 100.0 grams of salicylic acid produced 121.2 grams of aspirin. What was the % yield?

1 The RAMs are: C = 12, H = 1, O = 16.
   So the RMMs are: salicyclic acid, 138; aspirin, 180.

2 138 g salicylic acid   = 1 mole
   so 100 g = $\frac{100}{138}$ mole = 0.725 moles

3 1 mole of salicylic acid produces 1 mole of aspirin
   so 0.725 moles produce 0.725 moles of aspirin
   or 0.725 × 180 g = 130.5 g
   So 130.5 g is the **calculated mass** for the trial.

4 But the **actual mass** obtained was 121.2 g.
   So % yield = $\frac{121.2}{130.5}$ g × 100 = **92.9%**

This is a high yield – so it is worth continuing with those trials.

% purity

An impure product has impurities mixed with it.
You can work out its % purity like this:

$$\% \text{ purity of a product} = \frac{\textbf{mass of the product}}{\textbf{total mass of the mixture}} \times 100\%$$

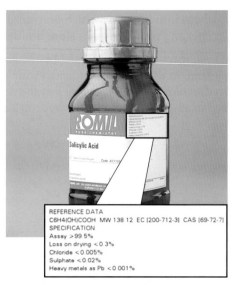

Chemicals can now be purified to 'parts in a trillion' – equivalent to a leaf in a forest!

Example 1 The aspirin obtained in the trial on page 304 was not pure. 121.2 g of solid was obtained, but analysis showed that only 109.2 g of this was aspirin. The rest was impurities such as unreacted salicylic acid.

% purity of the aspirin obtained $= \dfrac{109.2}{121.2}$ g $\times 100\% = \mathbf{90.0\%}$

This level of purity is not acceptable for a drug.
So the aspirin was purified by **crystallization**. (See page 21.) In fact it was recrystallized several times. Each time, the product was a little purer. Finally an 80 g sample was obtained, containing 79.8 g of aspirin.

% purity of the aspirin $= \dfrac{79.8}{80}$ g $\times 100\% = \mathbf{99.75\%}$

> **Purity check!**
>
> You can check the purity of any product by measuring its melting and boiling points and comparing them with tables.
>
> (Impurities lower the melting point and raise the boiling point.)

It is now more acceptable for medical use. But there is a downside. Making the sample purer has reduced the final yield, from 121.2 g to 80 g. So the % yield is now only **61.3%**.

Example 2 Chalk is almost pure calcium carbonate. You can work out its purity by measuring how much carbon dioxide it gives off.

10 g of chalk from the South Downs was reacted with an excess of dilute hydrochloric acid. 2280 cm³ of carbon dioxide gas was collected at room temperature and pressure (rtp).

Chalk cliffs at Beachy Head in Sussex, UK

The equation for the reaction is:
$CaCO_3$ (s) $+ 2HCl$ (aq) $\longrightarrow CaCl_2$ (aq) $+ H_2O$ (l) $+ CO_2$ (g)

1 The RMM of $CaCO_3 = 100$ (RAMs: Ca = 40, C= 12, O = 16)
2 1 mole of $CaCO_3$ gives 1 mole of CO_2 and
   1 mole of gas has a volume of 24 000 cm³ at rtp (page 82).
3 So 24 000 cm³ of gas is produced by 100 g of calcium carbonate
   and 2280 cm³ is produced by $\dfrac{2280}{24\,000} \times 100$ g or **9.5 g**.

That means there is 9.5 g of calcium carbonate in the 10 g of chalk.
So its % purity $= \dfrac{9.5}{10}$ g $\times 100 = \mathbf{95\%}$.

The chalk in the sample is **95%** calcium carbonate.

---

1 100 g of aspirin was obtained from 100 g of salicylic acid. What was the % yield?
2 17 kg of aluminium was produced from 51kg of aluminium oxide (Al_2O_3) by electrolysis. What was the percentage yield for the extraction? (RAMs: Al = 27, O = 16)
3 Some sea water was evaporated. The salt obtained was found to be 86% sodium chloride. What mass of sodium chloride could be obtained from 200 g of sea salt?
4 A 5.0 g sample of dry ice (solid carbon dioxide) produced 2400 cm³ of gas at rtp. What was the percentage purity of the dry ice? (RMM of $CO_2 = 44$)

20.8 More about chromatography

What is chromatography?

Chromatography is a method of separating and identifying mixtures. You already met paper chromatography on page 23. There are many other kinds. *All* of them depend on the interaction between:

- a non-moving or **stationary phase** (such as paper) and
- a moving or **mobile phase** containing the mixture to be separated. It is carried in a solvent.

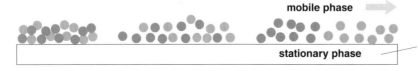

mobile phase can be liquid or gas

stationary phase can be paper, glass or plastic beads, or a coating on a glass plate

1 The mobile phase moves over the stationary phase.

2 Each substance in the mixture travels at a different speed.

3 So they eventually get separated.

Making use of chromatography

Chromatography is a really useful technique. You can use it to:
- separate complex mixtures of substances;
- purify a substance, by separating it from its impurities;
- identify substances.

For example, it is used to test for illegal drugs in urine and for impurities in food and medical drugs.

A scientist using gas chromatography to analyse a food sample

Example: Identifying amino acids by paper chromatography
You are given five aqueous solutions of amino acids, labelled A to E, obtained by the hydrolysis of proteins. Solution A contains an unknown mixture of amino acids. The others contain one amino acid each. Your task is to identify *all* the amino acids present.

1 Place a spot of each solution along a line on slotted filter paper, as shown in the diagram. (The slots keep the amino acids in each sample separate.) Label each spot in pencil at the *top* of the paper.

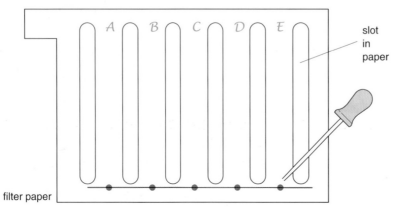

slot in paper

filter paper

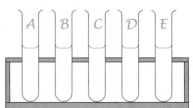

2 Place a suitable solvent (a mixture of ethanoic acid, butanol and water) in the bottom of a beaker.

3 Roll the filter paper into a cylinder and place it in the beaker. Cover the beaker.

4 The solvent rises up the paper. When it has almost reached the top, remove the paper.

5 Mark a line on it to show the level reached by the solvent. (But you can't see how far the amino acids have travelled because they are colourless.)

6 Place the paper in an oven to dry out.

7 Next spray it with a **locating agent** to locate the amino acid spots. A suitable one is **ninhydrin**. (You must use it in a fume cupboard.) After spraying, heat the paper in the oven for 10 minutes. The spots turn purple.

8 Mark a pencil dot at the centre of each spot. Now measure from the base line to each dot and to the mark showing the final solvent level.

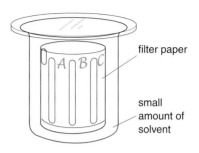

filter paper

small amount of solvent

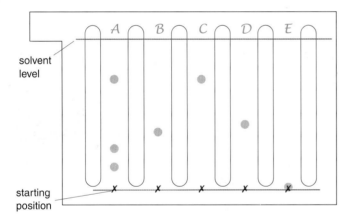

solvent level

starting position

9 Now work out the R$_f$ value for each amino acid. Like this:

$$R_f \text{ value} = \frac{\text{distance moved by amino acid}}{\text{distance moved by solvent}}$$

10 Finally, use R$_f$ tables to identify the amino acids. Part of an R$_f$ table for the solvent you used is shown on the right.

The R_f value can be used to identify the amino acids because:
the R_f value of a compound is always the same for a given solvent and conditions.

The filter paper with the separated spots is called a **chromatogram**.

R$_f$ values for amino acids
(for butanol ethanoic acid / water as solvent)

amino acid	R$_f$ value
glycine	0.26
alanine	0.38
valine 0.60 leucine	0.73
proline	0.43
serine	0.27
cysteine	0.08
lysine	0.14

1 Say what each term means in chromatography:
  a stationary phase b mobile phase
2 Explain in your own words how chromatography works.
3 a What do you think a *locating agent* is?
  b Why is one needed in the experiment above?

4 For the chromatogram above:
  a Were any of the amino acids in B – E also present in A? How can you tell at a glance?
  b Work out the R$_f$ values for the amino acids in A – E.
  c Now use the R$_f$ table above to name them.

20.9 Silicon and its compounds

The element silicon belongs to the same family as carbon. Do they have much in common?

Where silicon is found

Nearly all rocks contain silicon. In fact it accounts for over a quarter of Earth's crust (27%). It is the most common element in the crust after oxygen.

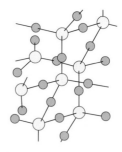

silicon(IV) oxide

It usually occurs combined with oxygen:

- as silicon(IV) oxide, which is also called silicon dioxide and **silica**. This has the formula SiO_2. It forms different giant structures.

  One example is **quartz**, which has the structure shown on the right. (And see the photo on page 160.) Sand is impure quartz; its colour is due to iron impurities.

- in complex **silicates**. Silicates contain ions such as SiO_4^{4-} and $Si_4O_{11}^{6-}$. One example is **mica**, a mineral found in clay and granite. Its formula is $K_2(MgFe)_6Si_3O_{10}(OH)_{12}$. Don't try to remember it!

Extracting silicon

It is made by heating sand and coke in an electric furnace:

$$SiO_2\ (s) + 2C\ (s) \longrightarrow Si\ (s) + 2CO_2\ (g)$$

The silicon(IV) oxide is reduced to silicon. Coke is the reducing agent.

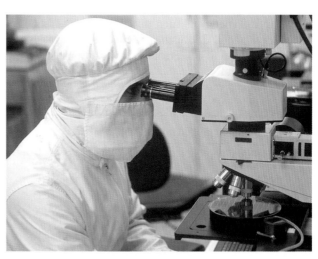

Inspecting a silicon chip under a microscope in a 'clean room'

The properties of silicon

- It is a hard grey solid.
- It has a high melting point (1410 °C).
- It is a **metalloid**. That means it behaves partly like a metal and partly like a nonmetal. (See page 38.)
- It can conduct electricity – but not as well as a metal. So it is called a **semiconductor**.

Uses of silicon and silicon compounds

Silicon is used to make:

- computer chips and solar cells, because it is a semiconductor. For these it must be ultra pure.
- **silicones**, which have hundreds of uses, for example, as waxes and polishes. See next page.

Silicon is extracted from sand. Sand is also used to make glass (page 58) and concrete and mortar (page 165).

White sandy beach of silica sand

Silicones

Silicon combines with organic carbon compounds to form a huge range of polymers called **silicones**.

In silicones, the 'backbone' is a long chain of silicon and oxygen atoms. Carbon atoms, or chains of carbon atoms, bond to the silicon atoms.

You can show it like this:

where □ represents carbon atoms or chains

Properties of silicones

- They may be oils, waxes, or plastics, depending on the length of the chains and the carbon groups attached.
  The plastics may be very hard, or flexible, or stretchy like rubber.
- They are unreactive and 'waterproof'.
- They do not break down on heating.

Overall, silicones are more stable than carbon-based polymers. They do not melt or burn so easily. This is because the Si – O bond is stronger than the C – C bond in carbon-based polymers.

Uses of silicones

Because of their properties, silicones have a wide range of uses.
For example:
- in varnishes, paints, polishes and adhesives
- as a coating for baking dishes
- to 'waterproof' material for raincoats and outdoor wear
- as a scratch resistant coating for car windshields
- as fireproof foams and sealants in buildings
- as waterproof sealants around baths and sinks
- as ingredients in cosmetics, shaving cream and suntan lotions.

Using silicones – sealing the edge of the bath

Using silicones – putting on lipstick

---

1 Silicon occurs naturally as silicon(IV) oxide and silicates.
  a Give two other names for silicon(IV) oxide.
  b What is a silicate?
2 a What is the main mineral in sand?
  b Write a formula for this mineral.
  c State four uses of sand.
  d Explain why sand is so hard. (Page 57 may help.)

3 a What are silicones?
  b Draw a diagram to show the general structure of a silicone.
  c Explain why silicones are so useful, and give three examples of their use around the home.
4 Silicon has the same structure as diamond (page 56). Explain why it has a high melting point.

Revision checklist

1 Particles in chemistry

1 Past ideas about atoms and elements (1.1)
2 Three pieces of evidence for the existence of particles in matter (1.1)
3 How matter changes from one state to another (1.2)
4 How melting and boiling points can be used to identify a substance (1.2)
5 The effect of an impurity on melting and boiling points (1.2)
6 How particles are arranged in a solid, liquid and gas (1.3)
7 Why a gas can be compressed (1.3)
8 The relationships between volume, pressure and temperature of a gas (1.4)
9 The link between the mass of the particles in a gas and its rate of diffusion (1.4)
10 Definitions of solute, solvent, solution, suspension, aqueous solution, immiscibility (1.5)
11 Examples of solvents other than water (1.5)
12 What a saturated solution is (1.6)
13 Examples of soluble and insoluble groups of compounds (1.6)
14 How solubility also depends on the solvent (1.6)
15 Definition of solubility (1.7)
16 Measuring and calculating solubility (1.7)
17 How to draw a solubility curve and what it tells you (1.7)
18 How the solubility of a gas changes with temperature and pressure (1.7)
19 Three methods of separating a solid from a liquid (1.8)
20 How to separate a mixture of two solids (1.8)
21 How to separate the solvent from a solution (1.9)
22 How to separate two liquids (1.9)
23 How to separate a mixture of substances using paper chromatography (1.9)

2 What are those particles?

1 The definition of an element (2.1)
2 The key differences between the metal and nonmetal elements (2.1)
3 What a compound is (2.1)
4 The structure of an atom (2.2)
5 Proton number and mass number and how they are linked (2.2)
6 How to draw diagrams to represent the atoms of the first twenty elements (2.3)
7 What isotopes are, with examples (2.3)
8 What a radioisotope is (2.3)
9 What the Periodic Table is and how it is set out and labelled (2.4)
10 Properties and trends for Group 0 (2.5)
11 Why Group 0 elements are unreactive and monatomic (2.5)
12 Properties and trends for Group 7 (2.5)
13 Why Group 7 elements are reactive and diatomic (2.5)
14 Properties and trends for Group 1 (2.6)
15 Properties and trends for Group 2 (2.6)
16 The general trends across Periods 2 and 3 (2.7)
17 The general properties of the transition elements (2.7)

3 Atoms combining

1 Why elements form compounds (3.1)
2 What an ion is, with examples (3.1)
3 What a compound ion is, with examples (3.1)
4 How to show an ionic bond on a diagram (3.2)
5 The structure of solid sodium chloride (3.2)
6 How to write the formula of an ionic compound (3.3)
7 What a covalent bond is, with examples (3.4)
8 How to show a covalent bond on a diagram (3.4)
9 What a molecule is (3.4)
10 The general properties of ionic solids (3.5)
11 The general properties of molecular solids (3.5)
12 How atoms are held together in a metal (3.6)
13 The general properties of metals (3.6)
14 The grain structure of metals (3.6)
15 Ways to make metals stronger (3.7)
16 What alloys are (3.7)
17 How the structures of diamond and graphite differ (3.8)
18 Why diamond and graphite have such different properties (3.8)
19 How their properties dictate their uses, with examples (3.8)
20 How the structure and properties of glass are linked (3.9)
21 How the structure and properties of plastics are linked (3.9)

4 The mole

1 What the relative atomic mass of an atom is (4.1)
2 How the RAM of an element is calculated (4.1)
3 How the formula mass or RMM is calculated (4.1)
4 A_r and M_r – other terms for RAM and RMM (4.1)
5 What a mole of atoms is and how many atoms it contains (4.2)
6 What Avogadro's number (or constant) is (4.2)
7 How to change mass to moles and vice versa (4.2)
8 What the Law of Constant Composition says (4.3)
9 How to calculate the percentage composition of a compound (4.3)
10 What the formula of a compound tells you (4.4)
11 What the empirical formula of a compound tells you (4.4)

5 Chemical equations

6 Chemical change

7 Electricity and chemical change

8 How fast are reactions?

6 How carbon dioxide contributes to global warming (11.3)
7 Why carbon monoxide is poisonous (11.3)
8 Ways to reduce pollution from fossil fuels (11.3)
9 What CFCs are and why they are harmful (11.3)
10 Tests for water (11.4)
11 The importance of water in the human body (11.4)
12 What the water cycle is and its stages (11.4)
13 How we obtain a supply of clean safe tap water (11.5)
14 How waste water is cleaned up at the sewage works (11.5)
15 What soft and hard water are (11.6)
16 Why hard water causes scum (11.6)
17 Where hard water gets its hardness from (11.6)
18 The difference between temporary and permanent hardness (11.6)
19 Some advantages and disadvantages of hardness in water (11.6)
20 Four ways to remove hardness from water (11.7)
21 How to compare the hardness in samples of water using soap solution (11.7)

12 The behaviour of metals

1 The properties of metals (12.1)
2 The main differences between metals and non-metals (12.1)
3 How different metals react differently with:
oxygen (12.2)
water (12.2)
dilute hydrochloric acid (12.3)
4 What happens when a metal is:
– heated with the oxide of a less reactive metal (12.3)
– placed in a solution of a compound of a less reactive metal (12.3)
5 What happens when carbon is mixed with a metal oxide and heated (12.3)
6 How metals can be arranged in a reactivity series, using the reactions in 4 – 6 above (12.4)
7 Seven key points to remember about the reactivity series (12.4)
8 What a cell is and why it produces electricity (12.4)
9 How the choice of metals affects the voltage of the cell (12.4)

13 Making use of metals

1 The stages in exploring for minerals (13.1)
2 How mining brings benefits but can also cause problems (13.1)
3 How the method of extraction depends on the metal's reactivity (13.2)
4 Examples of extraction methods for different metals (13.2)

5 Why extraction means the ore is reduced (13.2)
6 How extraction can cause pollution (13.2)
7 How the properties of metals dictate their uses (13.3)
8 How a metal is made more useful by turning it into an alloy (13.3)
9 Some common alloys and their uses (13.3)
10 How aluminium is mined and extracted (13.4)
11 Details of the electrolysis of aluminium (13.4)
12 Properties and uses of aluminium (13.4)
13 How iron is extracted in the blast furnace (13.5)
14 What cast iron is and what it is used for (13.5)
15 How iron is turned into different steels and their uses (13.5)
16 What corrosion is (13.6)
17 How both air and water are needed for rusting (13.6)
18 Ways to prevent rusting (13.6)
19 Why corrosion is limited for aluminium (13.6)
20 How aluminium is anodized and why (13.6)

14 Non-metals: hydrogen and nitrogen

1 What the nitrogen cycle is and its stages (14.1)
2 The properties and uses of nitrogen (14.1)
3 The properties and uses of hydrogen (14.2)
4 How hydrogen is made in industry (14.2)
5 What hydrogen fuel cells are (14.2)
6 Hydrogen's potential as a fuel (14.2)
7 How ammonia is made in the lab (14.3)
8 The properties of ammonia (14.3)
9 How ammonia is made in industry by the Haber process (14.4)
10 What the optimum conditions for making ammonia are and why (14.4)
11 The uses of ammonia (14.4)
12 How nitric acid is made in industry (14.4)
13 Uses of nitric acid (14.4)
14 The elements plants need (14.5)
15 What fertilizers are and why they are important (14.5)
16 Problems caused by fertilizers (14.5)

15 Non-metals: oxygen, sulphur, and the halogens

1 How oxygen is made in the lab (15.1)
2 The properties of oxygen (15.1)
3 What goes on during the respiration process (15.1)
4 How respiration and combustion are similar (15.1)
5 The test for oxygen (15.1)
6 What basic oxides are and which elements produce them (15.2)
7 What acidic oxides are and which elements produce them (15.2)

29 Vegetable oils as natural esters (17.5)
30 What fatty acids are (17.5)
31 What unsaturated oils are (17.5)
32 Hardening of oils and how it is carried out (17.5)
33 How soaps are made by hydrolysis of oils (17.5)
34 How soaps dissolve grease (17.5)

18 Rocks

1 The structure of Earth (18.1)
2 How temperature, pressure and density increase with distance below Earth's surface.
3 Why Earth has a magnetic field and how it flips (18.1)

4 The three types of rock in Earth's crust (18.1)
5 How igneous rock is formed (18.2)
6 The difference between magma and lava (18.2)
7 The difference between granite and basalt (18.2)
8 The effect of cooling on crystal size for igneous rock (18.2)
9 What plutons are with examples of different types (18.2)
10 Some uses of igneous rock (18.2)
11 What weathering is (18.3)
12 Three types of physical weathering (18.3)
13 What chemical weathering is, with examples (18.3)
14 How chemical weathering produces soil and ores (18.3)
15 How physical and chemical weathering work together (18.3)
16 What erosion and transport are (18.4)
17 How and why sediment is deposited (18.4)
18 What you can learn from examining sediment (18.4)
19 The three ways in which sedimentary rock is formed (18.5)
20 What fossils are, and how some can be used to date rock (18.5)
21 Some uses of sedimentary rock (18.5)
22 How metamorphic rock is formed (18.6)
23 What foliation is and what causes it (18.6)
24 Where metamorphic rock is usually found (18.6)
25 The difference between contact and regional metamorphism (18.6)
26 Some uses of metamorphic rock (18.6)
27 How rock gets weathered and eroded and eventually turned into new rock, in the rock cycle (18.7)
28 How to sketch the rock cycle (18.7)
29 The link between rock type and depth of burial (18.7)
30 Examples of clues gained from studying rocks (18.7)

19 The changing face of Earth

1 The general features of Earth's crust (19.1)
2 Wegener's theory of continental drift, and evidence he used to support it (19.2)
3 What sea floor spreading is and proof that it occurs (19.2 and 19.3)
4 What plates are and how fast they move (19.3)
5 What causes plates to move (19.4)
6 The three different types of plate movements (19.4)
7 The meaning of constructive boundary, destructive boundary, conservative boundary and subduction (19.4)
8 How faults and folds result from plate movements (19.5)
9 How fold mountains are formed by plate movements (19.5)
10 The connection between earthquakes, volcanoes and plate movements (19.3, 19.4 and 19.5)

20 Further topics

There is no revision checklist for Section 20 as you will choose your topics according to your syllabus.

Examination questions

The questions are taken from Higher tier papers of the various boards. They are arranged roughly in chapter order and provide comprehensive coverage of the chemistry syllabus. For some of the questions you will need to refer to the periodic table on page 330. You will need graph paper for some of the questions.

1 Barium is in Group II of the Periodic Table. The chemistry of this metal and of its compounds is very similar to that of calcium (relative atomic mass: C12, O16, Ba137).
   a Barium reacts vigorously with cold water.
      i Suggest the name of another metal in the same period that reacts with cold water. [I]
      ii Complete the word equation.

      barium + water $\longrightarrow$ .... + ... [2]

   b Use the information given below to predict the formula of barium sulphate and of barium phosphate.
      the formula of the barium ion is Ba^{2+}
      the formula of the sulphate ion is SO_4^{2-}
      the formula of the phosphate ion is PO_4^{3-} [2]

   c Complete the equations for the action of heat on barium carbonate and on barium nitrate.

      $BaCO_3 \longrightarrow$ .... + ... [1]

      $Ba(NO_3)_2 \longrightarrow$ .... + ... + ... [2]

   d Barium is used to extract the element americium from the compound americium(III) fluoride.

      $3Ba + 2AmF_3 \longrightarrow 2Am + 3BaF_3$

      i Complete the following equations by including the electron transfer.

      Ba $\longrightarrow Ba^{2+}$
      $Am^{3+} \longrightarrow Am$ [2]

      ii Which of these equations represents oxidation? [1]
   e An excess of hydrochloric acid was added to 1.23 g of impure barium carbonate. The volume of carbon dioxide collected at r.t.p. was 0.120dm³. The impurities did not react with the acid. Calculate the percentage purity of the barium carbonate.

      $BaCO_3 + 2HCl \longrightarrow BaCl_2 + CO_2 + H_2O$
      Molar gas volume at r.t.p. is 24 dm³.

      i The number of moles of CO_2 collected [1]
      ii The number of moles of $BaCO_3$ reacted [1]
      iii Mass of one mole of $BaCO_3$ (grams) [1]
      iv Mass of barium carbonate (grams) [1]
      v Percentage purity of the barium carbonate [1]
      Total marks: 16 CAM/IGCSE

2 This question is about metals and alloys.
   a These diagrams represent the relative sizes of magnesium and aluminium atoms.

      magnesium   aluminium

      An **alloy** of these two metals is stronger than pure magnesium or pure aluminium. Explain why, using the information above. A diagram may help your answer. [2]
   b Gwen tests two identical springy steel needles. The table shows her results.

Test	Result
1 Heat a needle until red-hot in a bunsen flame, and then cool it slowly above the flame.	The cooled needle is soft enough to bend easily.
2 Heat a needle until red-hot in a bunsen flame, and then plunge it into cold water.	The cooled needle is harder to bend than an unheated needle and is very brittle.

      The metallic properties of the two heat-treated needles are different. Explain why, in terms of grain size. [2]
   c Explain why a metal such as copper can conduct electricity. [2]
      Total marks: 6 MEG

3 a Describe a chemical test for chlorine. Give the result of this test. [2]
   b Chlorine is used to make antiseptics such as *TCP*. The label has been taken from a bottle of *TCP*.

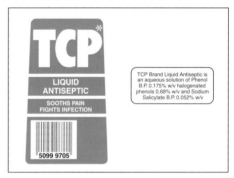

      i One of the active ingredients is halogenated phenol (known as *TCP*) made by treating phenol with chlorine. It contains 36.49% by mass of carbon, 8.10% oxygen, 1.53% hydrogen and 53.87% chlorine. Use this information to calculate the empirical formula of *TCP*.
      (Relative atomic masses: H = 1; C = 12; O = 16; Cl = 35.5) [3]
      ii The information on the label suggests that there is 0.68 g of *TCP* per 100 cm³ of solution. If the molecular formula is the same as the empirical formula, calculate the number of moles of *TCP* in 100 cm³ of solution. [2]
      Total marks: 7 NEAB

4 At present, most car bodies are made from either mild steel or plastic. A large German car manufacturer is investigating the use of an alternative material made from plant fibres for this purpose.
   a Plant fibres contain natural polymers, which are complex carbohydrates. These are made in green plants from simpler carbohydrates, such as glucose, by a process called *condensation polymerisation*.
      i Describe the formation of glucose by photosynthesis in a green plant. [4]
      ii Explain *condensation polymerisation*. [2]

iii Give the structure of a synthetic polymer that is made by condensation polymerisation. [2]

b Haematite, iron(III) oxide, is reduced to impure iron in a blast furnace. Two impurities in the iron are carbon and silicon. The impure iron is changed into mild steel using oxygen and powdered calcium carbonate.

i Complete the equation for the reduction of iron(III) oxide.

$Fe_2O_3 + \text{.....} CO \longrightarrow \text{...............} + \text{.................}$ [2]

ii How are the impurities removed when steel is made from impure iron? [3]

c Poly(phenylethene) is a synthetic polymer, usually called polystyrene. Its structure is given below.

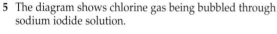

i Deduce the structure of the monomer of poly(phenylethene). [1]

ii Name the raw material from which most synthetic polymers are made. [1]

d i Suggest an environmental advantage of using the natural fibre material rather than a synthetic polymer. [1]

ii Suggest an advantage of using either the natural or a synthetic polymer rather than mild steel for car bodies. [1]

Total marks: 17 (CAM / IGCSE)

5 The diagram shows chlorine gas being bubbled through sodium iodide solution.

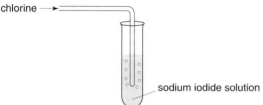

a i What safety precaution should be taken when carrying out this experiment? [1]

ii Why is this precaution necessary? [1]

b What colour is the sodium iodide solution at the start of the experiment? [1]

c This is the equation for the reaction between chlorine and the sodium iodide solution.

$Cl_2 + 2NaI \longrightarrow I_2 + 2NaCl$

Write the simplest balanced ionic equation for this reaction. [2]

d The reaction is described as a **redox reaction**. Explain, in terms of electron transfer, the meaning of the term redox reaction for this particular reaction.
Your answer should refer to what is oxidised, what is reduced, which is the oxidising agent and which is the reducing agent. [5]

Total marks: 10 MEG

6 a Ethanol can be made by the fermentation of glucose. Yeast is added to an aqueous solution of glucose. Carbon dioxide is given off and, after a while, the solution becomes warm because the reaction is exothermic.

$C_6H_{12}O_6 (aq) \longrightarrow 2C_2H_5OH (aq) + 2CO_2 (g)$

The graph below shows how the rate of reaction changed over several days.

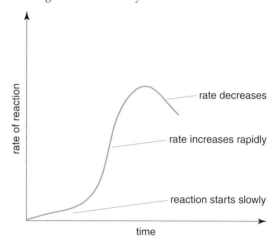

i Suggest a method of measuring the rate of this reaction. [2]

ii Suggest a reason why the reaction rate increases initially. [1]

iii Suggest a reason why the reaction rate eventually decreases. [2]

b Micro-organisms, different to those produced by yeast, change glucose into lactic acid. When lactic acid is heated, it decomposes to form acrylic acid and one other product. The structural formulae of these acids are shown below.

lactic acid   acrylic acid

i Give the empirical formula of lactic acid. [1]

ii Complete the word equation
lactic acid $\longrightarrow$ acrylic acid + ... [1]

iii Describe a test that would distinguish between these two acids. [3]

iv Other than using an indicator, describe a test that would show that both of these chemicals contain an acid group. [2]

v Suggest the name of the chemical that reacts with acrylic acid to form the ester ethyl acrylate. [1]

c Organic chemicals are made from petroleum as well as from natural materials such as glucose. The following steps are needed to make propanol from petroleum:

step 1 petroleum is cracked to make the suitable alkene;
step 2 this alkene reacts with steam to form propanol.

i Name the 'suitable alkene'. [1]
ii Give the structural formula of propanol. [1]
iii What type of reaction takes place between the alkene and steam? [1]
Total marks: 16 CAM / IGCSE

7 This question is about the insoluble salt, barium sulphate. Hospital patients are given a 'barium meal' before an X-ray of their stomach is taken. This 'barium meal' contains barium sulphate. It can be made as a white precipitate by reacting aqueous barium hydroxide with sulphuric acid.

a One mole of barium hydroxide reacts with one mole of sulphuric acid. Finish the equation for the reaction.

$Ba(OH)_2 + H_2SO_4 \longrightarrow$ .... + ... [2]

b The progress of the reaction between barium hydroxide and sulphuric acid can be followed by measuring electrical conductivity.

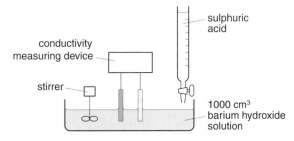

i The sulphuric acid is added 1 cm³ at a time. What is the name of the piece of apparatus containing the sulphuric acid? [1]
ii The mixture has to be stirred thoroughly as the acid is added. Explain why this is important in order to obtain reliable results. [1]

c An experiment was carried out, adding the sulphuric acid 1 cm³ at a time. After each addition of the sulphuric acid, the conductivity was measured. The results obtained are shown in the graph.

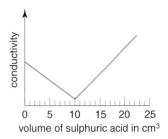

i What volume of sulphuric acid has exactly neutralised the 1000 cm³ of barium hydroxide solution? [1]
ii The concentration of the sulphuric acid is 1 mol/dm³. What is the concentration of the aqueous barium hydroxide? You **must** show how you work out your answer. [2]
iii Barium sulphate has to be very pure for medical use. Write down one reason why this method of preparation would be suitable. [1]

iv During the experiment there are changes in the number and types of ions present. Explain the shape of the graph using this information. [3]
Total marks: 11 MEG

8 This label has been taken from a can of *Tango Sparkling Apple Drink*.

INGREDIENTS:
CARBONATED WATER, APPLE JUICE, SUGAR, GLUCOSE SYRUP, MALIC ACID, PRESERVATIVE (SODIUM BENZOATE), ARTIFICIAL SWEETENER (SACCHARIN)

a The apple flavour is improved by the presence of malic acid, $C_4H_6O_5$. Malic acid is a *weak* acid. Explain what is meant by a *weak* acid. [1]
b It was found that 25.0 cm³ of the *Tango Sparkling Apple Drink* was neutralised by 15.0 cm³ of sodium hydroxide, of concentration 0.10 moles per litre.
The equation which represents this reaction is

$C_4H_6O_5 + 2NaOH \longrightarrow C_4H_4O_5Na_2 + 2H_2O$

Calculate the concentration of malic acid in:
i moles per litre. [2]
ii grams per litre.
(Relative atomic masses: H = 1; C = 12; O = 16) [2]
c Suggest why the concentration of malic acid in the Apple Drink may be less than that found by titration. [1]
Total marks: 6 NEAB

9 The table gives information about car exhaust emissions.

Pollutant	Petrol car no catalytic converter	Petrol car catalytic converter	Diesel car no catalytic converter	Diesel car catalytic converter
nitrogen oxides (NO$_x$)	xxx	x	xx	xx
carbon monoxide	xxx	xx	xx	x
hydrocarbons	xxx	xx	xx	x
particulates	xx	x	xxx	xx
carbon dioxide	xx	xxx	x	xx

Key: highest emissions xxx   intermediate xx   lowest emissions x
© *The Telegraph, plc, London, 1994.*

a i Particulates are very small particles of a black solid. Suggest the name of this black solid. [1]
ii Explain how the oxides of nitrogen are formed in car engines. [3]
iii Why does the catalytic converter increase the emission of carbon dioxide but decrease the emission of the other pollutants? [2]
b The oxides of nitrogen are one cause of acid rain. Acid rain increases the rate of rusting of steel.
i Name and describe the source of the other gas that causes acid rain. [3]
ii Explain why the rate at which steel rusts is lessened by 'sacrificial protection'. [2]

c Carbon dioxide is formed by the complete combustion of carbon-containing compounds. Another reaction that produces carbon dioxide is fermentation.

　i Complete the following equation for the fermentation of glucose.

$$C_6H_{12}O_6 \longrightarrow \underline{\hspace{2cm}} + \underline{\hspace{2cm}} \qquad [2]$$

　ii Give the conditions for this reaction. [1]

d Ozone is a serious air pollutant. It is formed by the following reaction.

$$NO_2 + O_2 \rightleftharpoons O_3 + NO$$

　i What type of reaction is this? [1]

　ii Predict the effect upon the concentration of ozone of increasing emissions of nitrogen monoxide, NO. [1]

Total marks: 16 CAM / IGCSE

10 A sample of mineral water has the following information on the label.

Ions present mg/l		
Positive ions	**Negative ions**	
calcium 27	fluoride	0.05
magnesium 9	chloride	9
sodium 6	nitrate	1
potassium 0.6	sulphate	6
	hydrogencarbonate	126

Total dry residue at 180 °C is 128 mg/l.

a Soap is sodium stearate. Predict the effect of shaking soap solution with this mineral water. Explain your prediction in terms of the ions involved. [3]

b Which of the negative ions decomposes when the water is evaporated? What solid product is formed? [2]

Total marks: 5 SEG

11 a Tin has been used for over three thousand years. Bronze, a copper/tin alloy, has been used for even longer.

　i Name another alloy that contains copper. [1]

　ii Suggest a reason why bronze might be used instead of pure copper. [1]

b The position of tin in the reactivity series is:
iron
tin
copper

　i The main ore of tin is tin(IV) oxide, SnO_2. By writing a word equation, suggest how this ore could be reduced to tin. [2]

　ii For each of the following decide if a reaction would occur. If there is a reaction, complete the equation, otherwise write 'no reaction'.

　　$Fe + SnCl_2 \longrightarrow$
　　$Sn + CuSO_4 \longrightarrow$
　　$Sn + Fe^{2+} \longrightarrow$ [4]

c Aqueous tin(II) sulphate is electrolysed using carbon electrodes. This electrolysis is similar to that of copper(II) sulphate using carbon electrodes.

　i What is the product at the negative electrode? [1]

　ii Write the equation for the reaction at the positive electrode. [2]

　iii Name the acid which is formed during the electrolysis. [1]

d The element tin can exist in two different solid forms. Grey tin has a diamond type structure and white tin has a metallic structure.

　i What type of chemical bond, ionic, covalent or metallic, is present in grey tin? [1]

　ii Describe how the carbon atoms are arranged in diamond. [2]

　iii Describe a typical metallic structure. [3]

　iv Which solid form of tin would be the better conductor of electricity? Explain your answer. [2]

Total marks: 20 CAM / IGCSE

12 Aluminium is extracted from bauxite.

a A sample of bauxite contained:

aluminium oxide	140 g
iron(III) oxide	44 g
silical	10 g
titanium(IV) oxide	6 g

Calculate the percentage of aluminium oxide in this sample of bauxite. [2]

b The bauxite is mixed with aqueous sodium hydroxide and heated. Balance the following equation for this reaction:

$$Al_2O_3 + \dots NaOH \longrightarrow 2NaAlO_2 + H_2O \qquad [1]$$

c Purified aluminium oxide is electrolysed in tanks lined with carbon.

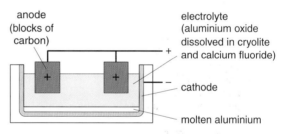

anode
(blocks of carbon)

electrolyte
(aluminium oxide dissolved in cryolite and calcium fluoride)

cathode

molten aluminium

　i How is the electrolyte kept molten? [1]

　ii What is the purpose of the cryolite? [1]

　iii Oxygen is produced at the positive electrode. The carbon blocks which form this electrode 'burn' away quickly. Write a symbol equation for this reaction. [2]

d Aluminium reacts with hydrochloric acid:

aluminium + hydrochloric → aluminium + hydrogen
　　　　　　　 acid　　　　 chloride

　i Aqueous aluminium chloride contains aluminium ions. What particles do aluminium atoms lose when they form aluminium ions? [1]

　ii Describe a test for hydrogen gas. [2]

　iii Describe what you would observe when aqueous sodium hydroxide is added to a solution containing aluminium ions. [2]

e Aluminium is used to make containers for soft drinks. Why should these containers be recycled? [1]

f The diagrams show the structures of three substances involved in the extraction of aluminium.

Use ideas about structure and bonding to explain the following facts about the following substances.

i Calcium fluoride is a compound. [1]

ii Calcium fluoride conducts electricity when molten but not when solid. [2]

iii Oxygen is a gas. [1]

iv Graphite has a very high melting point. [1]

v Graphite may be used as a lubricant. [2]

Total marks: 20 CAM / IGCSE

13 This question is about ammonia and fertilisers.
Ammonia is made from nitrogen and hydrogen in the Haber process. The equation is $N_2 + 3H_2 \rightleftharpoons 2NH_3$

a Write down the name of the catalyst in this process. [1]

b The diagram shows the energy change when ammonia is made.

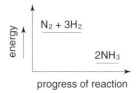

What can you conclude from this energy level diagram? [1]

c Ammonium nitrate is a fertiliser made from ammonia and nitric acid. Nitric acid is made from ammonia in three stages.

Stage 1
ammonia + oxygen $\xrightarrow{\text{platinum}}$ nitrogen monoxide + steam

Stage 2
nitrogen monoxide + oxygen $\longrightarrow$ nitrogen dioxide

Stage 3
nitrogen dioxide + water + oxygen $\longrightarrow$ nitric acid

i Ammonia is a raw material in this process. What are the other two raw materials? [2]

ii In Stage 1, the platinum has to be heated to 900 °C to start the reaction. Then the temperature of the catalyst stays at 900 °C without the need for further heating.
What does this tell you about the first stage? [1]

Total marks: 5 MEG

14 This question is about sulphuric acid. Sulphuric acid is manufactured from sulphur dioxide in a two-stage process.
In Stage 1, sulphur dioxide and oxygen are passed over vanadium(V) oxide at 400 °C. The equation for the reaction is:

$$2SO_2(g) + O_2(g) \rightleftharpoons 2SO_3(g)$$

In Stage 2, the sulphur trioxide produced is absorbed by concentrated sulphuric acid. This is then allowed to absorb water to produce 98.5% sulphuric acid.

a What is the job of the vanadium(V) oxide? [1]

b **Graph A** shows the yield of sulphur trioxide at different pressures. **Graph B** shows the cost of maintaining different pressures.

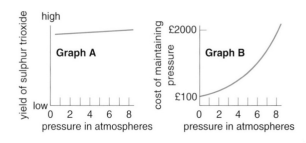

The production of sulphur trioxide is carried out at 1.5 atmospheres rather than at 8 atmospheres. Explain why, using the information in the graphs. [2]

c When dilute sulphuric acid is added to sodium carbonate, Na_2CO_3, bubbles of gas are readily given off. Write an equation for the reaction. [2]

Total marks: 5 MEG

15 The disposal of plastic waste presents problems. The waste usually contains a number of different polymers which are difficult to separate. The best way of disposing of plastic waste is to make useful products from it by recycling.

a Describe two environmental problems caused by the disposal of plastics. [2]

b Plastic waste could contain a mixture of addition and condensation polymers.

i For the following addition polymer, deduce the structure and name of the monomer.

[2]

ii A condensation polymer can be formed from the monomers $NH_2(CH_2)6NH_2$ and $HOOC(CH_2)8COOH$. Draw the structure of this polymer. [2]

c The first stage in recycling is to melt the plastic waste. If the plastic contains chlorine, hydrogen chloride is formed, which can be oxidised back to chlorine. Chlorine reacts with methane to give two useful products chloromethane and dichloromethane.

i What are the conditions for the reaction between methane and chlorine? [1]

ii Draw the structure of:
chloromethane [2]
dichloromethane [2]

d In the second stage, the plastic is heated to a higher temperature and it decomposes to a mixture that contains ethene, propane and naphtha. Naphtha is a mixture of liquid alkanes.

  i Suggest the name of the technique used to separate mixtures of liquid alkanes. [1]

  ii The alkanes in naphtha can be cracked. Construct an equation for the cracking of decane, $C_{10}H_{22}$.

  $$C_{10}H_{22} \longrightarrow \qquad\qquad [2]$$

e Alkenes are used to make a range of important organic chemicals.

  i Name the product of the reaction between butene and steam. [1]

  ii Write an equation and name the product of the reaction between propene and bromine. [3]

Total marks: 18 CAM /IGCSE

16 Hexadecane is a long chain hydrocarbon that can be broken down by cracking.

$$C_{16}H_{34} \longrightarrow H_2 + C_3H_{12} + C_5H_{10} + C_6H_{10}$$

a **i** Draw a structural formula for C_5H_{12}. [1]

  ii Draw a structural formula for C_5H_{10}. [1]

b **i** Explain with the aid of structural formulae how propene, C_3H_6, reacts with bromine to form 1, 2-dibromopropane. [4]

  ii Explain with the aid of structural formulae how propane, C_3H_8, reacts with blank bromine in the presence of UV light to form 1-bromopropane. [4]

c 20 cm³ of a hydrocarbon, C_xH_y, were mixed with 150 cm³ of excess oxygen. The mixture was then ignited. The volume of gas which remained was 120 cm³ which then decreased to 40 cm³ after passing through limewater. All volumes of gases were measured at the same temperature and pressure.

  i What volume of carbon dioxide, in cm³, was formed in the reaction? [1]

  ii What volume of oxygen, in cm³, reacted with the hydrocarbon? [1]

  iii Calculate the values of x and y. [3]

Total marks: 15 SEG

17 Here are the structures of some carbon compounds. They are labelled **A**, **B**, **C**, **D**, **E** and **F**.

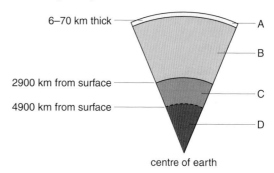

A **B** **C**

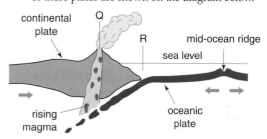

D **E** **F**

Choose from **A**, **B**, **C**, **D**, **E** or **F** to answer the following questions.

a **i** Which compound has the formula C_3H_8? [1]

  ii Which two compounds are alkenes? [1]

  iii Which two compounds are isomers of each other? [1]

b Look at compounds **A**, **B** and **C**.

  i Which **one** would you expect to have the highest boiling point? [1]

  ii Explain why you chose this compound. [2]

c What is the empirical formula of compound **B**? [1]

Total marks: 7 MEG

18 The diagram represents a section through the Earth, showing the layers which are labelled **A**, **B**, **C** and **D**.

6–70 km thick — A

— B

2900 km from surface — C

4900 km from surface — D

centre of earth

a Give the name of:

  i layer **A** [1]

  ii layer **B** [1]

b Give one difference between Layer **C** and Layer **D**. [1]

c The outer part of the Earth is made up of tectonic plates. These plates move relative to one another. Some of these plates are shown on the diagram below.

continental plate   Q   R   mid-ocean ridge   sea level   oceanic plate   rising magma

  i Volcanic activity is shown at area Q. Describe and explain the difference between igneous rocks which form within the Earth and those which form at the surface of volcanoes. [3]

  ii Explain what is happening at point R where the two plates meet. [2]

Total marks: 8 NEAB

19 The continents have continued to move apart very slowly over the last 200 million years.

Africa   South America

200 million years ago   Present day

a Give **two** pieces of geological evidence which show that the continents of South America and Africa were once joined. [2]

b The rigid outer layer of the Earth is made up of tectonic plates. Describe what causes these tectonic plates to move. [3]

Total marks: 5 SEG

20 One definition of an element is: 'A substance that cannot be broken down into simpler substances by chemical methods.'

The table below shows some of the 'substances' which Antoine Lavoisier thought were elements. He divided the 'substances' into four groups. He published these groups in 1789. The modern names of some of the 'substances' are given in brackets.

Acid-making elements	Gas-like elements	Metallic elements	Earthy elements
sulphur	light	cobalt mercury	lime (calcium oxide)
phosphorus	caloric (heat)	copper nickel	magnesia (magnesium oxide)
charcoal (carbon)	oxygen	gold platina (platinum)	barytes (barium sulphate)
	azote (nitrogen)	iron silver	argilla (aluminium oxide)
	hydrogen	lead tin	silex (silicon dioxide)
		manganese tungsten	
		zinc	

a Name **one** 'substance' in the list which is **not** a chemical element or compound. [1]

b i Name **one** substance in the list which is a compound. [1]

ii Suggest why Lavoisier thought that this substance was an element. [1]

c Dimitri Mendeleev devised a Periodic Table of the elements in 1869. Give two ways in which Mendeleev's table is more useful than Lavoisier's. [2]

Total marks: 5 NEAB

21 a Amino acids are important chemicals found in living organisms. The structures of three amino acids are shown.

Glycine   Cysteine   Alanine

i State **two** similarities in the three structures. [2]

ii Write down the molecular formula of alanine. [1]

iii Draw the structural formula of the product formed when glycine and alanine combine together. Label the peptide link. [2]

b The formula shows part of a protein molecule.

= the rest of
  the molecule

i What **two** conditions are needed to break down proteins into amino acids? [2]

ii What is the name of this breakdown process? [1]

iii Draw the structure of **one** of the amino acids formed when the protein shown breaks down. [2]

Total marks: 10 SEG

22 a Complete the table about atomic particles.

Atomic particle	Relative mass	Relative charge
proton		+1
neutron	1	0
electron	negligible	

b Read the following passage about potassium.
Potassium is a metallic element in Group I of the Periodic Table. It has a proton (atomic) number of 19. Its most common isotope is potassium-39, $[^{39}_{19}K]$. Another isotope, potassium-40, $[^{40}_{19}K]$, is a radioisotope.

i State the number of protons, neutrons and electrons in an atom of potassium-39. [2]

ii Explain why potassium-40 has a different mass number to potassium-39. [1]

iii What is meant by a *radioisotope*? [3]

iv Atoms of potassium-40 change into atoms of a different element. This element has a proton (atomic) number of 20 and a mass number of 40. Name, or give the symbol of, this new element. [1]

v Explain, in terms of atomic structure, why potassium-39 and potassium-40 have the same chemical reactions. [1]

c

Type of radiation	Change in	
	proton (atomic) number	mass number
alpha (α)	goes down by 2	goes down by 4
beta (β)	goes up by 1	no change
gamma (γ)	no change	no change

Use the table above, together with the Periodic Table, to help you answer the following questions.

i Name the type of radiation given out in part **b iv**. [1]

ii Give the name, or symbol, of the element formed when an atom of sodium-24 (proton number = 11) emits gamma radiation. [1]

iii State the Group number of the element formed when an atom of radon-220 (proton number = 86) emits an alpha particle. [1]

Total marks: 11 NEAB

23 Write a brief account of three of the following in the pages that follow. (Indicate your choice by writing the appropriate letter clearly at the beginning of each account.)

  a Outline the industrial production of ammonia starting with nitrogen and hydrogen. Discuss whether the advantages of using fertilizers made from ammonia outweigh the disadvantages.

  b Describe and explain how the composition of the Earth's atmosphere has changed over geological time.

  c Describe and explain in terms of their electronic structures the trends in reactivity in Groups I and VII of the Periodic Table.

  d Describe the differences in the structures and properties of thermoplastics and thermosets. How do these differences affect their uses?

$[3 \times 5$ marks $= 15]$ Welsh

Answers to numerical questions

1 e i 0.0050 moles   **ii** 0.0050 moles   **iii** 197 g
    iv 0.985   **v** 80.1 %

3 b i $C_6OH_3Cl_3$ **ii** 0.00344 moles

7 c i 10 cm^3 **ii** 0.01 mol/dm^3

8 b i 0.03 moles per litre   **ii** 4.32 grams per litre

12 a 70%

13 c ii 21.2%

16 c i 80 cm^3 **ii** 110 cm^3 **iii** x = 4 y = 6

Working with gases in the lab

Preparing gases in the lab

The usual method is to displace the gas from a solid or solution, using apparatus like that shown here. The table gives some examples.

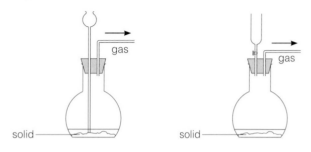

Apparatus for preparing gases. The dropping funnel in the right-hand apparatus is used for adding concentrated acids.

To make...	Place in flask...	Add...	Reaction
carbon dioxide	calcium carbonate (marble chips)	dilute hydrochloric acid	$CaCO_3(s) + 2HCl(aq) \longrightarrow$ $CaCl_2(aq) + H_2O(l) + CO_2(g)$
chlorine	manganese(IV) oxide (as an oxidizing agent)	conc hydrochloric acid	$2HCl(aq) + [O] \longrightarrow H_2O(l) + Cl_2(g)$
hydrogen	pieces of zinc	dilute hydrochloric acid	$Zn(s) + 2HCl(aq) \longrightarrow$ $ZnCl_2(aq) + H_2(g)$
hydrogen chloride	sodium chloride	conc sulphuric acid	$NaCl(s) + H_2SO_4(aq) \longrightarrow$ $NaHSO_4(aq) + HCl(g)$
oxygen	manganese(IV) oxide (as a catalyst)	hydrogen peroxide	$2H_2O_2(aq) \longrightarrow 2H_2O(l) + O_2(g)$

But to make ammonia, you just heat any ammonium compound with any strong alkali.

Collecting gases in the lab

Here are four ways to collect the gas you've prepared:

Method	Upward displacement of air	Downward displacement of air	Over water	Gas syringe
Use when...	gas is heavier than air	gas is lighter than air	gas is sparingly soluble in water	you want to measure the volume
Apparatus				
Examples	carbon dioxide, CO_2 chlorine, Cl_2 sulphur dioxide, SO_2 hydrogen chloride, HCl	ammonia, NH_3 hydrogen, H_2	carbon dioxide, CO_2 hydrogen, H_2 oxygen, O_2	any gas

Tests for gases

You have a sample of gas. You suspect it is carbon dioxide, but you're not sure. So you need to do a **test**. Below are some tests for common gases. Each is based on particular properties of the gas, including its appearance and sometimes its smell.

Ammonia, NH$_3$

Properties	Ammonia is a colourless alkaline gas with a pungent smell.
Test	Carefully smell the gas, and hold damp indicator paper in it.
Result	Recognizable odour, and indicator paper turns blue.

Carbon dioxide, CO$_2$

Properties	Carbon dioxide is a colourless, weakly acidic gas. It reacts with **lime water** (a solution of calcium hydroxide in water) to give a white precipitate of calcium carbonate:

$$CO_2(g) + Ca(OH)_2(aq) \longrightarrow CaCO_3(s) + H_2O(l)$$

Test	Bubble the gas through lime water.
Result	Lime water turns cloudy or milky.

Chlorine, Cl$_2$

Properties	Chlorine is a green poisonous gas which bleaches dyes.
Test	Hold damp indicator paper in the gas, *in a fume cupboard.*
Result	Indicator paper turns white.

Hydrogen, H$_2$

Properties	Hydrogen is colourless. It combines violently with oxygen when ignited. *Take care!*
Test	Collect the gas in a tube and hold a lighted splint to it.
Result	The gas burns with a squeaky pop.

Hydrogen chloride, HCl

Properties	Hydrogen chloride is a colourless, acidic gas with a choking smell. It reacts with the alkali ammonia to form the salt ammonium chloride:

$$HCl(g) + NH_3(g) \longrightarrow NH_4Cl(s)$$

Test	Hold a rod dipped in concentrated ammonia solution in the gas, *in a fume cupboard.*
Result	A white 'smoke' of ammonium chloride forms.

Oxygen, O$_2$

Properties	Oxygen is a colourless gas. Fuels burn much more readily in it than in air.
Test	Collect the gas in a test tube and hold a glowing splint to it.
Result	The splint immediately bursts into flame.

Sulphur dioxide, SO$_2$

Properties	Sulphur dioxide is a colourless acidic gas with a choking smell. It reacts with acidified potassium dichromate solution, reducing the orange dichromate ion, $Cr_2O_7^{2-}$, to the green chromium ion Cr^{3+}:

$$3SO_2(g) + 2H^+(aq) + Cr_2O_7^{2-}(aq) \longrightarrow 3SO_4^{2-}(aq) + H_2O(l) + 2Cr^{3+}(aq)$$

from the acid   orange solution   green solution

Test	Hold a piece of filter paper soaked in acidified potassium dichromate solution in the gas, *in a fume cupboard.*
Result	The paper turns from orange to green.

Testing for ions in the lab

You have a solid. You think it's sodium iodide, but you're not sure. Time for some detective work! Below are three methods of testing. Which would you use to test for sodium and iodide ions?

Method 1: Form a precipitate

If the salt is soluble in water, make a solution. Then try to precipitate an insoluble compound from it. You can use this method to test for any ion that forms an insoluble compound.

Testing for halide ions (Cl⁻, Br⁻, I⁻)
- Take a small amount of solution.
- Add an equal volume of dilute nitric acid.
- Then add silver nitrate solution.
- Silver halides are insoluble, so if halide ions are present a precipitate will form.

Precipitate	Indicates presence of ...	Reaction taking place
white	chloride ions, Cl⁻	$Ag^+ (aq) + Cl^- (aq) \longrightarrow AgCl (s)$
cream	bromide ions, Br⁻	$Ag^+ (aq) + Br^- (aq) \longrightarrow AgBr (s)$
yellow	iodide ions, I⁻	$Ag^+ (aq) + I^- (aq) \longrightarrow AgI (s)$

Note that iodide ions can also be identified using acidified lead(II) nitrate solution. A deep yellow precipitate of lead(II) iodide forms.

Testing for sulphate ions, SO_4^{2-}
- Take a small amount of solution.
- Add an equal volume of dilute hydrochloric acid.
- Then add barium chloride solution.
- Barium sulphate is insoluble, so if sulphate ions are present a precipitate will form.

Precipitate	Indicates presence of ...	Reaction taking place ...
white	sulphate ions, SO_4^{2-}	$Ba^{2+} (aq) + SO_4^{2-} (aq) \longrightarrow BaSO_4 (s)$

Testing for metal ions (Cu^{2+}, Fe^{2+}, Fe^{3+}, Al^{3+}, Zn^{2+}, Ca^{2+})
- Take a small volume of solution.
- Add a few drops of dilute sodium hydroxide solution.
- If a *coloured* precipitate forms, it indicates the presence of Cu^{2+}, Fe^{2+} or Fe^{3+} ions. These form insoluble coloured hydroxides:

Precipitate	Indicates presence of ...	Reaction taking place ...
pale blue†	copper ions, Cu^{2+}	$Cu^{2+} (aq) + 2OH- (aq) \longrightarrow Cu(OH)_2 (s)$
green	iron(II) ions, Fe^{2+}	$Fe^{2+} (aq) + 2OH (aq) \longrightarrow Fe(OH)_2 (s)$
red-brown	iron(III) ions, Fe^{3+}	$Fe^{3+} (aq) + 3OH- (aq) \longrightarrow Fe(OH)_3 (s)$

† This will redissolve in excess ammonia solution to form a deep blue solution (page 217).

- But a *white* precipitate indicates the presence of Al^{3+}, Zn^{2+} or Ca^{2+} ions. They form insoluble white hydroxides.
- To find out which of these ions it is, divide the solution containing the precipitate into two equal volumes.

- To one, add double the volume of sodium hydroxide solution. To the other, add double the volume of ammonium hydroxide solution. Check the result against this table:

If the precipitate ...	It indicates the presence of ...
dissolves again in sodium hydroxide solution	Al^{3+} ions
dissolves in both sodium and ammonium hydroxide solutions	Zn^{2+} ions
dissolves in neither	Ca^{2+} ions

Method 2: Carry out a reaction to produce a gas you can test

You can identify carbonate, ammonium, or nitrate ions in an unknown substance by carrying out a reaction that produces carbon dioxide or ammonia.

Carbonate ions (CO_3^{2-})
- Take a small amount of solid or solution.
- Add a little dilute hydrochloric acid.
- If the mixture bubbles and gives off a gas that turns lime water milky, the unknown substance contained carbonate ions.

The reaction that takes place is:
$$2H^+ (aq) \quad + \quad CO_3^{2-} (aq) \longrightarrow \quad CO_2 (g) \quad + \quad H_2O (l)$$
from the acid $\qquad\qquad\qquad\qquad$ carbon dioxide
$\qquad\qquad\qquad\qquad\qquad$ (turns lime water milky)

Ammonium ions (NH_4^+)
- Take a small amount of solid or solution.
- Add a little dilute sodium hydroxide solution and heat gently.
- If ammonia gas is given off, the unknown substance contained ammonium ions. (Ammonia gas has a pungent smell and turns red litmus blue.)

The reaction that takes place is:
$$NH_4^+ (aq) \quad + \quad OH^- (aq) \quad \longrightarrow \quad NH_3 (g) \quad + \quad H_2O (l)$$
$\qquad\qquad$ from sodium $\qquad$ ammonia
$\qquad\qquad$ hydroxide

Nitrate ions (NO_3^-)
- Take a small amount of solid or solution.
- Add a little sodium hydroxide solution.
- Add some aluminium powder or Devarda's alloy. Heat gently.
- If ammonia gas is given off, the unknown substance contained nitrate ions.

The reaction that takes place is:
$$8Al (s) + 3NO_3^- (aq) + 5OH^- (aq) + 2H_2O (l) \longrightarrow 3NH_3 (g) + 8AlO_2^- (aq)$$
$\qquad\qquad\qquad$ from sodium
$\qquad\qquad\qquad$ hydroxide

Method 3: Do a flame test

You can use a flame test to test for some metal ions.
- Make a paste of the solid in concentrated hydrochloric acid.
- Dip the end of a flame test wire (nichrome wire) in the paste, and then hold it in a Bunsen flame.

Colour of flame ...	Indicates presence of ...
yellow	sodium ions, Na^+
lilac	potassium ions, K^+
brick-red	calcium ions, Ca^{2+}
green-blue	copper ions, Cu^{2+}

Safety first!

The chemistry lab can be a dangerous place. But these rules will help you avoid accidents. How many of them do you follow already?

- Follow instructions precisely.
- Always wear your safety glasses when working in the lab. If chemicals get in your eyes they could blind you.
- Tie hair back.
- Don't fool around with Bunsen burners.
- Stand up when working with liquids.
- Don't splash chemicals around.
- Mop up spills carefully with a cloth for that purpose.
- If you have to smell a substance, do so carefully. Do NOT smell poisonous gases such as chlorine.
- Do NOT taste anything in the lab.
- Handle chemicals with the respect they deserve: especially acids, alkalis, and poisonous and flammable substances.
- Use a fume cupboard for experiments involving poisonous or highly reactive substances.
- Handle glassware with care.
- Find out *NOW* what to do in case of fire, or chemicals in your eye, or other kinds of accident. Talk the safety drill over with your teacher.
- Know where the fire extinguisher and fire blankets are, and how to use them.
- Know the symbols for chemical hazards. They're given below. Watch out for them on bottles and other chemical containers.

Toxic	These substances can kill. They may act when you swallow them, or breathe them in, or absorb them through your skin.	**Harmful**	These are similar to toxic substances but less dangerous.	**Highly flammable**	These substances catch fire easily. They pose a fire risk.
Examples:	chlorine gas mercury	Examples:	copper(II) sulphate lead oxide	Examples:	hydrogen ethanol

Oxidizing	These provide oxygen which allows other substances to burn more fiercely.	**Corrosive**	These attack and destroy living tissue, including eyes and skin.	**Irritant**	These are not corrosive but can cause reddening or blistering of skin.
Examples:	hydrogen peroxide potassium manganate(VII)	Examples:	conc sulphuric acid bromine	Examples:	calcium chloride zinc sulphate

Answers to numerical questions

page 19 2 11.6 g 3 a 27 g b 44 g 4 35 g

page 25 9 a i copper(II) sulphate 30 g, sodium chloride 36.5 g per 100 g of water ii copper(II) sulphate 40 g, sodium chloride 37 g per 100 g of water b i B ii A c 54 °C d 68 °C e No 10 a 35 g b 2 g

page 41 10 a The correct values for rubidium are: melting point 39 °C, boiling point 688 °C.

page 63 4 58 5 a 32 b 254 c 16 d 46 e 132

page 65 4 a 1 g b 127 g c 35.5 g d 71 g 5 a 32 g b 64 g 6 138 g 7 a 9 moles b 3 moles

page 67 1 60% 2 hydrogen 5%, oxygen 60% 3 80% 4 15 g

page 69 1 a 1 b 4 g 3 FeS 4 C_7H_8

page 71 4 C_2H_4 5 P_4O_6 6 CH_4

page 73 2 0.1 mol/dm^3 4 a 1 mole b 1 mole c 0.008 moles 5 a 2 mol/dm^3 b 0.05 mol/dm^3 c 0.06 mol/dm^3 6 a 0.5 dm^3 (500 cm^3) b 0.005 dm^3 (5 cm^3) 7 a 20 g b 0.5 g 8 a 0.5 mol/dm^3 b 0.25 mol/dm^3 9 1.7 g

page 74 1 a 64 g b 48 g c 48 g d 60 g e 355 g f 1.4 g g 4 g h 0.6 g i 24 g 2 a 2 g b 4 g c 32 g d 35.5 g e 248 g f 1024 g g 144 g 3 a 1 mole b 2 moles c 1 mole d 2 moles e 0.2 moles f 0.1 moles g 0.4 moles h 0.2 moles i 2 moles j 0.05 moles k 1000 moles 4 a 80 g of sulphur b 80 g of oxygen c 8 moles of chlorine atoms d 1 mole of oxygen molecules e 4 moles of sulphur atoms 5 a 18 g b 90 g c 160 g d 250 g e 34 g f 48 g g 30 g h 8 g i 15.8 g j 325 g 7 The missing numbers are: a 40, 16, 1, 56 b 1.6, 5.6 c 0.16 d 6, 3 e 71.4% 8 a 64 g b 4 moles c 2 moles d MnO_2 e 632.2 g 9 a 106.5 b 3 moles c 1 mole d $AlCl_3$ e 0.1 mol/dm^3

page 75 10 a N_2H_2 b C_2N_2 c N_2O_4 d $C_6H_{12}O_6$ 11 b CH_2 c A is C_3H_6, B is C_6H_{12} 12 a Zn_3P_2 b 27.2% 13 a P_2O_3 b 41.3 g c P_4O_{10} d P_4O_6 14 a 1 dm^3 of 2M sodium chloride b 1 dm^3 of 1M sodium chloride c 100 cm^3 of 2M sodium chloride d 1 dm^3 of 1M sodium hydroxide e 40 cm^3 of 1M sodium chloride 15 a 45.5 cm^3 b 41.7 cm^3 c 62.5 cm^3 d 25 cm^3 16 c The missing figures are: group 4, 0.19 g; group 5, 0.20 g e 0.16 g f 16 g g 2 moles h 1 mole i Cu_2O

page 81 2 b 2 moles c i 32 g ii 8 g 3 b $CuCO_3$, 124 g; CuO, 80 g; CO_2, 44 g c i 11 g ii 20 g

page 83 3 24 dm^3 4 a 168 dm^3 b 12 dm^3 c 0.0024 dm^3 or 24 cm^3 5 a 12 dm^3 b 2.4 dm^3 6 a 12 dm^3 b 12 dm^3 7 a 12 dm^3 b 6 dm^3

page 84 5 a 217 g b 20.1 g of mercury, 1.6 g of oxygen 6 a 32 g b sulphur c 11 g of iron(II) sulphide and 6 g of sulphur d 17.5 g 7 b 160 g c 2000 moles d 2 moles e 4000 moles f 224 kg 8 a 48 dm^3 b 12 dm^3 c 0.24 dm^3 d 7.2 dm^3 e 2.4 dm^3 f 0.072 dm^3 or 72 cm^3

page 85 9 b i 0.25 moles ii 0.125 moles iii 0.1 moles iv 25 moles v 0.05 moles vi 0.002 moles c i 0.5 g ii 4 g iii 2 g iv 700 g v 2.2 g vi 0.128 g 10 b i 2 moles

ii 168 g c i 1 mole ii 24 dm^3 or 24 000 cm^3 d i 12 dm^3 or 12 000 cm^3 ii 1.2 dm^3 or 1200 cm^3 11 b 0.5 moles c i 28 g ii 22 g iii 12 dm^3 or 12 000 cm^3 12 a 0.5 moles b 25 cm^3 c 75 cm^3 d 50 cm^3 13 a 4 moles b 19 moles c 4.75 moles d 114 dm^3 e 227 g f 502 dm^3 14 b i 34 g ii 0.88 moles per dm^3 or 0.88M iii 0.44 moles iv 10.6 dm^3 or 10 600 cm^3 15 a 0.1 moles b 8.4 g c 2.4 dm^3 d 9.5 g 16 a 1.4 g b 0.025 moles c 0.025 moles d Fe^{2+} e 0.6 dm^3

page 97 3 2100 joules

page 101 2 873.6 kJ/mol

page 102 2 b 0.1 moles c 0.1 moles 4 a 233 b 0.1 moles c 0.1 moles of each d 1 dm^3 6 b drop of 4 °C for NH_4NO_3, rise of 20 °C for $CaCl_2$ e i drop of 8 °C for NH_4NO_3, rise of 40 °C for $CaCl_2$ ii drop of 2 °C for NH_4NO_3, rise of 10 °C for $CaCl_2$ iii drop of 4 °C for NH_4NO_3, rise of 20 °C for $CaCl_2$ 7 d 100 cm^3

page 103 8 a 12 °C c i 50.4 kJ ii 2217.6 kJ 9 b i 30 kJ ii 280.4 kJ 10 c 55.6 kJ 12 c i 2220 kJ ii 2801 kJ d −581 kJ/mol g 1816 kJ 13 a 4 b none c 416 kJ f 330 kJ

page 115 3 26.8 amps 4 a 3600 coulombs b 0.0373 moles c 0.0187 moles d 1.194 g

page 112 11 b 3 moles c 201 451 g or 201.45 kg 12 c i 0.2 amps ii 0.119 g d Fe^{3+} 13 b ii 2 moles of electrons iii 193 000 coulombs c i 0.298 dm^3 or 298 cm^3

page 121 3 a i 29 cm^3 ii 39 cm^3 b i 0.65 minutes ii 1.5 minutes c i 5 cm^3 of hydrogen per minute ii 2 cm^3 of hydrogen per minute

page 123 1 a i 60 cm^3 ii 60 cm^3

page 125 1 a experiment 1, 0.55 g; experiment 2, 0.95 g b experiment 1, 0.55 g; experiment 2, 0.95 g c experiment 1, 0.33 g per minute; experiment 2, 0.5 g per minute

page 136 2 c i 14 cm^3 ii 9 cm^3 iii 8 cm^3 iv 7 cm^3 v 2 cm^3 d i 14 cm^3 per minute ii 9 cm^3 per minute iii 8 cm^3 per minute iv 7 cm^3 per minute v 2 cm^3 per minute e 40 cm^3 f 5 minutes g 8 cm^3 per minute

page 127 7 i 0.5 g

page 151 1 50 cm^3 2 1.6 mol/dm^3 3 75 cm^3

page 153 9 a 0.05 mol/dm^3 b 0.0005 moles e 0.1 mol/dm^3 f 15.0 cm^3 g 0.0015 moles h 3 10 b 0.014 moles c 0.007 moles d 0.742 g e 1.258 g f 0.7 moles g 10 moles

page 173 8 c i returnable 12.7 MJ, non-returnable 14.4 MJ ii twice 10 b 40% c ii 560 tonnes 11 b £5000

page 188 2 d 33 cm^3 e oxygen 33%, nitrogen 67% f oxygen 21%, nitrogen 78% 6 b i 2 sodium ions ii 2 sodium ions e 34.2 g

page 199 7 a iv the metal is iron (density 7.9 g/cm^3)

page 201 5 a 10 000 tonnes

page 213 5 e 1000 g of zinc ore

page 279 9 d i 18 000 years

The periodic table and atomic masses

Group

1   2

1 $^1_1$H hydrogen

2 $^7_3$Li lithium $^9_4$Be beryllium

3 $^{23}_{11}$Na sodium $^{24}_{12}$Mg magnesium

The transition metals

4 $^{39}_{19}$K potassium $^{40}_{20}$Ca calcium $^{45}_{21}$Sc scandium $^{48}_{22}$Ti titanium $^{51}_{23}$V vanadium $^{52}_{24}$Cr chromium $^{55}_{25}$Mn manganese $^{56}_{26}$Fe iron $^{59}_{27}$Co cobalt $^{59}_{28}$Ni nickel $^{64}_{29}$Cu copper $^{65}_{30}$Zn zinc

5 $^{85}_{37}$Rb rubidium $^{88}_{38}$Sr strontium $^{89}_{39}$Y yttrium $^{91}_{40}$Zr zirconium $^{93}_{41}$Nb niobium $^{96}_{42}$Mo molybdenum $^{98}_{43}$Tc technetium $^{101}_{44}$Ru ruthenium $^{103}_{45}$Rh rhodium $^{106}_{46}$Pd palladium $^{108}_{47}$Ag silver $^{112}_{48}$Cd cadmium

6 $^{133}_{55}$Cs caesium $^{137}_{56}$Ba barium $^{139}_{57}$La lanthanum $^{178.5}_{72}$Hf hafnium $^{181}_{73}$Ta tantalum $^{184}_{74}$W tungsten $^{186}_{75}$Re rhenium $^{190}_{76}$Os osmium $^{192}_{77}$Ir iridium $^{195}_{78}$Pt platinum $^{197}_{79}$Au gold $^{201}_{80}$Hg mercury

7 $^{223}_{37}$Fr francium $^{226}_{88}$Ra radium $^{227}_{89}$Ac actinium

$^{140}_{58}$Ce cerium $^{141}_{59}$Pr prae-sodimium $^{144}_{60}$Nd neodimium $^{147}_{61}$Pm promethium $^{150}_{62}$Sm samarium $^{152}_{63}$Eu europium $^{157}_{64}$Gd gadolinium $^{159}_{65}$Tb terbium

$^{232}_{90}$Th thorium $^{231}_{91}$Pa prot-actinium $^{238}_{92}$U uranium $^{237}_{93}$Np neptunium $^{242}_{94}$Pu plutonium $^{243}_{95}$Am americum $^{247}_{96}$Cm curium $^{247}_{97}$Bk berkelium

Relative atomic masses based on internationally agreed figures

Element	Symbol	Atomic number	Relative atomic mass	Element	Symbol	Atomic number	Relative atomic mass
Actinium	Ac	89		Erbium	Er	68	167.26
Aluminium	Al	13	26.9815	Europium	Eu	63	151.96
Americium	Am	95		Fermium	Fm	100	
Antimony	Sb	51	121.75	Fluorine	F	9	18.9984
Argon	Ar	18	39.948	Francium	Fr	87	
Arsenic	As	33	74.9216	Gadolinium	Gd	64	157.25
Astatine	At	85		Gallium	Ga	31	69.72
Barium	Ba	56	137.34	Germanium	Ge	32	72.59
Berkelium	Bk	97		Gold	Au	79	196.967
Beryllium	Be	4	9.0122	Hafnium	Hf	72	178.49
Bismuth	Bi	83	208.980	Helium	He	2	4.0026
Boron	B	5	10.811	Holmium	Ho	67	164.930
Bromine	Br	35	79.909	Hydrogen	H	1	1.00797
Cadmium	Cd	48	112.40	Indium	In	49	114.82
Caesium	Cs	55	132.905	Iodine	I	53	126.9044
Calcium	Ca	20	40.08	Iridium	Ir	77	192.2
Californium	Cf	98		Iron	Fe	26	55.847
Carbon	C	6	12.01115	Krypton	Kr	36	83.80
Cerium	Ce	58	140.12	Lanthanum	La	57	138.91
Chlorine	Cl	17	35.453	Lawrencium	Lw	103	
Chromium	Cr	24	51.996	Lead	Pb	82	207.19
Cobalt	Co	27	58.9332	Lithium	Li	3	6.939
Copper	Cu	29	63.54	Lutetium	Lu	71	174.97
Curium	Cm	96		Magnesium	Mg	12	24.312
Dysprosium	Dy	66	162.50	Manganese	Mn	25	54.9380
Einsteinium	Es	99		Mendelevium	Md	101	

Group

					0
3	**4**	**5**	**6**	**7**	$^{4}_{2}$He helium
$^{11}_{5}$B boron	$^{12}_{6}$C carbon	$^{14}_{7}$N nitrogen	$^{16}_{8}$O oxygen	$^{19}_{9}$F fluorine	$^{20}_{10}$Ne neon
$^{27}_{13}$Al aluminium	$^{28}_{14}$Si silicon	$^{31}_{15}$P phosphorus	$^{32}_{16}$S sulphur	$^{35.5}_{17}$Cl chlorine	$^{40}_{18}$Ar argon
$^{70}_{31}$Ga gallium	$^{73}_{32}$Ge germanium	$^{75}_{33}$As arsenic	$^{79}_{34}$Se selenium	$^{80}_{35}$Br bromine	$^{84}_{36}$Kr krypton
$^{115}_{49}$In indium	$^{119}_{50}$Sn tin	$^{122}_{51}$Sb antimony	$^{128}_{52}$Te tellurium	$^{127}_{53}$I iodine	$^{131}_{54}$Xe xenon
$^{204}_{81}$Tl thallium	$^{207}_{82}$Pb lead	$^{209}_{83}$Bi bismuth	$^{210}_{84}$Po polonium	$^{210}_{85}$At astatine	$^{222}_{86}$Rn radon

$^{162}_{66}$Dy dysprosium	$^{165}_{67}$Ho holmium	$^{167}_{68}$Er erbium	$^{169}_{69}$Tm thulium	$^{173}_{70}$Yb ytterbium	$^{175}_{71}$Lu lutecium
$^{251}_{98}$Cf californium	$^{254}_{99}$Es einsteinium	$^{253}_{100}$Fm fermium	$^{256}_{101}$Md mendelevium	$^{254}_{102}$No nobelium	$^{257}_{103}$Lr lawrencium

Approximate atomic masses for calculations

Element	Symbol	Atomic mass for calculations
Aluminium	Al	27
Bromine	Br	80
Calcium	Ca	40
Carbon	C	12
Chlorine	Cl	35.5
Copper	Cu	64
Helium	He	4
Hydrogen	H	1
Iodine	I	127
Iron	Fe	56
Lead	Pb	207
Lithium	Li	7
Magnesium	Mg	24
Manganese	Mn	55
Neon	Ne	20
Nitrogen	N	14
Oxygen	O	16
Phosphorus	P	31
Potassium	K	39
Silicon	Si	28
Silver	Ag	108
Sodium	Na	23
Sulphur	S	32
Zinc	Zn	65

Element	Symbol	Atomic number	Relative atomic mass
Mercury	Hg	80	200.59
Molybdenum	Mo	42	95.94
Neodymium	Nd	60	144.24
Neon	Ne	10	20.179
Neptunium	Np	93	
Nickel	Ni	28	58.71
Niobium	Nb	41	92.906
Nitrogen	N	7	14.0067
Nobelium	No	102	
Osmium	Os	76	190.2
Oxygen	O	8	15.9994
Palladium	Pd	46	106.4
Phosphorus	P	15	30.9738
Platinum	Pt	78	195.09
Plutonium	Pu	94	
Polonium	Po	84	
Potassium	K	19	39.102
Praseodymium	Pr	59	140.907
Promethium	Pm	61	
Protactinium	Pa	91	
Radium	Ra	88	
Radon	Rn	86	
Rhenium	Re	75	186.2
Rhodium	Rh	45	102.905
Rubidium	Rb	37	85.47
Ruthenium	Ru	44	101.07

Element	Symbol	Atomic number	Relative atomic mass
Samarium	Sm	62	150.35
Scandium	Sc	21	44.956
Selenium	Se	34	78.96
Silicon	Si	14	28.086
Silver	Ag	47	107.868
Sodium	Na	11	22.9898
Strontium	Sr	38	87.62
Sulphur	S	16	32.064
Tantalum	Ta	73	180.948
Technetium	Tc	43	
Tellurium	Te	52	127.60
Terbium	Tb	65	158.924
Thallium	Tl	81	204.37
Thorium	Th	90	232.038
Thulium	Tm	69	168.934
Tin	Sn	50	118.69
Titanium	Ti	22	47.90
Tungsten	W	74	183.85
Uranium	U	92	238.03
Vanadium	V	23	50.942
Xenon	Xe	54	131.30
Ytterbium	Yb	70	173.04
Yttrium	Y	39	88.905
Zinc	Zn	30	65.37
Zirconium	Zr	40	91.22

Index

*Where several page numbers are given and one is **bold**, look that one up first.*